POLITICAL GEOGRA

# POLITICAL GEOGRAPHY

## *A READER*

**Edited by**
**John Agnew**

A member of the Hodder Headline Group
LONDON • NEW YORK • SYDNEY • AUCKLAND

First published in Great Britain in 1997 by
Arnold, a member of the Hodder Headline Group
338 Euston Road, London NW1 3BH

Copublished in the US, Central and South America by
John Wiley & Sons, Inc., 605 Third Avenue,
New York NY 10158–0012



*British Library Cataloguing in Publication Data*
A catalogue entry for this book is available from the British Library

*Library of Congress Cataloging-in-Publication Data*
Political geography : a reader / edited by John Agnew.
p. cm.
Includes bibliographical references and index.
ISBN 0-470-23655-8 (pb.) 0-470-23656-6 (hb.)
1. Political geography. I. Agnew, John A.
JC319.P583 1996
320.9—dc20 96-21012

ISBN 0 340 67743 0 (Pb)
ISBN 0 340 67742 2 (Hb)

Typeset in 10/12 Times by Anneset, Weston-super-Mare, Somerset
Printed and bound in Great Britain by J W Arrowsmith Ltd, Bristol

# CONTENTS

**Acknowledgements** vii
**General Introduction** 1

**Section One: Approaching Political Geography**
Editor's Introduction 5
**1** Kevin R. Cox and David R. Reynolds, 'Locational Approaches to Power and Conflict' (1974) 10
**2** Peter J. Taylor, 'World-systems Analysis and Regional Geography' (1988) 17
**3** Derek Gregory, 'Areal Differentiation and Post-modern Human Geography' (1989) 25

**Section Two: The Spatiality of States**
Editor's Introduction 31
**4** Stein Rokkan, 'Territories, Centres, and Peripheries: Toward a Geoethnic–Geoeconomic–Geopolitical Model of Differentiation Within Western Europe' (1980) 37
**5** Michael Mann, 'The Autonomous Power of the State' (1984) 58
**6** Sankaran Krishna, 'Cartographic Anxiety: Mapping the Body Politic in India' (1994) 81

**Section Three: Geopolitics**
Editor's Introduction 93
**7** Alan K. Henrikson, 'America's Changing Place in the World: from "Periphery" to "Centre"?' (1980) 98
**8** Stuart Corbridge, 'Maximizing Entropy: New Geopolitical Orders and the Internationalization of Business' (1994) 122
**9** Gearóid Ó Tuathail, 'The Effacement of Place? US Foreign Policy and the Spatiality of the Gulf Crisis' (1993) 140

**Section Four: Geographies of Political and Social Movements**
Editor's Introduction 165
**10** Sari Bennett and Carville Earle, 'Socialism in America: a Geographical Interpretation of its Failure' (1983) 172
**11** Peter Osei-Kwame and Peter J. Taylor, 'A Politics of Failure: the Political Geography of Ghanaian Elections, 1954–1979' (1984) 198

**12** Paul Routledge, 'Putting Politics in its Place. Baliapal, India, as a Terrain of Resistance' (1992) 219

**Section Five: Places and the Politics of Identities**
Editor's Introduction 249
**13** Alexander B. Murphy, 'Linguistic Regionalism and the Social Construction of Space in Belgium' (1993) 256
**14** Loïc J. D. Wacquant, 'The New Urban Color Line: the State and the Fate of the Ghetto in PostFordist America' (1994) 269
**15** Benjamin Forest, 'West Hollywood as Symbol: the Significance of Place in the Construction of a Gay Identity' (1995) 287

**Section Six: Geographies of Nationalism and Ethnic Conflict**
Editor's Introduction 317
**16** Daniele Conversi, 'Reassessing Current Theories of Nationalism: Nationalism as Boundary Maintenance and Creation' (1995) 325
**17** Colin H. Williams, 'The Question of National Congruence' (1989) 336
**18** Nuala Johnson, 'Cast in Stone: Monuments, Geography, and Nationalism' (1995) 347

**Index** 365

# ACKNOWLEDGEMENTS

I would like to thank Stuart Corbridge, Alec Murphy, Joanne Sharp, and Graham Smith for their help with the 'concept' of this Reader. Laura McKelvie has been an enthusiastic and helpful sponsor of the project from the beginning. As I was engaged in the arduous task of moving from Syracuse, New York to Los Angeles, California she kindly prompted me to keep my interest in it. I have been fortunate over the years to work with a number of remarkable colleagues and graduate students on various facets of political geography. I am grateful to them for their stimulation and support. From an early age my parents encouraged an interest in the larger world beyond our village in the North of England. As a result, I owe much of my urge for understanding global political geography and much of what I have been able to accomplish in my own work to their stimulus and support. More recently my wife, Susan, and daughters, Katie and Christine, have given me the love and encouragement without which little can ever be accomplished with enduring satisfaction.

The editor and publisher would like to thank the following for permission to reproduce copyright material:
Blackwell Publishers for Gearóid Ó Tuathail, 'The Effacement of Place: US Foreign Policy and the Spatiality of the Gulf Crisis', *Antipode* 25(1) (1993), pp. 4–31, for the excerpts from Loïc J. D. Wacquant, 'The New Urban Color Line: The State and the Fate of the Ghetto in PostFordist America' from *Social Theory and the Politics of Identity* by Craig Calhoun (ed.) (1994), and for Colin H. Williams, 'The Question of National Congruence' from *A World in Crisis? Geographical Perspectives* by R. J. Johnston and P. J. Taylor (eds) (1989); Blackwell Publishers, Inc. for Peter Osei-Kwame and Peter J. Taylor, 'A Politics of Failure: The Political Geography of Ghanaian Elections, 1954–1979', *Annals of the Association of American Geographers* 74(4) (1984), pp. 547–89, and Peter J. Taylor, 'World-Systems Analysis and Regional Geography', *The Professional Geographer* 40 (1988), pp. 259–65; Frank Cass & Co. Ltd for Daniele Conversi, 'Reassessing Current Theories of Nationalism: Nationalism as Boundary Maintenance and Creation', *Nationalism and Ethnic Politics* 1(1) (1995), pp. 73–85. © Frank Cass & Co. Ltd; Elsevier Science Ltd for Sari Bennett and Carville Earle, 'Socialism in America: A Geographical Interpretation of its Failure', *Political Geography Quarterly* 2(1) (1983), pp. 31–55, and Paul Routledge, 'Putting Politics in its Place: Baliapal, India, as a Terrain of Resistance', *Political Geography* 11(6) (1992), pp. 588–611; Mouton de Gruyter, a division of Walter de Gruyter &

Co. Publishers for Alexander B. Murphy, 'Linguistic Regionalism and the Social Construction of Space in Belgium', *International Journal for the Sociology of Language* 104 (1993), pp. 49–64; Macmillan Press Ltd for Derek Gregory, 'Areal Differentiation and Post-Modern Human Geography' from *Horizons in Human Geography* by Derek Gregory and Rex Walford (eds) (1989); Michael Mann, 'The Autonomous Power of the State.' The text of this article has been previously printed in *European Journal of Sociology* XXV (1984), pp. 185–213. Reprinted with permission; Pion Ltd for Benjamin Forest, 'West Hollywood as Symbol: The Significance of Place in the Construction of a Gay Identity', *Environment and Planning D* 13 (1995), pp. 133–57, and Nuala Johnson, 'Cast in Stone: Monuments, Geography and Nationalism', *Environment and Planning D* 13 (1995), pp. 51–65; Lynne Rienner Publishers, Inc. for Sankaran Krishna, 'Cartographic Anxiety: Mapping the Body Politic in India', *Alternatives: Social Transformation and Human Governance* 19,4 (1994), pp. 507–21. © 1994 by Alternatives; Sage Publications, Inc. for Alan K. Henrikson, 'America's Changing Place in the World: From "Periphery" to "Centre"?', and for the excerpts from Stein Rokkan, 'Territories, Centres and Peripheries: Toward a Geoethnic–Geoeconomic–Geopolitical Model of Differentiation within Western Europe', both from *Centre and Periphery: Spatial Variation in Politics* by Jean Gottmann (ed.) (1980); Westview Press for Stuart Corbridge, 'Maximizing Entropy? New Geopolitical Orders and the Internationalization of Business' from *Reordering the World: Geopolitical Perspectives on the 21st Century* by G. J. Demko and W. B. Wood (eds) (1994); John Wiley & Sons, Inc. for Kevin R. Cox and David R. Reynolds, 'Locational Approaches to Power and Conflict' from *Locational Approaches to Power and Conflict* by Kevin R. Cox and David R. Reynolds (eds) (Halsted Press, 1974).

# GENERAL INTRODUCTION

Political geography concerns the processes involved in creating and the consequences for human populations of the uneven distribution of power over the earth's surface. This power is manifested geographically in the definition of boundaries between states or other political–territorial units, in the control exerted by powerful states and empires over less powerful ones, and in the material and emotional connections people make between themselves and the places or territories that they inhabit, thus limiting the access of others to them. Like so many other fields of study, political geography has been largely defined either by its most influential textbooks, such as, for example, those of Cox (1979) and Taylor (1989), or by those perspectives currently popular in the main academic journals. This is advantageous from the point of view of presenting a coherent perspective around which a course can be organized. Unfortunately, this approach tends to avoid both confronting theoretical controversies and giving a full presentation of the range of perspectives within the field. Intellectual liveliness is sacrificed for either succinct or fashionable presentation. Yet, the academic literature in political geography has flourished in recent years. After many years in the doldrums, there has been a major revival of academic interest in the field. This comes from both those who call themselves 'geographers' and from others, particularly in disciplines such as diplomatic history, political science, political sociology, international relations and literary studies. The revival dates from the late 1960s when a number of pioneers began to investigate questions concerning the spatial organization of politics and the political organization of space that had not excited much interest for many years. Initially, attention focused largely on place-to-place differences in the votes obtained by different political parties, the incidence of conflicts over local issues such as where to locate noxious facilities, and the history of the world political map. Since then, however, more theoretically informed approaches to empirical inquiry within the broad confines of the term 'political geography' have taken shape. In this context, the student of political geography needs to know how study is carried on in the field as much as about the substantive issues that are addressed from within the confines of a single framework. This Reader is designed to expose the student both to the major substantive areas of contemporary political geography and the main theoretical ways research has been addressed within them.

This Reader is structured around the five main areas into which research in political geography is now conventionally divided, although none of these is mutually exclusive of the others. The intention is to provide a set of readings which combine theoretical arguments with an empirical content that draws from situations and events in different parts of the world. In this way theoretical controversies can be identified and discussed with reference to concrete examples. The sections move from the classical topics of state spatiality and geopolitics to consider geographies of political and social movements, places and the politics of identities and, finally, geographies of nationalism and ethnic conflict. Three types of theoretical viewpoint have emerged to prominence within the field over the past 30 years. These are the spatial-analytic, the political-economic, and the postmodern. The spatial-analytic refers to perspectives in which terrestrial space is seen as having direct causal effects on the conduct of politics. In contrast, the political-economic perspectives tend to see the distribution of political activities as the outcome of institutional processes that inscribe themselves in space. Finally, the postmodern refers to a range of perspectives that question the possibility of 'objective' knowledge, seeing the social position of writers and the nature of language as not simply constraints but fundamental barriers to knowledge independent of social-political standpoints and linguistic conventions. From this point of view, the goal of research is attempting to recover the discourses governing the geography of political practices rather than searching for an ontologically independent 'reality' beyond the limits of theorizing.

Each of these designations is potentially problematic. The first, for example, is not restricted solely to perspectives that are strongly spatially determinist such as is often indicated by the term 'spatial analysis'. Indeed, there is very little of that in political geography. Rather, what I have in mind are those perspectives that give some causal role to spatial organization in political activities of one sort or another. Likewise with political-economic perspectives. They vary in the extent to which they focus on the political or the economic. The postmodern is the most problematic of all, encompassing such positions as poststructuralism (with its stress on language not just as the medium but as the message of communication), postcolonialism (with its emphasis on the politicization of knowledge in a world marked by major power disparities between people and places), and postmodernism in a narrow sense (with its tendency to savour the joy of multiple stories with no single one able to claim priority over others). Innovative research also plays around the margins of established positions, thus destabilizing neat typological boundaries by blurring genres (see, for very different examples of this, Enloe 1990; Billig 1995; Deudney 1995; Massey 1995). Notwithstanding the tendency of the bundling into three 'streams' of perspectives to tidy up a more complex theoretical picture, I have used them as a way of selecting the items that I have included in the Reader. Of the three articles and extracts in each of the six sections, one is selected to represent each of the three streams of theoretical perspectives. Running across the substantive or thematic sections, therefore, is a logical (if highly contestable) set of theoretical positions.

One important tendency in political geography has been to enlarge the scope of the 'political' beyond the bounds of the rather restricted focus on Great Powers and voting patterns that characterized the field in the past. Partly this is a response to disillusionment with politics-as-usual and a fear that political geography will be complicit with practices that are decried. But it is also the result of a growing sense that power is everywhere; not simply present in a realm of 'high' politics but very much part of the transactions of everyday life (Painter 1995). In this view, 'power' is not simply the capacity to coerce others (power *over*) but also the power *to* pursue interests and gain recognition or identity. This can lead to a questioning of the academic division of labour that designates a field as exclusively involved with the geography of power. Indeed, it is difficult to defend the idea that political geography should have a monopoly on studying the spatial organization of power. There is nevertheless still the need for an area of study of the more formalized and institutionalized arenas of power such as the spatiality of states, interstate geopolitics, the geographies of political and social movements, and territorial conflicts involving national and ethnic groups. This Reader has sections on each of these topics; but it also includes a section on places and the politics of identities, a topic of great interest to those who would like to extend the reach of the term 'political' beyond the limits of 'official' politics.

The specific items I have chosen for inclusion come from a wide range of sources: research articles and books published by people in a number of different academic disciplines. Political geography is not only the province of geographers. Whenever scholars engage with questions relating to the geographical deployment of power, then that is political geography. In an era when disciplinary boundaries are increasingly problematic, it is important to identify political geography as an emerging field of *interdisciplinary* inquiry. The corollary of this is that work is not included simply because it is associated with the discipline of geography. Geographical concepts such as territory, place, boundary, centre–periphery, location, and spatial context must be given a privileged position in the presentation for a work to be included. This reflects the view that there is a conceptual rather than a professional basis to the definition of a field (on this view, see the General Introduction to Agnew *et al.* 1996). I have chosen fewer longer articles and extracts so that the reader can follow the arguments of authors as fully as possible. The alternative of more shorter pieces would have meant sacrificing the goal of initiating readers in as fully a way as possible into the specific nature of the different theoretical perspectives and how they affect the interpretation of empirical information.

## References

Agnew, J. A., Livingstone, D. N. and Rogers, A. (eds) 1996: *Human geography: an essential anthology*. Oxford: Blackwell.

Billig, M. 1995: *Banal nationalism*. London: Sage.

Cox, K. R. 1979: *Location and public problems*. London: Methuen.

Deudney, D. 1995: Ground identity: nature, place, and space in nationalism. In Y. Lapid and F. Kratochwil (eds) *The return of culture and identity in IR theory*. Boulder Co: Lynne Rienner.

Enloe, C. 1990: *Bananas, beaches and bases: making feminist sense of international politics*. Berkeley CA: University of California Press.

Massey, D. 1995: *Space, place and gender*. Cambridge: Polity Press.

Painter, J. 1995: *Politics, geography and 'political geography'*. London: Arnold.

Taylor, P. J. 1989: *Political geography: world-economy, nation-state and locality*. London: Longman.

# SECTION ONE

# *APPROACHING POLITICAL GEOGRAPHY*

**Editor's Introduction**

The revival of political geography in the 1960s followed a long period in which the field, inside and out of geography, was without intellectual vitality. It had suffered from association with the efforts of previous generations to create a 'science' of geopolitics that was widely seen as having contributed in the form of Nazi expansionism to the onset of the Second World War. Contemporary human geography was largely apolitical in its understanding of 'how the world works'. In the United States and Western Europe political boundaries between states seemed set, the Cold War froze the world map into a seemingly fixed form, regional differences in political affiliations and living standards seemed in decline and political identities seemed very much to be synonymous with national ones. In short, there did not seem to be very much for political geography to study; except in the past. The field had limited present-day relevance. The onset of the Civil Rights movement in the United States (drawing attention to all kinds of inequalities between racial groups, particularly in the South, but also within cities between their central areas and their suburbs), changes in the world political map following from the independence of former colonies, the beginning or revival of separatist movements in Canada and Western Europe, the growth of the European Community and the expansion of 'movement politics' involving, first, organized labour and students and, later, the feminist and environmental movements, suggested, perhaps, that rumours of the 'death' of political geography were at least premature. A number of pioneers, including Jean Gottmann, Stein Rokkan, Jean Laponce, Paul Claval, Kevin R. Cox, Yves Lacoste, Ronald J. Johnston, David R. Reynolds, Richard Morrill and Peter J. Taylor, certainly came together as proponents of a revived political geography around much the same time. A coincidence, perhaps? Certainly, a tradition of political geography (associated with such historic figures as Friedrich Ratzel and Halford Mackinder) was available for expropriation. But the time was propitious too. Together these influences helped to establish a new political geography, building on the spatial concepts of the existing tradition but moving in largely new directions. Now there are a number of journals devoted to publishing research in political geography or closely allied areas, such as *Political*

*Geography, Society and Space* and *Space and Polity*. Multipurpose journals such as *Alternatives, Antipode, Annals of the Association of American Geographers, Transactions of the Institute of British Geographers, Ecumene, Environment and Planning A* and *C, International Organization, International Security, Review, Review of International Political Economy, Nations and Nationalism, Public Culture, Orbis, Politics and Society* and *Studies in American Political Development* also frequently contain articles on political-geographical topics.

The 1960s was the period in which geography underwent a 'spatial turn'. Space or distance was defined as geography's 'variable'. Older approaches, such as the regional-synthetic and the human-environmental, went into (temporary) eclipse. Not surprisingly, the revival of political geography, coinciding with this intellectual development, also took on something of a spatial-analytic cast. The search was on for distance-decay effects in the influence of voters on one another in the choice of political parties and externality-field effects from noxious facilities and 'undesirable' neighbours on the decision to become involved in neighbourhood political action. There was also considerable interest in the impact of districting methods on election outcomes, the modelling of distance-decay effects on the possibility of war breaking out between states and the spatial organization of local and regional governments. These interests and approaches have all persisted to a degree, although they are nowhere near as prevalent as once was the case. There has been a general retreat from a 'strong' spatial determinism but with a continuing commitment to analysing politics at a variety of geographical scales using spatial concepts.

An important statement of a spatial-analytic perspective in political geography is that provided by **Kevin R. Cox and David R. Reynolds (Reading 1)**. First published in 1974, this is an extract from a longer chapter that lays out the framework for a book entitled *Locational Approaches to Power and Conflict.* After a brief review of the neglect of geography in studies of power and conflict, the authors identify two factors which they see as leading towards an increased concern for space in political studies: the increasing impact of externalities (effects on others who are not parties to a transaction) on people in industrial societies and the adoption of 'systems' perspectives in political science which tended to increase the attention given to outcomes of the political process (who wins where) (also see Cox 1973). Although more recent work under this rubric departs from the specifics of the case made by Cox and Reynolds, nevertheless there is continuity in the focus on the spatial forms (maps of one kind or another) that give rise to political activity, on the one hand, or that are its outcome, on the other.

During the 1970s geography experienced something of a turning away from the dominance of spatial perspectives, an experience charted by Cox (1995) himself. In the context of an extended period of economic and political crisis in many Western countries, it is perhaps unsurprising that those dissatisfied with the explanatory adequacy of the conventional wisdom should turn towards a revived political economy. Usually drawing (if only nominally) from the Marxist tradition of political economy, a number of scholars have attempted to view a range of political-geographical phenomena as the result of processes of uneven development immanent within a capitalist world economy. One variety of this theorizing, that of the world-systems theory associated with the historian/sociologist Immanuel Wallerstein, has been particularly influential within political geography. It has been popularized by **Peter J. Taylor** in his textbook (Taylor 1989) and in numerous articles, such as the one reprinted here (**Reading 2**). Theoretically eclectic, drawing its essential tenets from the writings of such diverse figures as Fernand Braudel, Karl Marx and Karl Polanyi, this perspective tends to privilege the structural-geographical position occupied within the global division of labour (core, periphery, semi-periphery) in its explanation of everything else. Under this rubric, however, can be found a range of perspectives, some adhering more closely to an orthodox Marxism (e.g. Harvey 1982) and others exploring the independent powers of states (e.g. Skocpol 1994). What joins them is their view of space as a surface upon which political-economic processes (whatever the specifics) are inscribed and embedded but which is nevertheless essential to the outcome of the processes (providing a 'spatial fix' to the declining rate of profit, defining the spatial limits of state autonomy, etc.).

The 1980s saw a continuing sense of crisis, only this time it had even more profound intellectual consequences. A number of thinkers over the years had questioned the pretensions of 'grand theories' in the social sciences, others had suggested that knowledge was more a product of our language than based on 'facts' about the world. One reading (associated most with feminism and postcolonialism) emphasizes the partiality or 'situatedness' of knowing; knowledge is a function, at least in part, of the standpoint or 'subject position' at which the researcher/writer is located, particularly the historical experience of relative power in relation to others. Note the spatial metaphors applied to why there is doubt about the possibility of acquiring objective knowledge. The use of such metaphors can make these perspectives appear more *substantively* geographical than they sometimes are (Silber 1995). Another reading plays up the role of language in giving meaning. From this point of view, the world is written not discovered or explored (e.g. Barnes and Duncan 1992). Deconstructionist and poststructuralist positions have this emphasis. Finally, some would identify the tenuousness of all claims to tell 'stories' on behalf of others (see Alcoff 1991–92). Even 'emancipatory' narratives, stories told to benefit the interests and identities of others, involve a quest for transcendence that disciplines and limits the aspirations of presumed beneficiaries (for a critique of this logic see Harvey 1989). In this postmodern view respect for irony and ambiguity remain as the only guarantees against imposing order on others. To the extent that it is possible, one looks for the stories that groups share to understand their self-constructions. At the extreme one can *never* 'speak for others' (for a good critical review of these perspectives see Simpson 1996).

These are distinctive positions, therefore, and the use of the rubric 'postmodern' to cover all of them is problematic (see Appiah 1991; Krishna 1993; Duncan 1996). I use it, however, to convey the conviction common across all of them that knowledge production is both political and deeply compromised by the language and social conventions of academic fields and historical-geographical contexts. This is also the sense in which **Derek Gregory** uses the term in the extract reproduced here (**Reading 3**). These varied perspectives have made particular inroads into research on the spatiality of states, geopolitics and places and political identities. Some writers move uncertainly between the perspectives within their work. But the question of 'identity' – the relation of the self to larger groups and the world at large – bulks large in all of the perspectives. That this is inherently a geographical problem means that many of the substantive research areas in political geography are closely entangled with the concerns of postmodernism. Here is an intellectual two-way street.

The stimulus to the revival of political geography has come from a variety of directions. Initially largely under the influence of the 'spatial turn', political geography today is an intellectual zone of contestation between a much wider range of perspectives. None of the 'streams' or intellectual currents has yet dried up. Currently the third one is running the fastest. But who knows for how much longer? What is clear is that to be a student of political geography means knowing something about these varied perspectives and the way they address political-geographical problems. This Reader is designed to start the beginning student on that course.

## References

Alcoff, L. 1991–92: The problem of speaking for others. *Cultural Critique*, 20: 5–32.

Appiah, A. 1991: Is the post- in postmodernism the post- in postcolonial? *Critical Inquiry*, 17: 336–57.

Barnes, T. J. and Duncan, J. S. (eds) *Writing worlds: discourse, text, and metaphor in the representation of landscape*. London: Routledge.

Cox, K. R. 1973: *Conflict, power, and politics in the city: a geographic view*. New York: McGraw Hill.

Cox, K. R. 1995: Concepts of space, understanding in human geography, and spatial analysis. *Urban Geography*, 16: 304–26.

Duncan, N. 1996: Postmodernism in human geography. In C. Earle, K. Mathewson and M. Kenzer (eds) *Concepts in human geography*. Lanham MD: Rowman and Littlefield.

Harvey, D. 1982: *The limits to capital*. Chicago: University of Chicago Press.

Harvey, D. 1989: *The condition of postmodernity*. Oxford: Blackwell.
Krishna, S. 1993: The importance of being ironic: a postcolonial view on critical international relations theory. *Alternatives*, 18: 385–417.
Silber, I. F. 1995: Space, fields, boundaries: the rise of spatial metaphors in contemporary sociological theory. *Social Research*, 62: 323–55.
Simpson, D. 1996: *The academic postmodern and the rule of literature: a report on half-knowledge*. Chicago: University of Chicago Press.
Skocpol, T. 1994: *Social revolutions in the modern world*. Cambridge: Cambridge University Press.
Taylor, P. J. 1989: *Political geography: world-economy, nation-state and locality*. London: Longman.

# 1 Kevin R. Cox and David R. Reynolds, 'Locational Approaches to Power and Conflict'

Excerpt from: Kevin R. Cox and David R. Reynolds (eds), *Locational Approaches to Power and Conflict*. New York: Halsted Press (1974)

## Introduction

One might legitimately ask why yet another approach to the study of power and conflict – the traditional domain of political science – should even be contemplated, let alone developed. After all, the list of approaches is already impressive as an examination of almost any recent text in political science will reveal – those based on decision-making, modern political economy, micro-economics, political sociology, structural-functionalism, etc. In a trivial sense, it is quite true that the empirical sciences have always been concerned with locations for all the events and objects of interest in any empirical science must by definition occur at specifiable locations in space and time if they are to be empirical (i.e., capable of being observed and measured). However, spatial relations including physical distance, contiguity, distribution, and empirical and theoretical questions pertaining to geographical scale and the areal aggregation of locations have received only casual treatment by most modern analysts of conflict behavior.

Earlier in this century, on the other hand, locational factors played a considerably more central role in explanations of political and social phenomena. An examination of the works of geographers such as Ratzel and Mackinder and theoretical and empirical works of the sociologists of the "Chicago School" of human ecology attest to the once proud tradition of "spatial analysis" in social science. R. E. Park, one of the founders of human ecology, wrote in one of his seminal articles:

> Since so much that students of society are ordinarily interested in seems to be intimately related to position, distribution, and movements in space, it is not impossible that all we ordinarily conceive as social may eventually be construed and described in terms of space and the changes of position of the individuals within the limits of an area of competitive cooperation [Park, 1926].

The locational (spatial) approaches of students of early geopolitics were eventually abandoned because they resulted in overly simplistic, single-variable explanations, smacked of environmental determinism, or were not susceptible to empirical test. The spatial approach of the early human ecologists was rejected on similar but more complex grounds; it appeared to rest on an unfashionable biologistic view of social organization as well as appealing to a combination of environmental and economic determinism. Perhaps most important, however, the early human ecology school of sociology failed because its practitioners merely assumed that the spatial and social organization of society were interrelated rather than attempting to specify what forms the interrelationships took.

It should not be concluded that there was a wholesale abandonment of locational and spatial approaches in all of the social sciences. Indeed there were some significant exceptions. Most notable were the development of agricultural and industrial location theory in economics and the development of retail location theory by both economists and geographers. In the development of these theories the costs incurred as a result of spatial factors, particularly distance, rendered their consideration essential. Empirically, it could be, and was, demonstrated that industries based on geographically limited resources as well as many forms of agricultural activity were highly sensitive to transportation costs; in the case of retail establishments, it was empirically obvious that profitability was dependent on an establishment's geographical location vis-à-vis a dispersed set of consumers and its potential competitors (Tullock, 1967: 70). In fact, due to the apparent behavioral tendency of consumers to travel no further than necessary to purchase a given good (a tendency in perfect accord with consumer rationality posited in economic theory) the theory of monopolistic competition found empirical relevance in the problem of retail establishment location; it was found that the assumption of rationality coupled with a geographically dispersed set of consumers led to "natural" monopolies – monopolies that could not be explained without reference to factors that were explicitly spatial. Nevertheless, with regard to the problems in which other social scientists evinced an interest, the statement Kemeny made in 1961 appears to be accurate – "the social sciences may be characterized by the fact that in most of their problems numerical measurements seem to be absent and *considerations of space are irrelevant*" (italics added; Kemeny, 1961: 35).

To arrive at definitive statements which account for the apparently widespread contention in social science that considerations of space have tended (at least until recently) to be irrelevant would demand an excursion into the history and sociology of social science. Although such an exercise would no doubt be extremely valuable, we have not conducted an exhaustive one. Instead, we can tentatively suggest that one explanation is to be found in the paradigms under which and theories with which social scientists have labored. These paradigms and theories by accident or by disciplinary myopia have conditioned us to think that spatial considerations are irrelevant. In the exceptions discussed above spatial factors, such as distance and distribution, found reasonable, straightforward, economic interpretations as costs and hence did not require any radical conceptual retooling on the part of economists and economic geographers. We are led, then, to hypothesize that in other areas of social science spatial considerations tended to be viewed as relatively unimportant either because they found no such readymade counterparts in extant models and theories or because social scientists simply wished to avoid the predictable rounds of vociferous criticism that would greet any approach that seemed tainted with even the most remote vestige of the locational variant of environmental determinism.

Whatever the reasons for the general abandonment of locational approaches in social science in general and the study of politics in particular, it is important to stress that there are indications that the number of social scientists attempting to develop locational approaches is increasing in relative as well as absolute terms. Not unexpectedly, there is as yet no one generally accepted unified locational approach to the analysis of power and conflict, since after years of being imbued in aspatial analyses, social scientists seem only now to be developing theories and perspectives that could be classified as locational in a nontrivial sense. However, the exigencies of life in modern society, as we hope to make clear below, demand that we develop new ways of thinking about social, economic and political problems many of which will demand conceptualization in a locational context. There is some evidence that the basic foundations for the further elaboration of locational theories of power and conflict have already been laid. These we also hope to demonstrate in the remaining portions of this chapter.

## Factors associated with the revival of interest in the locational analysis of politics

Two major reasons can be advanced as to why there is growing dissatisfaction amongst students of politics with the view that considerations of space are irrelevant to their problems. The first stems from the increased realization that much, if not most, political conflict in an industrial-urban society is, at least in part, the result of geographical externalities; and the second is largely the product of the increased adoption of systems analytic perspectives and methodologies by students of politics.

### *The empirical imperative and locational conflict*

It has become common for social scientists to recognize that in modern society almost everything is in one manner or another related to everything else. Yet, it is equally apparent in their writings that they possess the essentially unshaken belief that despite the complexity of social phenomena, certain things are more related than others (or at least more directly related).

Clearly the things which a social science discipline perceives as more closely related will tend to be intrinsic to the particular discipline; this is a major justification and product of the academic division of labor. The view of the geographer at first appears the simplest and most naive of all. It can best be described as "everything is related to everything else, but near things are more related" (Tobler, 1970: 234).

The popular impression, on the other hand, is that whereas accessibility formerly may have been important in human affairs, modern communication technologies have rendered it unimportant. The advent of such technologies has no doubt revolutionized the form, content, and rapidity of interpersonal interaction – communication is more rapid (and hence can extend over greater distances in less time); "out-of-pocket" transport costs have decreased in general (at least for long distances); more individuals have more options in the mode of communication; and, not unimportantly, with this increase in options there has been a shift from primary interaction to more impersonal forms of communication. All this is granted. What is not granted is that locational factors including accessibility must of technological necessity be irrelevant in accounting for social phenomena.

The most obvious spatial concomitant of industrialization that has been observed is the massive redistribution (as well as growth) of population over space. It is perhaps too easily forgotten that industrialization does not take place without the development of a supporting pattern of settlement. Also, one suspects that the well-documented changes in the scopes of various levels of government (as reflected in increased budgets, range of activities undertaken, etc.) are not casually related to industrialization and to urbanization. The fact that more people are closer together in geographical space is perhaps one of the reasons why most social scientists feel compelled to lend lip-service to the above mentioned "everything-affects-everything-else" viewpoint. It is also well verified that changes in patterns and density of settlements and the distribution of industry have contributed to a host of problems with which political systems in societies have traditionally not had to cope. Examples are numerous: traffic congestion, environmental pollution, the social and psychological costs of economically enforced and socially sanctioned segregation by race, income and class the fiscal crisis of central cities.

Perhaps the most important consequence of the spatial aspects of urbanization from a political perspective is that there has been a rapid proliferation in, and awareness of, what the economist refers to as externalities (variously referred to as neighborhood effects, third party effects, free-rider costs and benefits). All of the problems listed above are problems because of their nature as externalities. The standard textbook case of an externality arises when a contract between two persons will have some effect on a third person (Tullock, 1970: 71–95). However, the number of parties involved in the contract and the actual number of unconsulted, yet affected, persons is irrelevant in a general definition of an externality. The important point is that an externality will exist whenever at least one person who is affected by a transaction is excluded from a decision-determining role in the group whose consent is necessary for the transaction to occur.

The importance of the increasing externalities associated with urbanization for the analysis of politics is implicit in James Buchanan's discussion of the applicability of the neo-classical interpretation of "economic man" as a hypothetical individual whose behavior is predicated on materialistic self-interest.

> The departures from behavior patterns based on narrowly materialistic utility functions seem to be almost universal only when *personal* externality relationships exist. This is to say, the argument against the narrow self-interest assumption applies fully only when the potential externality relationship is limited to a critically *small number of persons*. In large-number groups, by comparison, there may be little or no incorporation of the interests of "others" in the utility calculus of individuals. Here the individual really has no "neighbors," or may have none in any effective behavioral sense, despite the presence of "neighborhood effects" . . . the person

> who litters the non-residential street in the large city probably does not worry much about the effects of his actions on others [Buchanan, 1969: 80–81].

It is precisely with that often noted concomitant of urbanization, the "breakdown of 'community'," that individuals find themselves in large groups held together by impersonal modes of communication and comprised of shifting memberships in which it is extremely difficult, if not impossible, for them to accurately assess the costs that their actions may impose on others in the group. In the small primary groups typical of nonindustrial life, social sanctions or rewards could be and are readily applied such as to compel the potentially deviant individual to include reasonably accurate estimates of the costs or benefits he might inflict (bestow) on third parties as a result of an action. In an urban society, to whom or to what collectivity do the third parties turn for the resolution of the conflict engendered by the existence of negative externalities? Increasingly, it has been some level of government and increasingly to the national government!

As has been stressed by Mishan (1967), economic growth almost by definition generates negative (and unpriced) externalities; but, as argued above, they cannot be internalized, except at the risk of incurring prohibitive information collection and bargaining costs, through individual initiative or through informal social controls as urbanization and industrial growth continues. The only rational recourse is for the individual to behave as the self-interested economic man. This also entails appealing to the political system when it is in his self-interest.

A more general argument than that above, but one which also suggests the necessity of integrating locational views of politics with the more common aspatial views, is based on the not totally unrealistic assumption that political systems are directed primarily toward the provision of "public goods" (this is not a particularly novel view of politics and has achieved widespread attention in economics, in sociology, and in some political science circles. See, e.g., Coleman, 1970, Olson, 1968, and Tullock, 1970).

The two essential characteristics of a *pure* public good are nonexcludability and equal availability. If any person in some specified group consumes a good and it cannot feasibly be withheld from others in the group and also if additional consumption of the good by one person does not diminish the amount freely available to others, then the good is a pure public good. Common examples of public goods include police and fire protection and national defense (although whether even these are pure is a debatable question). If political systems were concerned exclusively with the provision of pure public goods, the political analyst would have less need to view political systems spatially. However, it is apparent that most, if not all, public goods are impure – impure for decidedly locational or spatial reasons.

First, the "friction of geographical space" renders the amount of a public good freely available to persons in the polity considerably less than equal. An impure public good is an externality (or spillover) generating good and typically such externalities are geographic, i.e., the amount of the good freely available increases or decreases (1) with increasing geographical distance out from the initial source or recipient, and (2) with geographical contiguity. Consider the consumption of the public good represented by a public park, for instance. Second, the degree to which a public good is nonexcludable is dependent not only upon its physical characteristics, but also on the manner in which geographical space has been organized for the production and provision of the public good in question and on a person's location in this spatial system. For example, in the United States persons not residing in middle-class suburbs (at least at the time of this writing) can be excluded from the generally high-quality, relative low-cost education provided there, whereas suburban residents cannot.

Nor does the spatially relevant analysis stop there. Many private goods have public effects and can therefore be regarded as providing localized benefits and costs or externalities to those who were not involved in the consumption decision: consider the costs imposed by the humble power lawnmower, for instance. In many ways, the impure public goods provided by political systems function as impure private goods and provide localized benefits and costs to adjacent systems. As a result of interstate mobility, for instance, a state cannot exclude adjacent states from the benefits provided by its production of educated manpower. This is a classic example of an externality or spillover problem.

Implicit in the production of public benefits and public costs or externalities is an allocation problem: how should scarce public resources be allocated to diverse public ends? It is precisely this allocation problem engendering conflicts between those with diverse preferences which political systems are intended to solve.

Given the localized costs and benefits associated with public goods, it follows that ensuing conflicts are likely to be locational in character (e.g., conflict between central city and suburb, Quebec and the rest of Canada, ghetto and non-ghetto, etc.). Such conflicts are the result of allocations regarded as more or less inequitable by some localized groups and feed in to affect the next round of allocations. In short, locational change and social change may well be the proverbial "two sides of the same coin."

### *Growth of systems perspectives*

The growth of systems perspectives in social science in general and in political science in particular is resulting in a heightened concern with the outcomes of political processes and their impacts on future inputs to a political system and not just with the process itself. Schick (1971: 144), in writing about the analytical predilections of political scientists in the 1950s and 1960s, states that "rather than showing concern about political outcomes, they were preoccupied with celebrating an *ancien regime* that exhibited few signs of the traumas developing within it." The *ancien regime* to which Schick refers is that of process politics as viewed by the pluralists – wherein government is sometimes a representative of special interests and sometimes an arbitrator of conflicts between interest groups but always an elaborate structure and set of procedures whose primary function is not that of promoting some overarching public interest but merely that of keeping the process going and managing conflict. The process approach of the pluralists

> offered a convenient escape from difficult value questions. A decisional system that focuses on the outcomes and objectives of public policy cannot avoid controversy over the ends of government, the definition of the public interest, and the allocation of core values such as power, wealth, and status. But the pluralists by-passed these matters by concentrating on the structure and rules for choice, not on the choices themselves. They purported to describe the political world as it is, neglecting the important normative implications of their model. The pluralists scrupulously avoided interpersonal comparisons and the equally troublesome question of whose values shall prevail. Instead, they took the actual distribution of values (and money) as Pareto optimal, that is, as the best that could be achieved without disadvantaging at least one group [Schick, 1971: 141].

With the growth of a systems perspective in political science starting with Easton's *The Political System* (1953), the attention of researchers began to focus on the outputs of (the allocation of values), as well as on the inputs to, the political process (although Easton himself concentrates almost exclusively on inputs). Empirical studies on comparative state politics in which policy outcomes were correlated with economic and political characteristics over sets of political jurisdictions within states called into question the pluralist assumption that the group process produces representative and desirable outcomes (see, e.g., Dye, 1966, and Sharkansky, 1968). In brief, the view of politics as a positive sum game in which almost everyone wins or at least comes out ahead did not mesh well with the political reality in which there are losers and in which power itself is a scarce resource. Studies such as Dye's (1966) not only indicated that there are losers in political systems, but that there were jurisdictional and hence locational biases in their distribution as well. The development of a systems perspective in political analysis, therefore, has brought into refocus the realization that political systems, however conceptualized, whether in mechanistic input-output terms or more organic or functional terms, do something other than perpetuate themselves, i.e., as pointed out in the last section, they perform work or are action or allocational systems, and this allocation is either explicitly or implicitly locationally biased.

A locational consideration of basic importance in the analysis of political systems concerns the feedback relationships between the allocational outputs (location decisions) of political systems and the generation of new conflicts, demands and expectations with which the system must cope. In brief, the basic question is again one of geographic spillovers or externalities – what is the political impact of a benefit or cost at a given location on other locations or, phrased in a more macro sense, what are the dynamic relationships between a locational pattern of benefits and costs at time t and the locational pattern of conflicts, demands and expectations at time t + i?

The above two reasons, the first empirical and the second both empirical and theoretical, are in our view those most responsible for the recent development of more explicitly locational approaches to the analysis of politics. There are of course other factors which have also contributed to this development, e.g., the growth of the urban planning profession, quantification, and social engineering. All of these, however, are related to the reasons cited above.

**Selected references**

Buchanan, J. M. 1969: *Cost and Choice*. Chicago: Markham.

Coleman, J. S. 1970: "Political money," *American Political Science Review* 64: 1074–1087.

Dye, T. R. 1966: *Politics, Economics, and the Public: Policy Outcomes in the American States*. Chicago: Rand McNally.

Easton, D. 1953: *The Political System*. New York: Alfred A. Knopf.

Kemeny, J. G. 1961: "Mathematics without numbers," in D. Lerner (ed.) *Quantity and Quality*. New York: Free Press.

Mishan, E. J. 1967: *The Costs of Economic Growth*. London: Praeger.

Olson, M. 1968: *The Logic of Collective Action*. New York: Schocken.

Park, R. E. 1926: "The urban community as a spatial pattern and a moral order," in E. W. Burgess (ed.) *The Urban Community*. Chicago: University of Chicago Press.

Schick, A. 1971: "Systems politics and systems budgeting," in L. L. Roos, Jr. (ed.) *The Politics of Ecosuicide*. New York: Holt, Rinehart and Winston.

Sharkansky, M. 1968: *Spending in the American States*. Chicago: Rand McNally.

Tobler, W. R. 1970: "A computer movie simulating urban growth in the Detroit region." *Economic Geography* 46: 234–240.

Tullock, G. 1970: *Private Wants, Public Means*. New York: Basic Books.

Tullock, G. 1967: *Toward a Mathematics of Politics*. Ann Arbor: University of Michigan Press.

## 2 Peter J. Taylor, 'World-systems Analysis and Regional Geography'

Reprinted in full from: *Professional Geographer* 40(3), 259–65 (1988)

Now is an exciting time to be a geographer. The variety of approaches that has typified the last decade has resisted pressures to conform to a sterile applied geography. There is no applied geography paradigm; there is no paradigm in geography at all. As long as geography remains an intellectual pursuit, uncertainty will be endemic. Let us rejoice in our variety.

This essay considers three recent trends in modern geography: the revival of political geography (e.g., Editorial Board 1982), the rediscovery of the global scale (e.g., Johnston 1984), and the call for a new regional geography (e.g., Gregory 1978). I examine the relationship between these trends with an argument that derives a regional geography from a global approach that transcends the state. The framework adopted is world-systems analysis, which unites the three trends of modern geography into one theoretical perspective.

World-systems analysis means the attempt by Immanuel Wallerstein and his colleagues at the Fernand Braudel Center in Binghamton, New York, to produce a historical social science (Taylor 1986a, 1986b; Wallerstein 1979). It views the modern world as consisting of a single entity, the capitalist world-economy, which evolved from about 1500 to encompass the whole world by about 1900. This entity rather than the state is the prime object of analysis. The mechanisms generating and sustaining this modern world-system are expressed geographically as a core-periphery pattern, with a semi-periphery in between; and historically as a cyclical pattern with related secular trends. The result is an ever changing dynamic system, experienced today as growing material inequality between North and South in the downturn phase of the fourth Kondratieff long-wave cycle (Wallerstein 1984b).

World-systems analysis attracts geographers (e.g., Soja 1980, 222) with its explicit treatment of spatial pattern. Wallerstein's key concepts of core, periphery, and semi-periphery exclude a placeless social science. But where precisely does geography, and in particular regional geography, fit into this scheme? Wallerstein may superficially be "bringing geography back in," but many unanswered questions remain concerning the relations between world-systems analysis and geography. I address some of these queries in a four-part argument concentrating on regional geography. (1) World-systems analysis forces geographers to take a fresh, almost revolutionary, look at the data routinely used in comparative analyses. (2) The geographer's contribution should be to understand the places that make up the world-system. Enter regional geography. (3) The nature of the proposed regions is explored through the concept of "historical region." (4) World-systems analysis is distanced from the regional geography classification analogy under the heading "breaking all the rules!" No new regional taxonomy is proposed.

## New geographies, new statistics

Incorporating geography into world-systems analysis is much more than rediscovering a global perspective. It involves, among many other features, critiquing the statistics available for studying the world. The "data are . . . conceived of as social products: statistics are not collected but produced" and "their production is a social process which is carried out for specific reasons, and in specific ways" (Irvine *et al.* 1979, 3). The state is at the heart of that social process as the prime producer of statistics in the modern world. This social process of data production is particularly worrying from a world-systems perspective that tries to transcend state-centric thinking. The "official data" that are the life blood of so much social science are even more problematic for world-systems analysts than for other critical scientists. Not only are the social categories that the state defines unlikely to fit the needs of the analysis, but the very territorial basis of state-generated information is challenged. A new data production is needed with a recursive relation between new statistics and new geographies.

Tasks undertaken by researchers at the Fernand Braudel Center illustrate needs to collect, collate, and construct the necessary kinds of information (Hopkins and Wallerstein 1977, 141). The first task simply involves "the *process* of constituting boundaries, that is, learning to identify what is to be bounded" (Hopkins and Wallerstein 1977, 141). Nothing can be said quantitatively about world-system trends and fluctuations until this task is carried out. The second task is to map the continually changing division of labor within the newly defined bounds. This task specifies the economic geography of the world-economy through identification of commodity chains. The third task is to define the cyclical movements of the world-economy and their relation to the economic geography. The fourth task involves developing sets of indicators for different types of production processes to be classified "according to time and place of occurrence" (Hopkins and Wallerstein 1977, 143). The fifth task is to measure net capital accumulation in order to specify the contribution of both state and non-monetized economic processes for accumulation and disaccumulation across zones. The sixth and final task involves the mapping of labor force formation in different zones of the world-economy. All these tasks require the definition of new categories, the production of new statistics and the creation of new maps. For instance the sixth task considers degrees of proletarianization among households to define new household types which can then be mapped for different places at different times. The enormity of this task is exactly the point: a world-systems approach requires us to recast our categories for analysis and that means many new geographies.

The research tasks do not accord special privileges to geography as a discipline. Hopkins and Wallerstein (1977) define empirical tasks, and this concrete emphasis brings them into geographical fields of investigation. But we should not take this argument too far. Wallerstein's (1983) *Historical Capitalism* is avowedly concrete and hence historical: it is equally geographical on even the most cursory reading. All social relations must occur in space and time, however, negating the argument that the new geographies are the property of the discipline of geography any more than, for instance, the world political map of state territories is the sole preserve of political geography.

Hall's (1986) contribution to the first task, bounding the system, clearly illustrates this point. Hall tackled this most geographical question by reconsidering the nature of the category periphery in the light of Wolf's (1982) concern to give greater weight to non-core peoples in the making of our modern world. Hall defined four categories to replace Wallerstein's periphery and external arena. In addition to the dependent or "full blown" periphery he identified the marginal periphery as a region of refuge incorporated into the world for future expansion but left in limbo, and the contact periphery as part of Wallerstein's external arena. For Hall, the influence of the core on the contact periphery has radically transformed it, although not yet incorporated it into the world-system. The external arena is now restricted to zones of no contact. These categories of space are derived from a detailed study of New Mexico through its Spanish, Mexican, and American phases. This first modification of world-systems concepts through careful empirical study, although superficially geographical, is actually based on a new conception of social relations by a sociologist. The point is not that geographers cannot do such work, but rather that these concrete analyses are of general concern to all historical social scientists. One field of inquiry, however, invites geographers to use their particular skills to contribute to historical social science. Whereas new geographies are of general concern, new regional geographies are a specific geographical contribution.

### Understanding places

Geographers have always been distinguished from other scientists by their detailed knowledge of places. Although such studies have been relatively neglected in recent years, geographers still have more empirical knowledge about different places across the world than any other group of scholars. The problem is that this human resource has been poorly harnessed with the demise of traditional regional geography. No new theoretical basis has replaced the implicit physical determinism/possibilism of traditional regional geography. As a result regional geography has largely survived as a handmaiden of regional planning for the state. Regional synthesis has given way to regional science, a narrow applied economic geography. World-systems analysis offers a new theoretical basis for a new regional geography. The position of geographers today is analogous to that of the founders of modern geography, as Hartshorne (1939) defined it, in the period 1750 to 1840. He argued that geographers in that period moved away from practical utility – geography for statecraft – and reacted against the ordering of data through the political map, particularly during the Napoleonic period when the political map was so unstable. They attempted to develop a *reine* (pure) *geographie* based upon a physical division of the world. Although Hartshorne argued that by 1830 the debate had discarded fixed physical boundaries, the idea of natural boundaries nevertheless entered modern geography and influenced the regional geography produced in the first half of the twentieth century. Most regional divisions in textbooks, for instance, continued to be based on physical features. Geographers eventually appreciated that physical divisions were decreasingly important bases for socio-economic activities, and increasingly dismissed regional geography as old-fashioned, leaving the state unchallenged, once again, as the spatial framework for geographical studies (Cole 1981, 271).

Any analogy drawn across two centuries must not be taken too far. Although the world political map is challenged as the prime geographical basis for global analysis, no new *reine geographie* is proposed. To claim that the new regional geography is not state-centric is not to claim that it is pure in the sense of being non-political and objective. It simply provides a rationale for regional construction within the overarching structure of the capitalist world-economy.

Wallerstein's (1979, 69) tripartite concepts of core, periphery, and semi-periphery can be viewed as a first order division of the world, which he considers necessary for a proper functioning of the world-system. Wallerstein hypothesizes that the proportion of the world's population within each of these zones remained roughly constant throughout the history of the world-economy (Wallerstein 1984a, 7). This demographic stability does not imply a constant geographical pattern, however. The actual places that constitute each zone will change over time with rises and falls approximately compensating each other. For instance, incorporation of new areas into the periphery enables the system to support a larger core as happened in the late nineteenth century. Regional geographies as second order divisions of the world can aid in understanding such changes within and between the first order zones.

Core, periphery, and semi-periphery delineate a hierarchal spatial division of labor within the world-system. Each zone or category includes broadly similar production processes, as well as important differences within a category. Variations in resource bases, economic mixes, political structures, and other features facilitate movement of places between categories. This variety is not random but evolves historically to differentiate places within each category and to produce identifiable regions. By studying these regions one can begin to see how the larger zones are constituted and changed. Such a regional geography indispensably serves world-systems analysis.

**Historical regions**

The traditional geographers' natural regions were seemingly eternal. The world-systems analyst's regions, in contrast, are historical: they are created, exist for some period of time, and then come to an end. From their rise to their demise, regions are the particular outcomes of general mechanisms that incessantly reproduce the capitalist world-economy as core, periphery, and semi-periphery. These mechanisms operate through agents of change, individuals, and institutions that function across the world within regions and within zones. Acting through the operation of states, households, social classes, political movements, economic enterprises, and many other organizations, individuals have created, reproduced, and destroyed regions.

A concrete example will help to fix these ideas. The greater Caribbean region extending from northeast Brazil to Maryland was a peripheral region from the seventeenth to the nineteenth centuries. Wallerstein (1980, 157–175) described this region as a product of the second major phase of the capitalist world-economy, the stagnation period from 1600 to 1750. In the rest of the periphery this was a time of retrenchment, reorientating economies to local markets. In the Caribbean, however, there was expansion. A new peripheral region developed which concentrated on some of the growth commodities of the period, such as sugar and cotton. This process began with the conversion of the original imperial powers, Spain and Portugal, to semi-peripheral status. Although the Caribbean islands originally served as centers for contraband in this transfer of power to northwest Europe, the coming of legitimate trade generated the new region. Plantation America used indentured servants and slaves to serve the new tastes of the European core for sugar and tobacco. The relations between planter, banker, and merchant secured most of the benefits of this enterprise to the core in northwest Europe. The greater Caribbean region, more than any other section of the periphery, sustained the core in the crucial and difficult consolidation stage of the capitalist world-economy.

During the next growth phase of the system, however, the greater Caribbean fared poorly and much of the region became surplus to the core's requirements. The region declined under the destructive effects of the Napoleonic Wars and the development of an alternative crop, the sugar beet, in Europe. The region disintegrated in the nineteenth century. The northern mainland section converted from European periphery to United States periphery. Most of the region stagnated and even reverted to a backwater status like its position before its expansion. The greater Caribbean region was no more.

The greater Caribbean example illustrates three characteristics of regions in the world-system. Initially and obviously, it shows the rise and demise of a region: no region is eternal. It locates the region within the wider context of happenings in the system as a whole: no region is autonomous. It also shows a region comprising several political sovereignties: no region has to be a political unit, although regions can be political units. The discussion now returns to the question of the state and regional geography.

**Choosing political boundaries**

I have argued against automatically accepting the state as the basis for comparative analyses but the argument does not preclude choosing political boundaries for regions (e.g., Agnew 1987). Political boundaries may coincide with important economic boundaries, or else, in a particular context, political boundaries may be more important than the economic boundaries. The United States as a region illustrates both these cases.

The nineteenth-century territory of the United States could be viewed as covering all three zones of the world-economy: an increasingly core-like Northeast, a peripheral South, and an emerging semi-peripheral West. A view of the United States as one region enables the monitoring of relations between the zones in one polity to see how the "power-container" organizes, protects, and uses its particular combination of zones. This polity was successful: all the major sections of the United States became core-like by the second half of the twentieth century. While this outcome justifies the use of the United States as a region, the boundary with Canada does separate two core-like territories. The designation of two regions as states (United States and Canada) is a rejection of a single Anglo-American region. Despite the empirical evidence for the importance of this political boundary (e.g., Goldberg and Mercer 1986; Reitsma 1987) there is no simple right or wrong decision to be made here. Different emphases in studies explaining the rise of this first non-European sector of the core zone may suggest alternative boundary decisions. The same choice, however, seems far less problematic for the United States–Mexico boundary. This boundary so definitely coincides with a zonal boundary that nobody confuses Mexico City and Chicago as North American cities in the way Goldberg and Mercer (1986) criticize confusion over Toronto and Chicago. Anglo-America possibly is a region, but North America is not.

Political processes need not be neglected by a world-systems analysis, but they should be evaluated alongside other processes in determining regions. The only clue to the likely propensity of political boundaries in regions in world-systems analysis is Wallerstein's (1979, 117) suggestion that conflict is heightened in the semi-peripheral zone, where political processes are consequently more important.

**Breaking all the rules**

If most regions are not to be politically defined, how are geographers to identify them? This age-old geographical problem of region definition illustrates one final point concerning the study of places. How far should the regionalization-classification analogy be taken? Regional construction has always sat uneasily with the strict rules of classification. Although regional taxonomy was produced during the quantitative revolution, geographers lost more than they gained by trying to be more "scientific." Places are not like species: understanding places is a different type of knowledge from taxonomy.

In world-systems analysis each region is shaped by a singular combination of general mechanisms. Definition should proceed, therefore, by identifying the combination of mechanisms that characterizes a particular region, enumerating their expected outcomes, and then mapping their range. Such an exercise will inevitably produce an inner area containing most of the expected outcomes and an outer area where a smaller proportion appear. Other regional theories have noted this pattern. One should not be overly concerned to produce rigid boundaries, which are important primarily for precise quantitative and comparative analysis between regions and over time. But one must recognize generally that adjacent regions do not meet at a simple boundary but rather merge together across border zones. The fluidity of regions and outcomes demands flexibility. The outcomes will be stable enough to depict patterns, but the patterns are by no means fixed.

Translating flexibility into the greater Caribbean region example indicated the Caribbean sugar islands developed by the major core powers typified plantation America, whereas both mainland adjuncts constituted outer areas of the region. The northern outer area, for example, always had important differences with the islands and these increased with the new political processes set in train by US independence. The set of mechanisms and their outcomes that had generated a new distinctive periphery region became severely disrupted and the region finally disintegrated. The greater Caribbean example also illustrates the varying utility of political boundaries. The political processes on either side of the US southern maritime boundary became of vital importance, but the different and changing sovereignties of the sugar islands became much less relevant. The fact that Jamaica was British and Guadeloupe was French does not assign them to different regions, even though differences between English and French forms of imperialism cannot be ignored.

One of the implications of this treatment of the greater Caribbean region is that it can share places with other regions. The Florida to Maryland area obviously functioned within both the greater Caribbean and U.S. regions. The definition of a single set of regions covering the world is not proposed. This admission of overlap is anathema to regional taxonomy. But then the purpose of this new regional geography is not the production of some sort of ultimate geography as a complete description of the world, but rather an understanding of places within the wider compass of the world-economy. Overlapping regions contradict only the belief in a simple world of neat divisions. The world is a mosaic, to be sure, but it is not a stone one. It depicts the variability of institutions, individuals, and their mechanisms in particular places that reproduce the overall world-economy.

## Conclusion

The implications of this essay for the study of a contemporary region are some new questions to be asked and some old questions to be resurrected. For instance, whereabouts on the route from rise to demise can a region be located? What was the role of outside forces in the creation of the region? How does the region fit into the global division of labor? How will the demise of the region affect the status of its territories? Today the southern Africa region more than any other highlights the importance of such questions (Martin 1986). Regions may be neither eternal nor autonomous, but their analysis remains vital to understanding our world. Not only must we know the location of a region in the world-economy to fully understand it, but in order to properly understand the world-economy we must know the places that constitute its whole.

## References

Agnew, J. 1987: *The United States in the World-Economy: A Regional Geography*. Cambridge: Cambridge University Press.

Cole, J. P. 1981: *The Development Gap*. New York: Wiley.

Editorial Board. 1982: Political geography: Research agendas for the nineteen eighties. *Political Geography Quarterly* 1: 1–18.

Goldberg, M. and Mercer, J. 1986: *The Myth of the North American City*. Vancouver: University of British Columbia Press.

Gregory, D. 1978: *Ideology, Science and Human Geography*. London: Hutchinson.

Hall, T. D. 1986: Incorporation in the world-system: Toward a critique. *American Sociological Review* 51: 390–402.

Hartshorne, R. 1939: *The Nature of Geography*. Washington, D.C.: Association of American Geographers.

Hopkins, T. K. and Wallerstein, I. 1977: Patterns of development of the modern world-system. *Review* 1: 111–45.

Irvine, J., Miles, I. and Evans, J. 1979: *Demystifying Social Statistics*. London: Pluto.

Johnston, R. J. 1984: The world is our oyster. *Transactions of the Institute of British Geographers*, new series 9: 443–59.

Martin, J. G. 1986: Southern Africa and the world-economy. *Review* 10: 99–120.

Reitsma, H. A. 1987: Agricultural changes in the American-Canadian border zone, 1954–78. *Political Geography Quarterly* 7: 1–16.

Soja, E. 1980: The socio-spatial dialectic. *Annals of the Association of American Geographers* 70: 207–25.
Taylor, P. J. 1986a: World-systems analysis. In *The Dictionary of Human Geography*, 2nd edition, ed. R. J. Johnston, 527–29. Oxford: Blackwell.
Taylor, P. J. 1986b: The world-systems project. In *A World in Crisis?*, ed. R. J. Johnston and P. J. Taylor, 269–88. Oxford: Blackwell.
Wallerstein, I. 1979: *The Capitalist World-Economy*. Cambridge: Cambridge University Press.
Wallerstein, I. 1980: *The Modern World System II*. New York: Academic Press.
Wallerstein, I. 1983: *Historical Capitalism*. London: Verso.
Wallerstein, I. 1984a: *The Politics of the World-Economy*. Cambridge: Cambridge University Press.
Wallerstein, I. 1984b: Longwaves as capitalist process. *Review* 12: 559–75.
Wolf, E. R. 1982: *Europe and the People without History*. Berkeley: University of California Press.

# 3 Derek Gregory, 'Areal Differentiation and Post-modern Human Geography'

Excerpt from: Derek Gregory and Rex Walford (eds), *Horizons in Human Geography*. London: Macmillan (1989)

> Searching for an epigraph to his *Philosophical Investigations*, Ludwig Wittgenstein considered using a quotation from *King Lear*: 'I'll teach you differences.' 'Hegel,' he once told a friend, 'always seems to me to be wanting to say that things which look different are really the same. Whereas my interest is in showing that things which look the same are really different.'
>
> Terry Eagleton, *Against the Grain*

## Post-modernism

If my title seems strange, so much the better. In this essay I want to explore some fragments of the contemporary intellectual landscape and to suggest some of the ways in which they bear upon modern human geography: and all of this will, I suspect, be unsettling. (Or, at any rate, if I can convey what is happening successfully then it ought to be unsettling.)

I use 'post-modernism' as a short-hand for a heterogeneous movement which had its origins in architecture and literary theory. The relevance of the first of these for human geography must seem comparatively straightforward – especially if the interpretative arch is widened to span the production of the built environment or, wider still, the production of space[1] – but the second is, as I will seek to show, every bit as important for the future of geographical inquiry. The converse may also be true: Frederic Jameson, one of the most exhilarating literary critics writing today, claims that 'a model of political culture appropriate to our own situation will necessarily have to raise spatial issues as its fundamental organizing concern'.[2]

Post-modernism is, of course, much more than these two moments. It has spiralled way beyond architecture and literary theory until it now confronts the terrain of the humanities and social sciences *tout court*. But whatever its location, I shall argue that post-modernism raises urgent questions about place, space and landscape in the production of social life.

Post-modernism is, in its fundamentals, a critique of what is usually called 'the Enlightenment project'. The European Enlightenment of the eighteenth century provided one of the essential frameworks for the development of the modern humanities and social sciences. It was, above all, a celebration of the power of reason and the progress of rationality, of the ways in which their twin engines propelled modernity into the cobwebbed corners of the traditional world. Both 'reason' and 'rationality' were given highly specific meanings, however, and a number of thinkers have been disturbed by the triumph of the particular vision of knowledge which those terms entailed. The more radical of them have sought to overthrow its closures and its supposed certainties altogether. Their critique has, for the most part, been conducted at high levels of abstraction – the exchanges between Habermas and Lyotard are of just such an order[3] – but one of the most concrete illustrations of what is at stake has been provided by David Ley in a remarkable essay on the politico-cultural landscapes of inner Vancouver.

Ley contrasts two redevelopment projects on either side of False Creek. To the north, 'an instrumental landscape of neo-conservatism': high-density, high-rise buildings whose minimalist geometric forms provide the backdrop for the spectacular structures of a sports stadium, conference centre, elevated freeway and rapid transit system and the towering pavilions of Expo '86. To the south, an 'expressive landscape of liberal reform': low-density groupings of buildings, diverse in design and construction, incorporating local motifs and local associations and allowing for a plurality of tenures, clustered around a lake which opens up vistas across the waterfront to the downtown skyline and the mountain rim beyond. The north shore is a monument to modern technology, to the internationalization of 'rational' planning and corporate engineering: one of Relph's 'placeless' landscapes. The south shore, by contrast, is redolent of what Frampton calls a critical regionalism, a post-modern landscape attentive to the needs of people rather than the demands of machines and (above all) sensitive to the specificities of particular places.[4]

This contrast is, of course, emblematic of others, not least between different styles of human geography. But, as I must now show (and as the term itself suggests), post-modernism is no traditionalist's dream of recovering a world we have lost. It is a movement *beyond* the modern and, simultaneously, an invitation to construct our *own* human geographies. I will build my argument on three of its basic features.

Firstly, post-modernism is, in a very real sense, 'post-paradigm': that is to say, post-modern writers are immensely suspicious of any attempt to construct a system of thought which claims to be complete and comprehensive. In geography, of course, there have been no end of attempts of this kind, and many of those who have – in my view, mistakenly – made use of Kuhn's notion of a paradigm have done so *prescriptively*. They have claimed the authority of 'positivism', 'structuralism', 'humanism' or whatever as a means of legislating for the proper conduct of geographical inquiry and of excluding work which lies beyond the competence of these various systems. Others have preferred to transcend these, to them partial, perspectives and to offer some more general ('meta-theoretical') framework in which all these competing claims are supposed to be reconciled.

For over a decade this was usually assumed to be some kind of systems approach and now, apparently, it is the philosophy of realism (perhaps coupled with some version of Habermas's critical theory) which holds out a similar promise. But post-modernism rejects all of these manoeuvres. All of these systems of thought are – of necessity – incomplete, and if there is then no alternative but to pluck different elements from different systems for different purposes this is not a licence for an uncritical eclecticism: patching them together must, rather, display a sensitivity towards the differences and disjunctures between them.[5] And 'sensitivity' implies that those different integrities must be respected and retained: not fudged. The certainties which were once offered by epistemology – by theories of knowledge which assumed that it was possible to 'put a floor under' or 'ground' intellectual inquiry in some safe and secure way – are no longer credible in a post-modern world.[6]

Secondly, this implies, in turn, that post-modern writers are hostile to the 'totalizing' ambitions of the conventional social sciences (and, for that matter, those of the humanities). Their critique points in two directions. First, they reject the notion that social life displays what could be called a 'global coherence': that our day-to-day social practices are moments in the reproduction of a self-maintaining social system whose fundamental, so to speak 'structural' imperatives necessarily regulate our everyday lives in some automatic, pre-set fashion. These writers do not, of course, deny the importance of the interdependencies which have become such a commonplace of the late twentieth-century world, and neither do they minimize the routine character of social reproduction nor the various powers which enclose our day-to-day routines. (These are, on the contrary, some of the most salient foci of their work.) But they do object to the concept of totality which informs much of modern social theory, because it tacitly assumes that social life somehow adds up to (or 'makes sense in terms of') a coherent system with its own superordinate logic.

Second, and closely connected to this, these writers reject the notion that social life can be explained in terms of some 'deeper' structure. This was one of the premisses of structuralism, of course, and it still surfaces in some of the cruder versions of realism. It is largely through this opposition that post-modernism is sometimes identified with 'post-structuralism' and, put like this, I imagine that the post-modern critique will seem to echo the complaints of those who saw in structuralism a displacement of the human subject. In human geography as elsewhere, many commentators were dismayed by the way in which various versions of structuralism replaced the concrete complexities of human agency by the disembodied transformations of abstract structures.[7] But post-modernism is not another humanism. It objects to structuralism because its sharpened concept of structure points towards a 'centre' around which social life revolves, rather like a kaleidoscope or a child's mobile; but it objects to humanism for the very same reason. Most forms of humanism appeal to the human subject or to human agency as the self-evident centre of social life. And yet we are now beginning to discover just how problematic those terms are. Concepts of 'the person', for example, differ widely over space and through time and so, paradoxically, it is their very importance which ensures that they cannot provide a constant foundation for the human sciences. They are the *explanandum* not the *explanans*.

Thirdly, the accent on 'difference' which dominates the preceding paragraphs is a *leitmotif* of post-modernism. One of the distinguishing features of post-modern culture is its sensitivity to heterogeneity, particularity and uniqueness. To some readers this insistence on 'difference' will raise the spectre of the idiographic, which is supposed to have been laid once and for all (in geography at any rate) by the Hartshorne–Schaefer debate in the 1950s and by the consolidation of a generalizing spatial science during the 1960s. To be sure, the caricature of Hartshorne as a crusty empiricist, indifferent to the search for spatial order, blind to location theory and ignorant of quantitative methods could never survive any serious reading of *The Nature of Geography*. There were, as several commentators have emphasized, deep-seated continuities between the Hartshornian orthodoxy and the prospectus of the so-called 'New Geography'. But Schaefer's clarion call for geography as a nomothetic science, compelled to produce morphological laws and to disclose the fundamental geometries of spatial patterns, undoubtedly sounded a retreat from areal differentiation which was heard (and welcomed) in many quarters. Specificity became eccentricity, and the new conceptual apparatus made no secret of its confinement: it was, variously, a 'residual'; background 'noise' to be 'filtered out'; a 'deviation' from the 'normal'. And yet in the 1980s other writers in other fields have given specificity a wider resonance. In philosophy, Lyotard claims that 'post-modern knowledge . . . refines our sensitivity to differences'; in social theory De Certeau wants to fashion 'a science of singularity . . . that links everyday pursuits to particular circumstances'; and in anthropology Geertz parades 'the diversity of things' and seeks illumination from 'the light of local knowledge'.[8]

In geography too there has been a remarkable return to areal differentiation. But it is a return with a difference. When Harvey speaks of the 'uneven development' of capitalism, for example, or when (in a radically different vocabulary) Hägerstrand talks about 'pockets of local order', they – and now countless others like them – are attempting much more than the recovery of geography's traditional project. For they herald not so much the reconstruction of modern geography as its *deconstruction*. I mean this to be understood in an entirely positive and specifically technical sense: not as a new nihilism, still less as the enthronement of some new orthodoxy, but as the transformation of the modern intellectual landscape as a whole. I should admit at once that it is still barely possible to map that new landscape – not least because it is radically unstable – but in what follows I will try to put some preliminary markers around what Soja calls the 'post-modernization' of geography.[9]

Two disclaimers are immediately necessary. First, to work within disciplinary boundaries is obviously open to objection – and I am as uneasy about doing so as anyone else – but I have retained the conventional enclosures because I want to show that 'geography' has as much to contribute to post-modernism as it has to learn from it. In so far as social life cannot be theorized on the point of a pin, then, so it seems to me, the introduction of concepts of place, space and landscape must radically transform the nature of modern social theory. Second, to say that geography has re-opened the question of areal differentiation is to invite the response that, for many, it was never closed. I accept that it would be wrong to minimize the continuing power of traditional regional geographies which, at their very best, have always provided remarkably sensitive evocations of the particular relations between people and the places in which they live. And I insist on this not as a politeness to be pushed to one side as soon as possible. On the contrary, the 'problem of geographical description' with which so many of these writers struggle is part of a more general 'crisis of representation' throughout the contemporary human sciences. This realization, pregnant with consequences at once theoretical and practical, has also played its part in changing the modern intellectual landscape: so much so that we need new, theoretically-informed ways of conveying the complexities of areal differentiation if we are to make sense of the post-modern world.

The tragedy would be to treat the developments I have described here as symptomatic of yet another 'revolution', one more sea-change to roll with or roll back. These are, instead, ideas to think about and to work with – critically, vigilantly, constructively. One way of measuring the distance between them and modern human geography is, perhaps, to reverse one of the catch-phrases of spatial science: that there is more *disorder* in the world than appears at first sight is not discovered until that disorder is looked for.[10] That is more than mere word-play; it may be that one of the most ideological impulses of all – the 'commonsense' response to the complexity of the world – is to impose a coherence and a simplicity which is, at bottom, illusory.

Even so, I would not wish this essay to be taken as an unqualified manifesto for post-modernism. I am well aware that post-modernism can be read in a number of different ways, some of them acutely conservative as well as insistently radical; that there are all sorts of difficulties in its formulations which I have had no space to consider here; and that some of its own criticisms (of Habermas, for example) are wide of the mark. That said, the various themes which I have pulled together raise questions which, in my view, cannot be ignored. For, like Eagleton, I suspect that we are presently strung out between notions of social totality which are plainly discreditable and a 'politics of the fragment or conjuncture' which is largely ineffectual.[11]

And to go *beyond* these limitations, I suspect, we need, in part, to go *back* to the question of areal differentiation: but armed with a new theoretical sensitivity towards the world in which we live and to the ways in which we represent it. Whether we focus on 'order' or 'disorder' or on the tension between the two – and no matter how we choose to define those terms – we still have to 'look'. We are still making geography.

**Notes**

1 See Michael Dear, 'Postmodernism and planning', *Environment and Planning D: Society and Space*, 4 (1986) pp. 367–384.
2 Frederic Jameson, 'Postmodernism, or the cultural logic of late capitalism', *New Left Review*, 146 (1984) pp. 53–92.
3 See Richard Rorty, 'Habermas and Lyotard on Post-modernity', in Richard J. Bernstein (ed.) *Habermas and Modernity* (Cambridge: Polity Press, 1985) pp. 161–175; Peter Dews, 'From post-structuralism to post-modernity: Habermas's counter-perspective', *ICA Documents* 4 (1986) pp. 12–16.
4 David Ley, 'Styles of the times: liberal and neo-conservative landscapes in inner Vancouver, 1968–1986', *J. Hist. Geogr.*, 13 (1987) pp. 40–56; see also Edward Relph, *Place and Placelessness* (London: Pion, 1976) and Kenneth Frampton, 'Towards a critical regionalism: six points for an architecture of resistance', in Hal Foster (ed.) *The Anti-aesthetic: Essays on Postmodern Culture* (Port Townsend: Bay Press, 1983) pp. 16–30.
5 It is to the disclosure of these differences and disjunctures that 'deconstruction' is directed. I have found the following introductions particularly helpful: Terry Eagleton, *Literary Theory: An Introduction* (Oxford: Basil Blackwell, 1983) pp. 127–50; Christopher Norris, *Deconstruction: Theory and Practice* (London: Methuen, 1982); Michael Ryan, *Marxism and Deconstruction: A Critical Articulation* (Baltimore: Johns Hopkins University Press, 1982) Chapter 1.
6 This thesis is argued with a special clarity in Richard Rorty, *Philosophy and the Mirror of Nature* (Oxford: Basil Blackwell, 1980); see also Richard J. Bernstein, *Beyond Objectivism and Relativism: Science, Hermeneutics and Praxis* (Oxford: Basil Blackwell, 1983).
7 James Duncan and David Ley, 'Structural Marxism and human geography: a critical assessment', *Ann. Ass. Am. Geogr.*, 72 (1982) pp. 30–59.
8 Jean-François Lyotard, *The Postmodern Condition: A Report on Knowledge* (Manchester: Manchester University Press, 1984); Michel de Certeau, *The Practice of Everyday Life* (Berkeley: University of California Press, 1984); Clifford Geertz, *Local Knowledge: Further Essays in Interpretative Anthropology* (New York: Basic Books, 1983). I cite these three texts as examples, not because I accept their particular theses.
9 Edward Soja, 'What's new? A review essay on the postmodernization of geography', *Ann. Ass. Am. Geogr.*, 77 (1987) pp. 289–293. Cf. Michael E. Eliot Hurst, 'Geography has neither existence nor future', in R. J. Johnston (ed.) *The Future of Geography* (London and New York: Methuen, 1985) pp. 59–91.
10 The original phrase was Sigwart's, cited in P. Haggett and R. J. Chorley, 'Models, paradigms and the New Geography', in R. J. Chorley and P. Haggett (eds) *Models in Geography* (London: Methuen, 1967) p. 20.
11 Terry Eagleton, *Against the Grain: Selected Essays* (London: Verso, 1986) p. 5.

# SECTION TWO

# *THE SPATIALITY OF STATES*

**Editor's introduction**

Along with geopolitics this is the most established area of study in political geography. The focus is on the geography of state formation and the role of territoriality in state creation and administration. The spatiality of states refers, therefore, to both the external bounding and the internal spatial organization of states. The study of boundary claims, frontier disputes and territorial organization has long been one of the most active research areas in political geography (see, e.g., Kasperson and Minghi 1969; Prescott 1987; Sack 1986). There is also a considerable body of work on local governments as sub-state actors and the specific geographical features of federal systems (see, e.g., for a good overview, Paddison 1983). Although this kind of work continues to flourish, particularly with respect to local governments as agents of economic development and political activism (e.g. Hoggart 1991; Kirby 1993) or as hierarchical systems of government control (e.g. Whitney 1969; Vandergeest and Peluso 1995) and federalism as a means of managing inter-group conflicts (e.g. Smith 1995), increased attention is now being given to the general nature of the spatiality inherent in the modern state system of competitive sovereign territories claiming total jurisdiction over their inhabitants (e.g. Agnew 1994; Campbell 1992; Walker 1993; Inayatullah and Blaney 1995; Murphy 1996; Shapiro and Alker 1995).

This emphasis is in part probably a reflection of changing times. Rather than taking the modern territorial state for granted as a transcendental feature of political organization, the onset of globalization in finance and production, the explosion of migrant and refugee populations, the fiscal crisis of the Western welfare state, the collapse of 'strong' states in Eastern Europe and the former Soviet Union, the rise of supra-regional (as in the European Community) and global (as in the IMF and UN agencies) forms of governance, the 'disappearance' of central-state authority in many states under the onslaught of ethnic and other conflicts, all call into question the image of a 'fixed' territoriality and encourage the quest for understanding the historical roots of the territorial state as a form of governance and the possible limitation or displacement of state sovereignty as the governing principle of international relations (see, e.g., Cerny 1995; Hirst and Thompson 1996; Kofman and Youngs 1996). The increasing popularity of less state-centred understandings of power has also contributed to a questioning of established conventions about states and their 'hold' over people and territories (see, e.g., Bensel 1984; Agnew 1987; Walker 1993). In particular, the growth of regional tiers of government within states as important regulators of economic activities suggests that the concept of 'territory' itself should no longer be restricted to the geographical scale of the state alone (Andreucci and Pescarolo 1989; Trigilia 1991; Putnam 1993). Finally, there is a growing sense that statehood in the past was more complicated than dominant accounts have made it. Thomson (1994), for example, has pointed out that state monopoly over the means of violence has always been more partial and limited than writing about 'the state' has made it. Deudney (1995) suggests that there has been a variety of state-forms, even within the category of territorial states, that cannot be reduced to the 'real-state' of much political theory.

The idea that there is a necessary connection between political community and territory is an old one in Western thought (Gottmann 1973). Aristotle, for example, was of the view that governance could only be practised satisfactorily within a discrete and well-defined expanse of territory. To later thinkers the Roman experience suggested that this was all too true. Expanding into an empire, the Romans sacrificed political community for territory. Only with the rise of the modern territorial state in Europe between the sixteenth and nineteenth centuries was there finally a close affiliation between co-residence in a territory and political community. The transfer of sovereignty from the person of the monarch or prince to the people within a given territory (symbolized above all by the French Revolution of 1789 but also very much at the heart of the American Revolution shortly before) finally joined citizenship to territory (Walker 1993). It was in this context that ideas of 'good government' and 'democracy' became tied to the establishment and maintenance of the territorial state. To this day discussions of democracy largely presume a territorial state as the main 'target' of political opportunity (Held 1995; Pateman 1996).

Until recently much of the debate in political theory and political geography about the 'nature' of the state has emphasized sovereignty as a realized ideal and turned to questions about the character of the state apparatus or political institutions associated with different 'kinds' of states (usually using polarities such as strong–weak, capitalist–socialist, authoritarian–democratic, etc.) (see, e.g., Carnoy 1984; Kazancigil 1986; Clark and Dear 1984; Johnston 1982). There is much to be said on behalf of this approach, if only because questions about citizenship rights, access to institutions, local government and the role of states in legitimizing social divisions are given critical attention. This *internalist* orientation, however, can lead to the neglect of the geographical underpinnings of the sovereign state itself and the critical importance of the internal/external distinction that is a key attribute of modern statehood. The clear bounding of territory by state is one of the major differences between modern European-origin political organization and the types of political organization that prevailed in nomadic, clan, imperial, absolutist and feudal societies around the world in the past.

The achievement of the territorial form of statehood in Europe out of older types of rule was the main research interest for a number of years of the late Norwegian political sociologist **Stein Rokkan**. One of the founders of a revived political geography, Rokkan tried to show through an analysis of European history from the collapse of the Roman Empire down to the twentieth century the geographical process by which the modern European states came into existence. The extract reprinted here (**Reading 4**) provides a summary of a lengthier presentation available in Rokkan and Urwin (1983). Rokkan's purpose is to show the various spatial elements involved in state-creation, particularly the emerging core areas of states and the consequences of urbanization and relative location within Europe (central versus fringe) for later state development. The experience of particular states, therefore, is placed within a Europe-wide frame of reference.

More recent research by Tilly (1990) and others (Tilly and Blockmans 1994) examines hypotheses about the roles of military capacities and the rise of merchant and industrial capitalism to suggest that Rokkan's spatial-analytic approach misses out important causes of the rise of the European states. Their political-economic approach draws attention to the relative autonomy of state-building elites in processes of state formation. Others such as Wallerstein (1974) and Anderson (1974) have tended to give much more importance to economic factors, particularly the growth of trade and rise to power of new classes. The emphasis on state autonomy is best represented by the massive research project of **Michael Mann**, like Rokkan a political sociologist by discipline. In the article included here (**Reading 5**), Mann summarizes the argument he presents at greater length in a number of publications (Mann 1986, 1993; also see Mann 1988, 1990). After a very helpful definition of the state, his main theme is the necessity of territoriality for states in both agrarian and industrial societies because the state penetrates into the life of social groups as these groups give up powers to the state in return for various favours. In turn it is the very territoriality of the state that guarantees the state a degree of autonomy in relation to society. Its powers cannot be reduced to those of any one group. Mann identifies the *infrastructural* power of the state (its capacity to carry out functions impossible to any one group) as the major secret of its success. He distinguishes this from the *despotic* power exerted by state elites which cannot give much autonomous power to the state because it is usually precarious and cannot insinuate itself like infrastructural power into the routines of social life.

Of course, the success of the state as an autonomous actor depends on the degree to which its powers are vested with legitimacy by its population. From a postmodern point of view, an argument like Mann's serves (even if inadvertently) such a purpose. It naturalizes and normalizes 'the state' as a legitimate actor in peoples' lives (Shapiro 1994). A postcolonial perspective would add to this the violence that statehood brings in its train. The experience of 'independence' for the former colonies of European empires serves to bring attention to the arbitrary nature of the bounding process involved in statehood and the difficulties boundary drawing entails for those living at the borders. People must choose to be on one side or the other; there is no recognized borderland identity, only competing state ones. **Sankaran Krishna** (**Reading 6**) uses the metaphor of 'cartographic anxiety' to convey how discourses about an Indian 'nation' are used to define the borders of the Indian state. The 'body politic' of India is defined in terms of a physical map that tries to conjure up a 'historical original', a 'homeland' that never existed prior to British colonialism. So, not only those at the borders are caught up in an exercise in spatial self-definition that is the essence of statehood: abstracting from history a set of stable, legitimate boundaries that fix the history of the state in place and guarantee it a place in future (also see Winichakul 1994 on Thailand).

This brings us back to the dilemma created by the fact that Western thinking about governance in general and democracy in particular has usually centred around the state (Wolin 1960). But state spatiality is based fundamentally on exclusion from concerns about the 'outside' and with penetration of the state *into* society. As Machiavelli taught in *The Prince*, politics are possible only within state boundaries, *raison d'état* operates beyond them (Viroli 1992). The spatial attributes of modern statehood, therefore, have more than passing interest (Connolly 1993). They are at the heart of debates over the possibility of democratic governance in a world in which many decisions affecting us all on a day-to-day basis increasingly emanate from distant seats of power outside the reach of state authorities.

## References

Agnew, J. A. 1987: *Place and politics: the geographical mediation of state and society*. London: Allen and Unwin.

Agnew, J. A. 1994: The territorial trap: the geographical assumptions of international relations theory. *Review of International Political Economy*, 1: 53–80.

Anderson, P. 1974: *Lineages of the absolutist state*. London: New Left Books.

Andreucci, F. and Pescarolo, A. (eds) 1989: *Gli spazi del potere. Aree, regioni, stati: le coordinate territoriali della storia contemporanea*. Florence, Italy: Usher.

Bensel, R. F. 1984: *Sectionalism and American political development, 1880–1980*. Madison WI: University of Wisconsin Press.

Campbell, D. 1992: *Writing security: United States foreign policy and the politics of identity*. Minneapolis MN: University of Minnesota Press.

Carnoy, M. 1984: *The state and political theory*. Princeton NJ: Princeton University Press.

Cerny, P. 1995: Globalization and the changing logic of collective action. *International Organization*, 49: 595–625.

Clark, G. L. and Dear, M. 1984: *State apparatus: structures and language of legitimacy*. Boston MA: Allen and Unwin.

Connolly, W. E. 1993: Democracy and territoriality. In F. M. Dolan and T. L. Dumm (eds) *Rhetorical republic: governing representations in American politics*. Amherst MA: University of Massachusetts Press.

Deudney, D. H. 1995: The Philadelphian system: sovereignty, arms control, and balance of power in the American states-union, circa 1787–1861. *International Organization*, 49: 191–228.

Gottmann, J. 1973: *The significance of territory*. Charlottesville VA: University Press of Virginia.

Held, D. 1995: *Democracy and the global order: from the modern state to cosmopolitan governance*. Cambridge: Polity Press.

Hirst, P. and Thompson, G. 1996: *Globalization in question: the international economy and the possibilities of governance*. Cambridge: Polity Press.

Hoggart, K. 1991: *People, power and place*. London: Routledge.

Inayatullah, N. and Blaney, D. L. 1995: Realizing sovereignty. *Review of International Studies*, 21: 3–20.

Johnston, R. J. 1982: *Geography and the state: an essay in political geography*. New York: St. Martin's Press.

Kasperson, R. E. and Minghi, J. V. 1969: *The structure of political geography*. Chicago: Aldine.

Kazancigil, A. (ed.) 1986: *The state in global perspective*. Aldershot: Gower.

Kirby, A. 1993: *Power/resistance: local politics and the chaotic state*. Bloomington IN: Indiana University Press.

Kofman, E. and Youngs, G. (eds) 1996: *Globalization: theory and practice*. London: Pinter.

Mann, M. 1986: *The sources of social power. Vol. I: A history of power from the beginning to A.D. 1760*. Cambridge: Cambridge University Press.

Mann, M. 1988: *States, war and capitalism*. Oxford: Blackwell.

Mann, M. (ed.) 1990: *The rise and decline of the nation-state*. Oxford: Blackwell.

Mann, M. 1993: *The sources of social power. Vol. II: The rise of classes and nation-states, 1760–1914*. Cambridge: Cambridge University Press.

Murphy, A. B. 1996: The sovereign state system as political-territorial ideal: historical and contemporary considerations. In T. Biersteker and C. Weber (eds) *State sovereignty as social construct*. Cambridge: Cambridge University Press.

Paddison, R. 1983: *The fragmented state: the political geography of power*. Oxford: Blackwell.

Pateman, C. 1996: Democracy and democratization. *Political Science Review*, 17: 5–12.

Putnam, R. D. 1993: *Making democracy work: civic traditions in modern Italy*. Princeton NJ: Princeton University Press.

Prescott, J. R. V. 1987: *Political frontiers and boundaries*. London: Allen & Unwin.

Rokkan, S. and Urwin, D. 1983: *Economy, territory, identity: politics of West European peripheries*. London: Sage.

Sack, R. D. 1986: *Human territoriality: its theory and history*. Cambridge:

Cambridge University Press.
Shapiro, M. J. 1994: Moral geographies and the ethics of post-sovereignty. *Public Culture*, 6: 479–502.
Shapiro, M. J. and Alker, H. (eds) 1995: *Challenging boundaries: global flows, territorial identities*. Minneapolis MN: University of Minnesota Press.
Smith, G. (ed) 1995: *Federalism: the multiethnic challenge*. London: Longman.
Thomson, J. 1994: *Mercenaries, pirates and sovereigns*. Princeton NJ: Princeton University Press.
Tilly, C. 1990: *Coercion, capital and European states: AD 990–1992*. Oxford: Blackwell.
Tilly, C. and Blockmans, W. P. (eds) 1994: *Cities and the rise of states in Europe, A.D. 1000 to 1800*. Boulder CO: Westview Press.
Trigilia, C. 1991: The paradox of the region: economic regulation and the representation of interests. *Economy and Society*, 20: 306–327.
Vandergeest, P. and Peluso, N. 1995: Territorialization and state power in Thailand. *Theory and Society*, 24: 385–426.
Viroli, M. 1992: *From politics to reason of state: the acquisition and transformation of the language of politics, 1250–1600*. Cambridge: Cambridge University Press.
Wallerstein, I. 1974: *The modern world-system: capitalist agriculture and the origins of the European world-economy in the sixteenth century*. New York: Academic Press.
Walker, R. B. J. 1993: *Inside/outside: international relations as political theory*. Cambridge: Cambridge University Press.
Whitney, J. B. R. 1969: *China: area, administration and nation building*. Chicago: University of Chicago, Department of Geography, Research Paper, Number 123.
Winichakul, T. 1994: *Siam mapped: a history of the geo-body of a nation*. Honolulu: University of Hawaii Press.
Wolin, S. 1960: *Politics and vision: continuity and innovation in Western political thought*. Boston MA: Little, Brown.

# 4 Stein Rokkan, 'Territories, Centres, and Peripheries: Toward a Geoethnic–Geoeconomic–Geopolitical Model of Differentiation Within Western Europe'

Excerpt from: Jean Gottmann (ed.), *Centre and Periphery: Spatial Variation in Politics*, Chapter 9. London: Sage (1980)

Originally trained in political philosophy and political sociology, I have been increasingly influenced over the last few years by the theoretical revolution in geography. In my early work on the development of mass politics in Norway, I concentrated much of my analytical effort on the deciphering of the marked contrasts between central and peripheral communities and tried to reach some understanding of the modes of interaction between what I called the territorial and the functional dimensions: I studied these processes of interaction as they manifested themselves both in the differential successes of the early waves of political mobilization and in the structuring of regional party systems (Rokkan, 1962). In my later work on the comparative history of mass politics in Western Europe, I combined these two dimensions in an abstract model and tried to show how this model might help to account for variations in the timing of extensions of political rights and in the structure of electoral and organizational alternatives. In the first version of this model I dealt with each territorial polity in isolation from its immediate context: the thrust of the analysis was essentially *typo*logical (Rokkan *et al.*, 1970: Chapter 3). In a later set of articles (Rokkan, 1975, 1973), I tried to recast the model within a *topo*logical framework: I made an effort to locate each case within a geoeconomic–geocultural–geopolitical space and constructed what I called a *conceptual map* of Western Europe. I worked out what amounted to a baseline model of the system of territorial differentiations characteristic of Western Europe and used this model as an engine for the generation of hypotheses about the sources of differences in internal political development case by case. My primary dependent variables were (1) the strength of representative institutions during the phase of absolutist rule from the sixteenth to the nineteenth century, (2) the sequencing of steps toward full suffrage democracy, (3) the structuring of alternatives for the mass citizenry as expressed in the system of parties, and (4) the vulnerability of each system of mass politics to disruption by monolithic movements during the crisis years from 1918 to 1940. I have not yet been able to present a systematic review of all the hypotheses generated within this framework, and I have done even less to assess in any orderly fashion the wealth of evidence for or against the hypotheses.[1] I am still far from satisfied with the baseline model and want to develop it in further depth before I proceed to a detailed review of its value as an engine for the generation of hypotheses about sources of differences in internal political development.

In this chapter I shall set out, as concretely as I can within the space allotted me, the historical and geographical underpinnings of the basic modelling effort. I shall go beyond the original formulation on one point: I shall discuss the three core components of the baseline model within their *geoethnic context.* The baseline model ignored the concrete ethnic composition of each cell in the geoeconomic–geocultural–geopolitical map of Europe. In this chapter I shall present a first effort to link up a set of geoethnic variables with the geoeconomic–geocultural–geopolitical core components and suggest the contours of an overarching model combining all these elements. This is perhaps a foolhardy enterprise: I am not at all sure that I am on the right track. I publish this early version in the hope that it will inspire others to search for parsimonious ways of systematizing this extraordinary wealth of information. I also entertain the perhaps vain hope that the reactions of my colleagues within political science and geography will force me to recast my model still further and, if possible, make it a more useful tool in the design of comparative analyses across Europe.[2]

## The core elements of the model

The model is developmental but not unilineal: it brings out the crucial significance of discontinuities, retrenchments, and recombinations of elements. The process of development is analyzed from the vantage point of an isolated primordial community: a closely knit, kinship-regulated local unit covering only a small territory and commanding only elementary technologies of communication. The model posits three part-processes of peripheralization under increasingly powerful systems of long-distance communication and control: one *military–administrative*, one *economic*, and one *cultural*. For each of these processes of territorial aggregation, the model posits a distinctive set of centralizing agencies. These need not control separate physical locations, but may in some cases be found together in close fusion in one dominant centre. Figure 4.1 spells out this baseline model in further detail.

But the model does not only serve as a tool for the comparative analysis of large-scale efforts of territorial aggregation: it has proved much more directly useful in the study of processes of *fragmentation, retrenchment, and reorganization* of territorial structures. Figure 4.2 shows how the model can be used to study the combinatorics of processes of breakdown in one concrete case: the disintegration of the western Roman Empire. The midpoints on each of the three core vectors suggest three distinctive modes of disintegration: feudalization, vernacularization, and centre formation on the periphery of the fallen empire. What proved crucial in the Western European case was that these three processes *got out of phase with each other* and that these differences in the timing and impact of the processes produced very different configurations from south to north and from west to east. These contrasts are spelled out in the discussion of the "conceptual map" in sections that follow. The gist of this typological–topological scheme can be stated in two sentences. The emergence of the city belt from south to north in Europe *stopped* the process of feudal fragmentation and produced a new and powerful thrust of

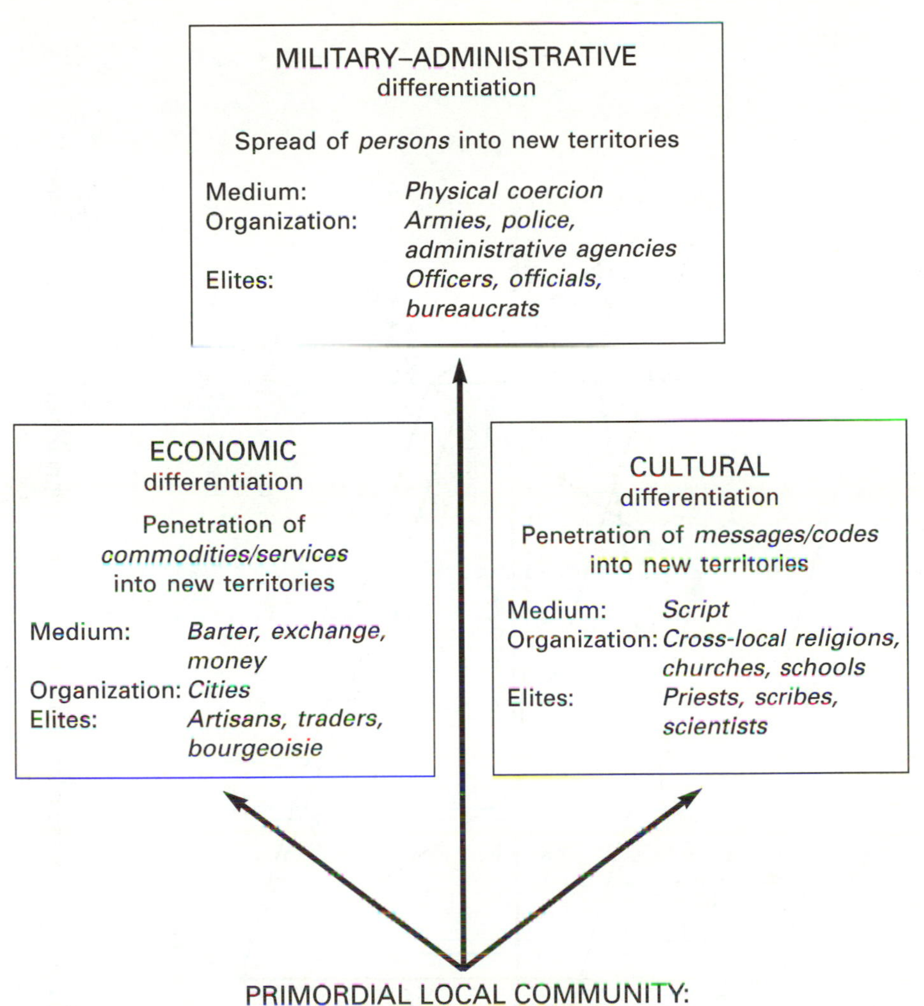


**Fig. 4.1** Three directions of differentiation in large-scale territorial systems

long-distance communication and boundary transcendence, while the strengthening of vernacular cultures and the development of major territorial centres at the edges of the empire *accelerated* the break-up of the old system, consolidated new sets of boundaries, and set the stage for the development of a range of highly distinctive political systems within Western Europe. The breakthrough toward merchant capitalism produced a world network of economic transactions and undermined established boundaries, while the emergence of strong nation-states tended to mark off clear-cut boundaries and accentuate territorial identity and citizenship. This is the great paradox of Western European development: the model was designed as a tool for systematic research on the sources and consequences of this paradox.

In this original formulation the model was clearly too abstract: it not only

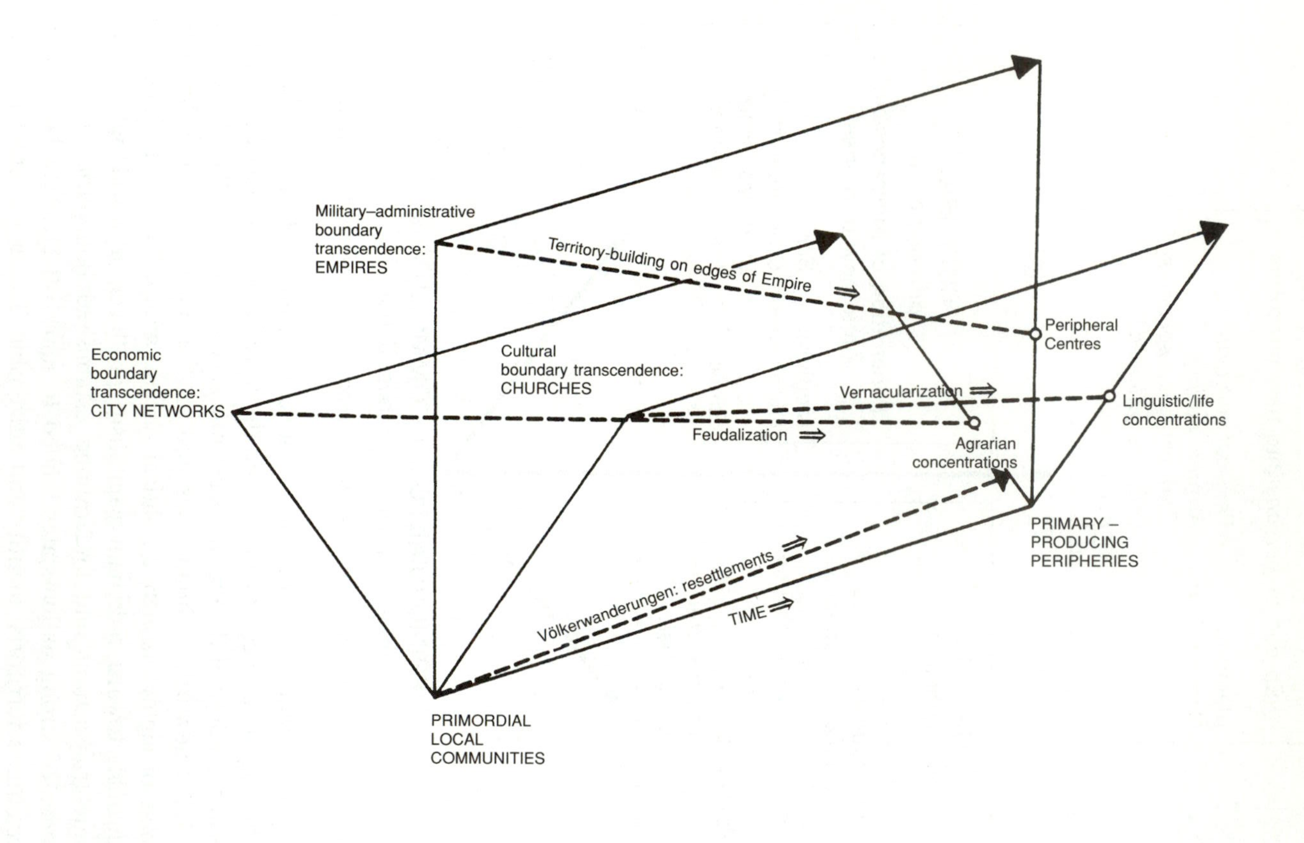


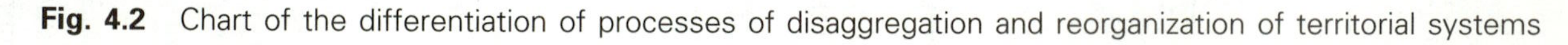

**Fig. 4.2** Chart of the differentiation of processes of disaggregation and reorganization of territorial systems

ignored details of social, economic, and political history, it also ran roughshod over differences in ethnic legacies and traditional affinities among local and regional cultures within and across the politically and economically defined boundaries. To put it bluntly: the model left out of account the complex ethnic configurations produced by the successive *Voelkerwanderungen*. The distintegration of Western Europe and the subsequent attempts at territory-building toward the North, West, and East were only too obviously affected by these large-scale movements of ethnically distinctive populations. These differed not only in their languages and customs but also in their ideas of governance, their styles of centre-building, and their resistance to peripheralization. The time-arrows at the bottom of Figure 4.2 suggest one set of consequences of these extensive movements of migration and resettlement: increased integration of peripheral communities within one or more of the three territory-controlling networks, but at the same time increased potential for peripheral protest and for some level of politicization of the peripheral predicament.

On the pages that follow we shall start out our discussion of sources of differentiation within Western Europe with a review of this set of *geoethnic* variables. We are very far from satisfied with the fit between our geoethnic chart and the originally constructed conceptual map, but we hope that this essay will offer some incentive for further efforts of model integration. We are thoroughly convinced that these geoethnic components must be incorporated within a broader model for the explanation of variations across the territories of Western Europe. The importance of efforts on these lines should come out with greatest clarity in our discussion of variations in modes of territorial consolidation and of sources of peripheral protest.

## The peopling of Western Europe

We cannot get anywhere toward an explanation of the successive changes in the territorial structure of Western Europe without some knowledge of the many waves of migration, conquest, and occupation that have layered the ethnic–linguistic landscape since the Early Iron Age. We can distinguish a total of *seven* major waves. Let us review these in chronological order.

First, the *Celtic* expansion: the Celts moved out from their heartland between the Rhine and the Danube from the sixth century B.C. and occupied large tracts of Gaul, Iberia, Britain, and even Greece.

Second, the long series of *Roman* conquests: the empire moved westward into southern Gaul and Iberia, northward toward the *limes* on the Rhine and the Danube, and then into Britain all the way up to the massive walls built up against the aggressive Picts in Caledonia, what was later called Scotland.

Third, the multiple invasions of the *Germanic* tribes into the crumbling western empire during the fourth and fifth centuries A.D.: the Ostrogoths, Visigoths, and Vandals all covered long stretches of territory on their way and ended up around the Mediterranean; the Lombards settled in northern Italy; the Burgundians in eastern Gaul; the Franks in northern Gaul; and the Jutes, Angles, and Saxons in parts of Britannia.

Fourth, the eighth century wave of *Arab* conquest northward across Iberia and for a brief spell even far into Gaul: this great thrust of Islamic forces was countered by a number of Christian counterthrusts, first into Spain and, later, with the Crusades, across the length of the Mediterranean.

Fifth, the succession of *Viking* raids and conquests: beyond all the plunder and devastation, these produced lasting settlements in Normandy (911), Ireland (until 1014), England (1013–1042 and again after 1066), and even Sicily and southern Italy (1072–1194).

Sixth, the westward drift of the *Slavs* and the *Finno-Ugric* peoples into the territories to the landward side of the Germans: the most spectacular consequences were the founding of Bohemian, Polish, Hungarian, and Serbian kingdoms during the tenth and eleventh centuries, and the beginnings of a Russian Empire centred on Kiev.

And finally, the *eastward expansion of the Germans* from the twelfth century onwards: this was part of the great drive to Christianize the rest of Europe, but was accompanied by well-planned efforts to colonize and improve poorly used agricultural land. The result was a thorough penetration of German settlers, religious orders, and merchants far into Slavic and Baltic lands, and a long history of conflict between the Marchland rulers and the kingdoms to the east.

These successive waves of conquest and occupation, penetration and retrenchment, produced a complex distribution of ethnic–linguistic groupings across Western Europe. Simplifying in the extreme we can reduce this geoethnic map to the configuration in Table 4.1.

Starting from the seaward fringe we can distinguish four sets of ethnic groupings along a west–east gradient. First, an *Atlantic periphery* made up of the Celtic and Basque lands and, after the collapse of the early Norwegian North Sea Empire, even West Norway, the Orkneys, Shetland, the Faeroes, and Iceland. Second, the *Western coastal plains,* the heartland of the early seaward kingdoms (the Danish, the Anglo-Saxon, the Frankish, and, considerably later and in a different context, the Iberian). Third, the *central plains* between the Meuse–Rhône line and the Elbe, the heartland of the German-Roman Empire. And fourth, the *landward periphery* caught in the cross-pressure between German and Swedish empire-building thrusts and the resistance of the Slavs, the Magyars, and the Finns.

Each of these four west–east slices can in turn be divided into at least three distinctive layers from north to south. First, the *lands beyond the reach of the Roman Empire*: Ireland, Scotland, northern Germany, Scandinavia, Poland, and the Baltic. Second, the *imperial lands north of the Alps*: England and Wales, France, Switzerland, southern Germany and Austria, and Hungary. Third, the *Mediterranean* lands, the territories most heavily imprinted by Latin institutions and least influenced by the Germanic invaders.

These territorial distributions provided the ethnic–linguistic infrastructures for the institutional developments of the High Middle Ages: the first steps toward the consolidation of centralized monarchies, the early leagues of cities, and the first consociational structures. In the next round, the distributions of ethnic identities and affinities determined the character and the cost of

**Table 4.1** A crude geoethnic map of Europe before the High Middle Ages

| | *Atlantic periphery* | *Coastal plains* | *Central plains and Alpine Territory* | *Landwards marshlands* |
|---|---|---|---|---|
| Beyond the reach of the Roman Empire | Icelanders<br>Faeroese<br>West Norse<br>Celts: Scotland, Ireland | East Norse<br>Danes | Swedes | Finns<br>Balts<br>Prussians<br>Poles, Lithuanians<br>Moravians, Czechs |
| Territory of the northern empire | Celts: Wales<br>Cornwall<br>Brittany | Angles, Saxons<br>Friesians, Jutes<br>West Franks/<br>Gallo–Romans<br>Normans | Germanic Tribes:<br>Burgundians East Franks<br>Saxons Thuringians<br>Alemannians Bavarians<br><br>Rhaetians | Hungarians<br><br>Bavarian settlers<br><br>Tirolians |
| Mediterranean territories | Basques | Occitans<br>Catalans<br>Corsicans<br>Castilians<br>Portuguese | Lombards<br>Italians<br>Sardinians<br>Sicilians | Slovenes<br>Croats<br>Serbs |

linguistic standardization within each of these territorial structures. The development of such central standards was accelerated by the invention of printing and the religious conflicts of the Reformation, and it put the peripheries under heavy pressure to accept the norms set by the territorial centres.

### The collapse of the Roman Empire and the formation of new centres[3]

To understand these processes of centre-building and peripheralization we have to refer back to Figures 4.1 and 4.2. Figure 4.1 distinguishes three dimensions of differentiation in the development of large-scale territorial systems; Figure 4.2 specifies the corresponding processes of "retrenchment" to smaller systems covering shorter distances.

At its height the Roman Empire maximized communications and controls in all three dimensions: economic, military–administrative, and cultural. It controlled a vast network of cities around the Mediterranean and at the same time built up a strong centre for the conquest of territories still at a low level of economic development. And, what was equally important, the Roman Empire also became the essential vehicle of penetration into new territories for a major script religion: Christianity.

In this way, the Roman Empire for some time drew strength from all of the three basic processes of differentiation. The three developments reinforced each other for a while, but generated separate organizational structures with resources of their own. The western empire collapsed as a military–administrative structure in the fifth century, but the city network was still there, as was the Roman Church and the tradition of long-distance communication via alphabetic script. The empire broke up as a political system of territorial control, but much of the economic as well as the cultural infrastructure for long-distance communication was left intact and in fact strengthened after the four or five centuries of conflict with Islam, the other empire-building religion of the Mediterranean.

To account for such processes of break-up and reorganization, we posit three processes of periphery build-up within the territory of disintegrating empires: feudalization, vernacularization, and state-building. The location of these processes can be represented graphically at some halfway point along each of the paths of development (see Figure 4.2). In the history of the territorial structuring of political systems, it is as important to analyze the processes of retrenchment as it is to study the phases of expansion. The system of states emerging in Europe from the twelfth to the twentieth century can only be understood against the background of the Roman inheritance and the reduction in the range and scope of cross-territorial communications that gradually took place in the wake of the fall of Rome.

The rise of feudal structures is a much-discussed theme in comparative history. Within our simple paradigm, what matters is the rise of intermediary power holders controlling resources in the primary economy. This "parochialization" of economic and political power was widespread in the territories of the old empire. But the process was far from uniform throughout

Europe: there were few signs of such a process in the territories north of the old Roman territory, and there were also variations depending on the level of agricultural development and the degree of exposure to the onslaughts of nomadic raiders and armies. What proved important in the later development of territorial units was the level of resource generation within such concentrated agrarian structures and the strategies adopted by the owners and controllers of land in their dealings with the urban bourgeoisie and with the military–bureaucratic agencies of state-building. This is the central thrust of Moore's (1966) important work on the development of large-scale territorial units in Europe and Asia.

Developments were much slower on the cultural front: the Roman Church established itself as the central spiritual authority across all of Western Europe and proved able to maintain its two languages, Greek and Latin, as the dominant standards of elite communication for centuries after the fall of the western empire. But, as Goody (1977) has shown in his penetrating analysis of the social consequences of script, there is a fundamental difference between empires built on ideographic communication and empires using alphabets. In China it was possible to keep the literati and the gentry integrated across a vast territory varying greatly in its local vernaculars. The ideographs had no direct relation to speech and could be pronounced in all kinds of ways even though conveying the same message. In the Roman Empire, Greek and Latin were maintained for centuries as vehicles of elite communication across Europe, but the alphabetic script opened up possibilities for the direct expression of vernacular languages. There was already a strong flourishing of such vernacular literatures in the Middle Ages, but the decisive break with the Greek and Latin standards came with the invention of printing and with the Reformation. These developments opened the flood-gates for the mass reproduction of messages in vernaculars and set the stage for the establishment of a variety of national standards of communication in an increasingly fragmented Europe. Gutenberg created an essential technology for the building of nations: the mass reproduction of books, tracts, and broadsheets made it possible to reach new strata within each territorial population and at the same time to confine communication within the limits of the particular vernacular. The Reformation reinforced this process in northern Europe. It meant much more than a break with Rome in matters of theological doctrine; it strengthened the distinctiveness of each territorial culture by integrating the priesthood into the administrative machinery of the state and by restricting priests to the confines of the given vernacular. In Hirschman's (1970) terms, the Reformation built up a wall against cultural "exits" into other territories. This wall was not only an important strategy in legitimizing the new territorial state but in the longer run was also a crucial step in preparing the broader population for the use of "voice" within their system.

This process was not uniform, however. One of the paradoxes of European development is that the strongest and the most durable systems emerged at the *periphery* of the old Empire; the heartlands and the Italian and German territories remained fragmented and dispersed until the nineteenth century.

To reach some understanding of this paradox we have to reason in several steps:

(1) The heartland of the old western empire was studded with cities in a broad trade route belt stretching from the Mediterranean to the east, as well as west of the Alps northward to the Rhine and the Danube.
(2) This "city belt" was at the same time the stronghold of the Roman Catholic Church and had a high density of cathedrals, monasteries, and ecclesiastical principalities.
(3) The very density of established centres within this territory made it difficult to single out any one as superior to all others; there was no geographically-given core area for the development of a strong territorial system.
(4) The resurrection of the Holy Roman Empire under the leadership of the four German tribes did not help to unify the territory. The emperors were prey to shifting electoral alliances, many of them were mere figureheads, and the best and the strongest of them expended their energies in quarrels with the Pope and with the Italian cities.
(5) By contrast, it proved much easier to develop effective core areas at the *edges* of the city-studded territories of the old empire; in these regions, centres could be built up under less competition and could achieve command of the resources in peripheral areas too far from the cities in the central trade belt.
(6) The earliest successes in such efforts of system-building at the edges of the old empire came on the *coastal plains to the west and the north*, in France, England, Scandinavia, and later in Spain. In all these cases the dynasties in the core areas were able to command resources from peripheral territories largely beyond the reach of the cities of the central trade belt.
(7) The second wave of successful centre-building took place on the *landward side*: first the Habsburgs, with their core area in Austria; then the eastern march of the German Empire; next the Swedes; and finally, and decisively, the Prussians.
(8) The fragmented middle belt of cities and petty states was the scene of endless onslaughts, countermoves, and efforts of reorganization during the long centuries from Charlemagne to Bismarck. First, the French monarchs gradually took over the old Lotharingian–Burgundian buffer zone from Provence to Flanders and incorporated such typical trade cities as Avignon, Aix, and Lyons. Second, the key cities to the north of the Alps managed to establish a defense league against all comers and gradually built up the *Swiss Confederation*: similar leagues were established along the Rhine and across the Baltic and North Seas, but they never managed to establish themselves as sovereign territorial formations. Third, the Habsburgs made a number of encroachments both on the west and the east of the belt and for some time controlled the crucial territories at the mouth of the Rhine, triggering the next successful effort of consociational confederation: the *United Netherlands*. Finally, in

the wake of the French Revolution, Napoleon moved across the middle belt both north and south of the Alps and set in motion a series of efforts of unification that ended with the successes of the Prussians and the Piedmontese in 1870.

**Peripheries at the interface between major core territories**

Let us proceed to discuss these propositions in further detail. The Roman Empire was symbolically reestablished by the Franks in 800, but this construction remained throughout its history a loose federation of kingdoms, principalities, free cities, and ecclesiastical territories. The division of the empire in 843 proved fateful: the western territory was gradually consolidated into a separate kingdom, and the territory in the middle, Lotharingia, was caught in a protracted conflict between the claims of the expanding Frankish kingdom and the different constellations of resource holders within the empire.

For a while it looked as if the eastern boundary of the French kingdom would remain stuck along the Rhône–Saône–Meuse line, but new opportunities opened up with the rapid decline of the internal cohesion of the empire after the extinction of the Hohenstaufen dynasty in the thirteenth century. The constitutional settlement called the Golden Bull (1356) increased the particularist powers of the princes against the Emperor, and the attempts of the Burgundians to create a strong "northern Lotharingia" with its core in what was later called Belgium created increasing tension along the eastern borders of the French territory. The first eastward expansions across the Rhône–Saône line occurred in the South: the Dauphiné came into French hands as early as 1349 and Provence was united to the French crown as an autonomous administrative unit in 1489. The later boundary changes came with the failure of the Burgundian effort to build up a strong state between the French kingdom and the old empire. The Burgundians straddled the Saône and had established great strength both within the French monarchy and within the empire. The western territory, the *Duchy*, was reorganized as the premier peerage of the kingdom, while the eastern territory, the *Franche Comté*, was a highly valued part of the empire. These two territories were brought together under the same dynasty in 1369 and a great effort of state-building got under way. This "Lotharingian" state was composed of the two Burgundies to the south, the Ardennes, the valley of the Meuse, and the Netherlands to the north. The effort collapsed, however, with the great empire-building thrust of the Habsburgs in the fifteenth century. The heiress of the last great duke married the Austrian archduke Maximilian in 1477 and Flanders and Holland, as well as the Franche Comté, came under the sway of the Habsburgs. The sixteenth and the seventeenth centuries saw protracted conflicts between the French and the Austrian-Spanish Habsburgs over these Lotharingian lands. The Thirty Years' War brought Metz and a number of other towns on the northeastern frontier further into French hands. Louis XIV continued these expansionist policies with great vigour and conquered the Franche Comté, the Alsace, and parts of Flanders in 1678, as well as Strasbourg in 1681. Progress was much slower in the Lorraine. After a brief period under the former King of Poland, Stanislaw I, the old duchy was incorporated into the French monarchy in 1766. But this did not stabilize the northeastern frontier for very long. The Lorraine and Flanders were to become the great battlefields between France and the consolidated German Empire in 1870, in 1914–1918, and in 1940, and parts of the Lorraine as well as Alsace changed hands three times over a period of 75 years.

We have gone into such details of territorial history to develop a general thesis: the great efforts of state-building in Western Europe since the High Middle Ages did not only produce a number of subject peripheries in the Celtic-Atlantic West and on the eastern marches, they also generated a number of marginal territories, problem peripheries, *between* the great state-building cores. We could call these *interface* peripheries: they were caught in the cross-fire between dominant centres and were never fully integrated into either of the blocs.

Starting from the North Sea we can identity these buffer territories and problem peripheries between France and different parts of the old empire. First of all, *Flanders* and *Wallonia*, territories straddling the age-old boundary between the Gallo-Roman-Frankish language and various Germanic dialects. Second, *Luxembourg*, a border territory with its own dialect but accepting both French and German as linguistic standards. Third, the *Lorraine* and the *Alsace*, both heavily Germanic-speaking but strongly oriented toward French culture and French political traditions. Fourth, the *Bernese Jura*, once a part of Burgundy, later a territory under the Prince Bishop of Basle, from 1815 incorporated into a German-speaking canton of the Swiss Confederation. Fifth, the *Savoie*, once a duchy within the Holy Roman Empire, later the core territory of the expanding dynasty that was to unify Italy, but was ceded to Napoleon III as the price for his consent to the formation of the Italian state in 1860. Sixth, the *Val d'Aosta*, a French-speaking enclave within Italian territory. And seventh, the city of *Nice*, a major tourist centre with strong Italian affinities, also ceded to the French by the House of Savoy in 1860.

We could have added to this list the entire *Occitanian* region, the region of the southern Gallo-Roman language, the *Langue d'Oc*, but this clearly could only be classified as an interface periphery by stretching that concept beyond recognition. We *could*, of course, make a case for interpreting the failure of the state-builders in Toulouse as a parallel to the Burgundian case, but, first, Occitania was part of the western section of the empire from the start, there was only for a brief period an alternative bloc competing effectively for the control of this territory. Occitania could better be classified, with Scotland, Catalonia, and Bavaria, as "failed-centre" peripheries: territories that *might* have built up their own core structures but were the victims of more effective drives of incorporation from other centres.

We could continue our tour around the entire territory of the old German-Roman Empire and identify a number of further "interface" peripheries. But the character of the cultural and political cross-pressures varied with the blocs confronting each other.

At its height the old empire could count on large chunks of Italy as well as the Germanic lands to the north. The standard literary language of the northern empire was *hochdeutsch*; the standard south of the Alps was increasingly the Florentine variant of Italian. But there remained great variations among dialects and in a few cases even distinctive languages. The most important of these linguistic "isolates" within the old empire were the *Rhaeto-Roman* populations in the *Grisons* in Switzerland and in *Friulia* on the border between Italy and what is now Yugoslavia. We might call these "enclave" peripheries since they do not fit into the dominant culture of the surrounding territory.

By contrast we could classify as "interface" peripheries the territories caught in cross-pressure between two major blocs. Between Italy and the Germanic-speaking territory, the most important such case is the *Alto Adige/South Tirol*. In their efforts to extend their powers across the old empire and beyond, the Habsburgs had acquired Tirol in 1363, Milan and Lombardy in 1535, and Venice and Venetia with the Congress of Vienna in 1815. The territories with clear cut Italian majorities were brought into the unified kingdom in 1859 and in 1866, but there remained an important Italian-speaking minority (as well as a small group of Rhaetians called the Ladins) in what was then called South Tirol, the area stretching south of the Brenner down to Lake Garda. This territory was promised to the Italians in the negotiations that led to their entry into World War I on the side of the Western Alliance in 1915. When the peace treaty was negotiated, the Italians succeeded in imposing a *strategic* boundary along the Alps, but this did not correspond in any way to the *cultural* boundary between German-speakers and Italians. There was a considerable German population south of the Brenner and Mussolini did not succeed in his efforts to "Italianize" the region through accelerated industrialization and public works. The result was a protracted conflict. A new round of negotiations between Italy and Austria took place in 1947 and, after a period of terrorist attacks and counterattacks, a *Proporzpaket*, a detailed package of proportionalizing measures, was signed in 1972.

The other interface peripheries to the landward side of the old empire all confront Italians or Germans with Slavs or Hungarians. Starting from the south, these are the major areas of tension:

(1) the *Italian–Slovene* interface in the Trieste-Fiume provinces
(2) the *Austrian–Slovene* interface in Carinthia
(3) the *Austrian–Hungarian* interface in the Burgenland
(4) the extensive *German–Czech* interface in the Sudetenland
(5) the equally extensive *German–Polish* interface, once pushed far to the east toward the Lithuanian border, since World War II placed on the Oder-Neisse line.

We could again add to this list a couple of enclave peripheries: the most important would be the two *Sorbian/Lusatian* territories in the Dresden and Cottbus *Bezirke* of what is today East Germany.

On the northern front of the old empire we can identify two major interfaces: the *German–Danish* in Slesvig and the *Friesian* straddling the

Netherlands and Lower Saxony. The Danish kings were active in imperial politics for centuries and held important fiefs south of the linguistic border. These were conquered by the Prussians, but the question of the exact tracing of the military–administrative boundary remained an intractable issue at least until 1920. There were no corresponding conflicts on the German–Dutch border, as the Friesians adjusted peacefully to their predicament.

We can go further north to identify interface peripheries within the extensive territories well beyond the reach of the German–Roman as well as the Roman Empire. Two cases deserve particular attention:

(1) the *Danish–Swedish* interface in Scania, Halland, and Blekinge
(2) the *Swedish–Finnish* interface in southeast Finland

Sweden conquered the Danish land east of the Sound in 1645 and 1658 and was able to integrate these territories remarkably quickly into its system. One important factor was the establishment of the University of Lund in 1672. This periphery has remained firmly in Swedish hands ever since and there has been no trace of serious irredentism.

By contrast, the Swedish settlement in Finland generated a long history of political conflict. Finland was an early territory of conquest for the kings of Sweden, and the urban elite was predominantly Swedish-speaking until well into the nineteenth century. The extraordinary mobilization of Finnish identity from the 1850s onward set the stage for a series of conflicts, but these were gradually settled through a policy of central equalization and local majority rule for the two language groups.

With the Swedish–Finnish case we have in fact moved into the analysis of a third category of peripheries: we could call these *external* peripheries. We have already reviewed the most important of these in Table 4.1. The left column in that geoethnic chart lists the Celtic-Atlantic peripheries as "external" to the great state-building endeavours on the coastal plains. We shall revert to these peripheries in our subsequent discussion of unitary-federative structures.

## A conceptual map of Europe[4]

The long sequences of migration, centre-building, cultural standardization, and boundary imposition produced an extraordinary tangle of territorial structures in Europe: some large, some small, some highly centralized, others made up of differentiated networks of self-reliant cities. The alphabet and the city decided the fate of Europe. The emergence of vernacular standards of communication prepared the ground for the later stages of nation-building at the mass level, and the geography of trade routes made for differences in the resources for state-building between East and West.

The essentials of these differences in the conditions of polity-building can be set out in a conceptual map: a simple two-dimensional typology interpreted within the framework of the overall topology of Europe.

In this map (Table 4.2) the West–East axis differentiates the economic resource bases of the state-building centres: surpluses from a highly

**Table 4.2** A typology of the political systems of twentieth-century Western Europe

| *System characteristics* | | | | | | | |
|---|---|---|---|---|---|---|---|
| *Size* | *City structure* | *Linguistic structure* | *Seaward peripheries: Sovereign after 1814* | *Seaward nation-states: Retrenched empires* | *City-state Europe: Early consociations or late unification* | *Landward nation-states: Retrenched empires* | *Landward buffers: Sovereign after 1918* |
| Larger | Monocephalic | Integrated<br><br>Endoglossic peripheries | | France (but some resistance among Bretons, Alsatians, Occitans) United Kingdom (Welsh, some Gaelic) | | (DDR) | |
| | Polycephalic | Integrated | | | German F.R. and Italy | | |
| | | Endoglossic peripheries | | Spain | | | |
| Smaller | Monocephalic | Integrated | Iceland (Danish never strong) | Denmark | Luxembourg (but exoglossic standards) | Sweden (near polycephalic) Austria | |
| | | Divided | Norway (two standards, one once close to Danish)<br><br>Ireland (Gaelic now very weak) | | | | Finland (one endoglossic, one exoglossic, standard) |
| | Polycephalic | Integrated | | Portugal | Netherlands (but religious *Verzuiling*) | | |
| | | Divided: Exoglossic standards | | | Belgium (Flemish, French) Switzerland (German, French, Italian) | | (Yugoslavia) |

monetized economy in the West; surpluses from agricultural labour in the East. The North–South axis measures the conditions for rapid cultural integration: the early closing of the borders in the Protestant North; the continued supraterritoriality of the Church in the Catholic South.

This conceptual map reflects the fundamental asymmetry of the geopolitical structure of Europe: the dominant city network of the politically fragmented trade belt from the Mediterranean toward the North, the strength of the cities in the territories consolidated to the seaward side of this belt, and the weakness of the cities in the territories brought together under the strong military centres on the landward Marchland.

The West–East contrast is the underlying dimension in Moore's (1966) analysis. He does not discuss the middle belt, but his contrast between the seaward powers, England and France, and the landward powers, Prussia and Russia, is directly reflected in the West–East gradient in the map. Essentially this was a contrast in the levels of monetization reached at the time of the decisive consolidation of the territorial centres: England and France during the sixteenth century, Prussia and Russia during the seventeenth and eighteenth centuries. In the West, the great surge of commercial activity made it possible for the centre builders to extract resources in easily convertible currency. In the East, the cities were much weaker partners and could not offer the essential resource base for the building of the military machineries of the new centres at the periphery of the old empire. The only alternative partners were the owners of land, and the resources they could offer were food and manpower: crofters, tenants, and smallholders in Sweden; serfs in Austria, Prussia, and Russia. This contrast in the resource bases for political consolidation goes far to explain the difference between the Western and the Eastern systems in their internal structure and in the character of the later transition to mass politics. This is the thrust of Moore's detailed analysis. It does not explain all the cases, however, and does not pinpoint the sources of variations on each side. There are important variations both on the seaward and the landward sides, and these can only be understood through an analysis of the other dimensions of the polities, quite particularly the cultural.

In the conceptual map of Europe, the West–East axis differentiates conditions of *state*-building, the South–North axis the conditions of *nation*-building. In the underlying model of development, the Reformation is interpreted as the first major step toward the definition of territorial nations. Lutherans and Calvinists broke with the supraterritoriality of the Roman Church and merged the ecclesiastical bureaucracies with the secular territorial establishments. This led to a closing of "exit options" on the cultural front and an accentuation of the cultural significance of the borders between territories. The Reformation occurred only a few decades after Gutenberg; the state churches of the Protestant North became major agencies for the standardization of national languages and for the socialization of the masses into unified national cultures. In Catholic Europe the Church remained supraterritorial and did not to the same extent prove an agency of nation-building. True enough, the Catholic Church played a major role in the development of peripheral nationalisms in some of the territories of Counter-

Reformation Europe, but these were much later developments; they occurred in the aftermath of the French Revolution and took the form of alliances between the Church and nationalist or secessionist leaders against the rulers at the centre, whether Protestant (Belgium before 1830, Ireland from the 1820s onward), Orthodox (Poland, Lithuania), or simply secularizing (the Carlist wars in Spain). Even in the most loyal of the Counter-Reformation states, the churches remained supraterritorial in outlook and never became central agencies of nation-building in the way the Protestant churches did in the North. Whether Protestant or Catholic, the churches had to work with populations varying widely in their openness to efforts of standardization.

The *Voelkerwanderungen* and the struggles of the Middle Ages had produced very different conditions for linguistic unification in the different territories of Europe. The result was a variety of conflicts between claims of territorial control and claims of national identity. There was nowhere a complete fit between the "state" and the "nation," and the conflicts between the two sets of claims were particularly violent in the central trade route belt and in Catholic Europe.

In the northern territories the processes of state-building and nation-building tended to proceed *pari passu*, but even in these systems at the edge of the old empire, claims for territorial control often clashed with claims for separate identities: the English versus the Celts, the Danes versus the Norwegians and the Icelanders, the Swedes versus the Finns. Trade-belt Europe inherited strong linguistic standards from the ancient and medieval empires: Italian in the South, German in the North. But there was no corresponding development at the political level; in these territories national identity came first, political unification only much later. In the "Lotharingian–Burgundian" zone, between the German *Reich* and the French monarchy, the linguistic borders hardly ever followed the territorial frontiers and a variety of developments took place. The Swiss quickly accepted several exoglossic literary standards and never built up a "linguistic nation" of their own; Alsace-Lorraine maintained its Germanic dialect but identified politically with France; Luxembourg also kept its dialect but veered between allegiance to Germany and to France; and Belgium was split between a Flemish-speaking North and a Walloon-speaking South. In the rest of Catholic Europe there were dominant languages at the territorial centres but strong movements of cultural resistance in the peripheries. France went furthest in linguistic standardization but had to keep on "building the nation" in such peripheries as Brittany and Occitania throughout the nineteenth century. The *levée en masse* for the great wars and the nationalization of public education were the decisive vehicles of cultural unification. In Spain the Castilians were never able to build up a unified national culture; the Basque country and Catalonia remained strongholds of regional resistance. Austria tried for centuries to make German the dominant language of Southeast Europe, but the drive never succeeded; with the rise of national ideologies this vast empire soon fell apart into a number of territorial fragments, but even these fragments were still torn by cultural conflicts. The contrast between the Austrian and the Prussian strategies is striking: Austria extended the domain of its state apparatus far beyond the borders of the Germanic language community and acquired a multilingual empire; Prussia had also built up its strength on the eastern marches but in the end turned westward, into the core areas of the ancient German nation. The Catholic power stuck to the supraterritorial idea; the Protestant power endeavoured to acquire territorial control over the one linguistic community. The struggle between *kleindeutsche* and *grossdeutsche* strategies was a struggle over conceptions of the state and of the nation, a struggle between a political and a cultural conception of territorial community.

**Federal versus unitary structures**

We have discussed a variety of sources of diversity within politically defined territories: linguistic minority–majority contrasts, religious divisions, and competitive relations among urban centres. We shall now proceed to review some of the *institutional* expressions of such diversity: the structures of representation, decision-making, and administration set up to coordinate the conflicting interests within the total territory.

Territorial institutions are normally classified on some continuum from loose confederations to highly centralized unitary structures. Riker (1974: 101) defines a federal political organization as one dividing the activities of government between regional units and a central government so as to ensure that each level of government has the final decision in *at least some* of the fields of activity.

Given a definition on these lines, structures will vary along a continuum depending on the ranges and the weights of activities left for final decision at the regional versus the central level.

Such classifications obviously make sense only if the *contents* of the decisions reserved for the lower-level units are specified. The most frequent solution is to reserve *cultural* matters to the regional units and all matters of *defence* and *foreign policy* to the central unit of government.

In this minimal sense, even *empires* are federative in character. In fact, an often-quoted distinguishing mark of an empire is that it allows autonomy in cultural affairs to a number of subject territories or populations while reserving for the central authority the control of economic transactions and the extraction of surplus resources for military–administrative needs. The *millet* system of the Ottoman Empire is perhaps the most extreme example: the imperial rulers allowed a number of religious communities to keep up their distinctive institutions, languages, and rituals against payment of tribute. There were equally pronounced elements of pluralism and particularism in the German-Roman Empire which was, from the start, a federation of the four original *Staemme*, and the King-Emperor never acquired extensive powers for himself and his court beyond the symbolic and ceremonial ones and the control of appointments, particularly those within the Church. This federal tradition in fact weighed heavily in the history of state-building within the central South-North belt of Europe. Two of the polities within that belt are explicitly federal; one was established as a federation but has moved toward a unitary structure; and two were established as unitary states but have been moving toward new levels of federalization. By contrast, we find hardly any examples of explicit federalization outside this central city belt. Let us look at the cases in turn.

The earliest federal structures emerged *within* the Holy Roman Empire and were essentially alliances of peasant communities or, more frequently, of cities against the encroachments of the stronger dynastic units competing for power within that system. A number of *leagues* were created along the south–north trade routes from the twelfth century onward: the Lombard, the Burgundian, the Alsatian, the Swabian, the Rhine Leagues and, of course, the Hanse, the largest and most wide-ranging of them all. All the purely urban leagues lost out in the conflict with territorial princes during the fifteenth and sixteenth centuries. Only two of the defensive alliances survived the great onslaughts of the territorial powers: the *Swiss Confederation* and the *United Netherlands*. Both these structures built up strength at crucial transition points in the trade route system: the Swiss controlled several important passes in the Alps, the Dutch the estuary of the Rhine. These "consociational" polities stood in marked contrast to the unitary nation-states: Switzerland and the United Provinces were the first to practise religious toleration in the wake of the fierce struggles of the Thirty Years' War. But there were important differences: the Dutch territory was much more compact, the Swiss more divided; the Dutch developed a standard language of their own (an "endoglossic" standard), the Swiss did not develop one national language but accepted external ("exoglossic") standards for their three major languages of written communication. These differences in geopolitical position and geocultural structure clearly affected the constitutional reorganizations after the Napoleonic Wars. The Netherlands developed a much more unitary structure; Switzerland retained its federal structure while increasing the powers of the central government, particularly after 1848.

The other territories of the old empire were either incorporated into the stronger nation-states to the west and east or remained a congeries of particularist principalities, cities, and bishoprics until Napoleon showed that unification was not only possible but worth fighting for. There followed two great movements of unification: one within Germany, the other in Italy. Germany was unified in several steps: first a *Zollverein*, a customs union, later, in 1848, a frustrating experience of "consensual" unification under Liberal leadership, and then finally, in 1866, the establishment of the *Norddeutsche Bund* under the military command of the Prussians. Bismarck did not, however, create a unified empire. He was fully aware of the strong particularist loyalties and allowed the principalities and free cities some rights of self-government within a federal structure. This German federalism was reaffirmed in the Weimar Constitution of 1919 and again after the break-up of the old territory between West and East in the Bonn constitution of 1949. Developments in Italy took a different path. There the constitution of the state-building unit, the Kingdom of Sardinia, was extended to the entire territory after the annexations of 1859, 1860, 1866, and 1870. The majority alliance of Piedmontese and Tuscan leaders insisted on the maintenance of a centralized system of government, particularly because of their interest in controlling developments in the southern periphery. The great difference between the German and the Italian unification processes reflected this contrast in the balance between central and peripheral forces. In Germany the highly urbanized western regions were unified from a military periphery: Prussia. In this situation, federalization was the best strategy of unification. In Italy, the unifying power had its base in the highly urbanized North and the state-builders did not need to resort to a federalizing strategy to gain control of the southern periphery. Some exceptions had to be made, however, because of strong cultural resistance to centralized rule. The Constitution of 1948 gave separate powers to five autonomous regions: Sicily, Sardinia, and three linguistically divided territories, Valle d'Aosta (French-speaking), Trentino-Alto Adige (large German minority in the South Tirol province of Bozen/Bolzano), and Friulia-Venezia-Giulia (Friulian and Slovene minorities). Steps were also taken to give the other fifteen regions greater powers. From the early 1970s the regional councils took control over the activities of the smallest units of local administration: the communes. There was still considerable tension in the system, however, since there was no fit between the economic structure, centred on the northern triangle of Milan-Turin-Genoa, and the political structure with its centre in Rome.

We can summarize processes of unification versus federalization in a schematic table (see Table 4.3). In this we have combined three sets of variables: the first economic, the second cultural, and the third political–administrative. The first dimension in the table groups territories by the *centralization of their city networks*: polycephalic versus monocephalic. The second dimension represents the degree of *ethnic–linguistic unification* of the territory. The third groups the countries by their *institutional policies* from the expressly federative to the most markedly unitary.

There is clearly no direct fit between economic centralization, cultural homogeneity, and institutional structure. In one case, we find a monocephalic city network within a federally structured state; in two other cases, we find polycephalic networks within unitary systems. In one case, Germany, we find a federal structure and a high level of ethnic–linguistic homogeneity within the territorial population; in another, Italy, we find a markedly more heterogeneous population within a near-unitary institutional structure. Practically all the possible combinations have occurred historically, but some of the combinations have clearly made for more peaceful interactions between centres and peripheries than others. Switzerland and Sweden are particularly revealing cases. The Swiss federal structure has no doubt served that multicultural population better than any unitary apparatus. The current conflict within the largest and most "centralist" of the cantons, Berne, seems likely to find a classical solution through territorial division. In the Swedish case, once the Danish and Norwegian territories conquered during the seventeenth century had been culturally integrated, the unitary structure of government certainly served the people well. It is not difficult to see from the review of cases in Table 4.3 that the greatest strains on territorial equilibria tend to occur in three situations of "poor fit" between economic-cultural distributions and institutional structure: first, in centralized regimes with culturally distinctive populations (e.g., France, Belgium, Finland); second,

**Table 4.3** Combinatorics of economic, cultural, and politico-administrative diversity

| *City network structure* | *Ethnic linguistic homogeneity* | *Institutional structure* | *Timing of state formation* | *Case* | *Particulars* |
|---|---|---|---|---|---|
| Monocephalic | High | Unitary | Medieval | Denmark | German minority in Slesvig |
| M | H | U | Med. | Sweden | City network less centralized; Lappish minority in North |
| M | H | U | 20th ct. | Iceland | |
| M | H | Federal | Med. | Austria | Slovene minority in Carinthia |
| M | Medium | U | Med. | France | Breton, Flemish, Alsatian, Corsican, Occitan, Catalan minorities |
| M | M | U | Med. | U.K. | Federalizing tendencies: N. Ireland, Scotland, Wales |
| M | M | U | 19th ct. | Norway | Two linguistic standards, but no clear-cut ethnic opposition |
| M | M | U | 20th ct. | Ireland | English dominant, but Irish-speakers in West |
| M | Low | U | 20th ct. | Finland | Swedish-speaking minority in Southwest |
| M | L | U→F | l9th ct. | Belgium | Flemish–Walloon opposition; federal-communal constitution under discussion |
| Polycephalic | H | U | 17th ct. | Netherl. | Friesian minority; elements of provincial autonomy |
| P | H | U | Med. | Portugal | |
| P | H | F | 19th ct./split 20th ct. | FRG | Friesian minority; Bavarian particularism |
| P | M | U→F | 19th ct. | Italy | Francophone, Germanic, Friulian minorities |
| P | L | U→F | Med. | Spain | Basque, Catalan minorities; devolved provincial government under discussion |
| P | L | F | Med. | Switzerland | Four linguistic groups |

in centralized regimes with a markedly polycephalic economic network (e.g., Italy and Spain); third, in "mixed strategy" regimes practising one policy for core areas, another for historically defined communities (e.g., the U.K.).

### Notes

1 A brief overview of the model and its primary ramifications was published in a *Festschrift* for Karl W. Deutsch in 1979: *Territories, Nations, Parties.*
2 An expanded version of this text was published in 1979 as part of a "data workbook" on *Centre–Periphery Structures in Western Europe*.
3 This section reproduces a number of passages from Rokkan (1973).
4 This section also reproduces passages from Rokkan (1973).

### Selected references

Goody, J. (1977) The *Domestication of the Savage Mind*. New York: Cambridge University Press.
Hirschman, A. O. (1970) *Exit, Voice and Loyalty*. Cambridge, MA: Harvard University Press.
Moore, B. (1966) *Social Origins of Dictatorship and Democracy*. Boston: Beacon.
Riker, W. H. (1974) "Federalism," pp. 93–172 in F. I. Greenstein and N. W. Polsby (eds) *Handbook of Political Science*. Reading, MA: Addison-Wesley.
Rokkan, S. (1975) "Dimensions of state formation and nation-building," pp. 562–600 in C. Tilly (ed.) *The Formation of National States in Western Europe*. Princeton, NJ: Princeton University Press (1973).
Rokkan, S. (1973) "Cities, states, and nations," Chapter 2 in S. N. Eisenstadt and S. Rokkan (eds.) *Building States and Nations, Vol. I*. Beverly Hills, CA: Sage.
Rokkan, S. (1962) "The mobilization of the periphery," in S. Rokkan (ed.) *Approaches to the Study of Political Participation*. Bergen: Michelsen Institute.
Rokkan, S. *et al.* (1970) *Citizens, Elections, Parties*. Oslo: Universitetsforlaget.

# 5 Michael Mann, 'The Autonomous Power of the State'

Reprinted in full from: *European Journal of Sociology* 25, 185–213 (1984)

This essay tries to specify the origins, mechanisms and results of the autonomous power which the state possesses in relation to the major power groupings of 'civil society'. The argument is couched generally, but it derives from a large, ongoing empirical research project into the development of power in human societies. At the moment, my generalizations are bolder about agrarian societies; concerning industrial societies I will be more tentative. I define the state and then pursue the implications of that definition. I discuss two essential parts of the definition, centrality and territoriality, in relation to two types of state power, termed here *despotic* and *infrastructural* power. I argue that state autonomy, of both despotic and infrastructural forms, flows principally from the state's unique ability to provide a *territorially centralized* form of organization.

Nowadays there is no need to belabour the point that most general theories of the state have been false because they have been reductionist. They have reduced the state to the pre-existing structures of civil society. This is obviously true of the Marxist, the liberal and the functionalist traditions of state theory, each of which has seen the state predominantly as a place, an *arena*, in which the struggles of classes, interest groups and individuals are expressed and institutionalized, and – in functionalist versions – in which a General Will (or, to use more modern terms, core values or normative consensus) is expressed and implemented. Though such theories disagree about many things, they are united in denying significant autonomous power to the state. But despite the existence of excellent critiques of such reductionism (e.g. by Wolin 1961) and despite the self-criticism implied by the constant use of the term 'relative autonomy' by recent Marxists (like Poulantzas 1972 and Therborn 1978), there has still been a curious reluctance to analyse this autonomy.

One major obstacle has been itself political. The main alternative theory which *appears* to uphold state autonomy has been associated with rather unpleasant politics. I refer to the militarist tradition of state theory embodied around the beginning of the century in the work of predominantly Germanic writers, like Gumplowicz (1899), Ratzenhofer and Schmitt. They saw the state as physical force, and as this was the prime mover in society, so the militaristic state was supreme over those economic and ideological structures identified by the reductionist theories. But the scientific merits of these theories were quickly submerged by their political associations – with Social Darwinism, racism, glorification of state power, and then Fascism. The final (deeply ironic) outcome was that militarist theory was defeated on the battlefield by the combined forces of (Marxist) Russia and the (liberal democratic and functionalist) Western allies. We have heard little of it directly since. But its indirect influence has been felt, especially recently, through the work of 'good Germans' like Weber, Hintze (1975), Rüstow (1982) and the anarchist Oppenheimer (1975), all influenced to one degree or another by the German militarist tradition, and all of whose major works have now been translated into English.

I am not advocating a return to this alternative tradition, even at its scientific level. For when we look more closely, we see that it is usually also reductionist. The state is still nothing in itself: it is merely the embodiment of physical force in society. The state is not an arena where domestic economic/ideological issues are resolved, rather it is an arena in which military force is mobilized domestically and used domestically and, above all, internationally.

Both types of the theory have merit, yet both are partial. So what would happen if we put them together in a single theory? We would assemble an essentially dual theory of the state. It would identify two dimensions: the domestic economic/ideological aspect of the state and the military, international aspect of states. In the present climate of comparative sociology, dominated by a Marxified Weberianism, domestic analysis would likely centre upon class relations. And as states would now be responding to two types of pressure and interest groups, a certain 'space' would be created in which a state elite could manoeuvre, play off classes against war factions and other states, and so stake out an area and degree of power autonomy for itself. To put the two together would give us a rudimentary account of state autonomy.

That is indeed precisely the point at which the best state theory has now arrived. It is exemplified by Theda Skocpol's excellent *States and Social Revolutions*. Skocpol draws upon Marx and Weber in about equal quantities. She quotes enthusiastically Otto Hintze's two-dimensional view of the determinants of state organization, 'first, the structure of social classes, and second, the external ordering of the states – their position relative to each other, and their over-all position in the world', and she then expands the latter in terms of military relations. These two 'basic sets of tasks' are undertaken by 'a set of administrative, policing and military organizations headed, and more or less well co-ordinated by, an executive authority' for whom resources are extracted from society. These resource-supported administrative and coercive organizations are 'the basis of state power as such'. This power can then be used with a degree of autonomy against either the dominant class, or against domestic war or peace factions and foreign states (Skocpol 1979: 29–31; Hintze 1975: 183). A very similar approach underlies Charles Tilly's recent work (e.g. 1981, Chaps. 5 & 8). And Anthony Giddens (1981) has argued in similar vein.

Now I do not wish to quite abandon this 'two-dimensional' model of the state – for I, too, have contributed a detailed analysis of English state finances in the period 1130–1815 starting from such a model (Mann 1980). All these works advance beyond reductionism. We can develop their insights considerably further, and so penetrate to the heart of state autonomy, its nature, degree and consequences. But to do this we must make a far more radical, yet in a sense peculiar and paradoxical, break with reductionism. I will argue in this paper that the state is merely and essentially an arena, a *place*, and yet *this* is the very source of its autonomy.

## Defining the state

The state is undeniably a messy concept. The main problem is that most definitions contain two different levels of analysis, the 'institutional' and the 'functional'. That is, the state can be defined in terms of what it looks like, institutionally, or what it does, its functions. Predominant is a mixed, but largely institutional, view put forward originally by Weber. In this the state contains four main elements, being:

a) a *differentiated* set of institutions and personnel embodying

b) *centrality* in the sense that political relations radiate outwards from a centre to cover
c) a *territorially-demarcated area*, over which it exercises
d) a monopoly of *authoritative binding rule-making*, backed up by a monopoly of the means of physical violence.
(See, for example, the definitions of Eisenstadt 1969: 5; MacIver 1926: 22; Tilly 1975: 27; Weber 1968: I, 64).

Apart from the last phrase which tends to equate the state with military force (see below), I will follow this definition. It is still something of a mixed bag. It contains a predominant institutional element: states can be recognized by the central location of their differentiated institutions. Yet it also contains a 'functional' element: the essence of the state's functions is a monopoly of binding rule-making. Nevertheless, my principal interest lies in those centralized institutions generally called 'states', and in the powers of the personnel who staff them, at the higher levels generally termed the 'state elite'. The central question for us here, then, is what is the nature of the power possessed by states and state elites? In answering I shall contrast state elites with power groupings whose base lies outside the state, in 'civil society'. In line with the model of power underlying my work, I divide these into three: ideological, economic, and military groups. So what, therefore, is the power of state elites as against the power of ideological movements, economic classes, and military elites?

### *Two meanings of state power*

What do we mean by 'the power of the state'? As soon as we begin to think about this commonplace phrase, we encounter two quite different senses in which states and their elites might be considered powerful. We must disentangle them. The first sense concerns what we might term the *despotic power* of the state elite, the range of actions which the elite is empowered to undertake without routine, institutionalized negotiation with civil society groups. The historical variations in such powers have been so enormous that we can safely leave on one side the ticklish problem of how we precisely measure them. The despotic powers of many historical states have been virtually unlimited. The Chinese Emperor, as the Son of Heaven, 'owned' the whole of China and could do as he wished with any individual or group within his domain. The Roman Emperor, only a minor god, acquired powers which were also in principle unlimited outside of a restricted area of affairs nominally controlled by the Senate. Some monarchs of early modern Europe also claimed divinely-derived, absolute powers (though they were not themselves divine). The contemporary Soviet state/party elite, as 'trustees' of the interests of the masses, also possess considerable despotic (though sometimes strictly unconstitutional) power. Great despotic power can be 'measured' most vividly in the ability of all these Red Queens to shout 'off with his head' and have their whim gratified without further ado – provided the person is at hand. Despotic power is also usually what is meant in the literature by 'autonomy of power'.

But there is a second sense in which people talk of 'the power of the state', especially in today's capitalist democracies. We might term this *infrastructural power*, the capacity of the state to actually penetrate civil society, and to implement logistically political decisions throughout the realm. This was comparatively weak in the historical societies just mentioned – once you were out of sight of the Red Queen, she had difficulty in getting at you. But it is powerfully developed in all industrial societies. When people in the West today complain of the growing power of the state, they cannot be referring sensibly to the despotic powers of the state elite itself, for if anything these are still declining. It is, after all, only forty years since universal suffrage was fully established in several of the advanced capitalist states, and the basic political rights of groups such as ethnic minorities and women are still increasing. But the complaint is more justly levelled against the state's infrastructural encroachments. These powers are now immense. The state can assess and tax our income and wealth at source, without our consent or that of our neighbours or kin (which states before about 1850 were *never* able to do); it stores and can recall immediately a massive amount of information about all of us; it can enforce its will within the day almost anywhere in its domains; its influence on the overall economy is enormous; it even directly provides the subsistence of most of us (in state employment, in pensions, in family allowances, etc.). The state penetrates everyday life more than did any historical state. Its infrastructural power has increased enormously. If there were a Red Queen, we would all quail at her words – from Alaska to Florida, from the Shetlands to Cornwall there is no hiding place from the infrastructural reach of the modern state.

But who controls these states? Without prejudging a complex issue entirely, the answer in the capitalist democracies is less likely to be 'an autonomous state elite' than in most historic societies. In these countries most of the formal political leadership is elected and recallable. Whether one regards the democracy as genuine or not, few would contest that politicians are largely controlled by outside civil society groups (either by their financiers or by the electorate) as well as by the law. President Nixon or M. Chaban-Delmas may have paid no taxes; political leaders may surreptitiously amass wealth, infringe the civil liberties of their opponents, and hold onto power by slyly undemocratic means. But they do not brazenly expropriate or kill their enemies or dare to overturn legal traditions enshrining constitutional rule, private property or individual freedoms. On the rare occasions this happens, we refer to it as a *coup* or a revolution, an overturning of the norms. If we turn from elected politicians to permanent bureaucrats we still do not find them exercising significant autonomous power over civil society. Perhaps I should qualify this, for the secret decisions of politicians and bureaucrats penetrate our everyday lives in an often infuriating way, deciding we are not eligible for this or that benefit, including, for some persons, citizenship itself. But their power to change the fundamental rules and overturn the distribution of power within civil society is feeble – without the backing of a formidable social movement.

So, in one sense states in the capitalist democracies are weak, in another they are strong. They are 'despotically weak' but 'infrastructurally strong'. Let us clearly distinguish these two types of state power. The first sense denotes power by the state elite itself over civil society. The second denotes the power of the state to penetrate and centrally co-ordinate the activities of civil society through its own infrastructure. The second type of power still allows the possibility that the state itself is a mere instrument of forces within civil society, i.e. that it has no despotic power at all. The two are analytically autonomous dimensions of power. In practice, of course, there may be a relationship between them. For example, the greater the state's infrastructural power, the greater the volume of binding rule-making, and therefore the greater the likelihood of despotic power over individuals and perhaps also over marginal, minority groups. All infrastructurally powerful states, including the capitalist democracies, are strong in relation to individuals and to the weaker groups in civil society, but the capitalist democratic states are feeble in relation to dominant groups – at least in comparison to most historical states.

From these two independent dimensions of state power we can derive four ideal-types in Figure 5.1.

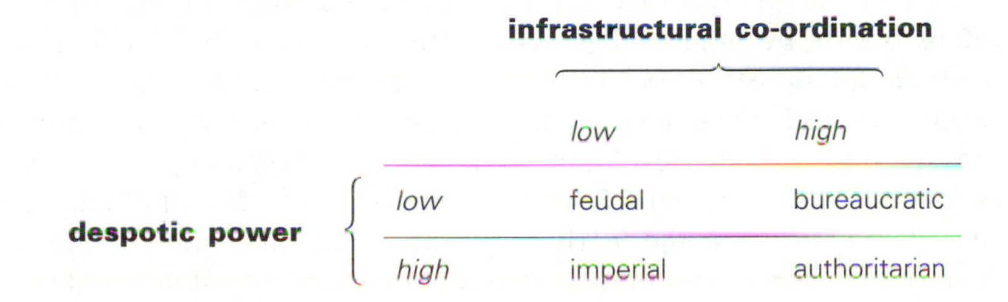

| | | infrastructural co-ordination | |
|---|---|---|---|
| | | *low* | *high* |
| **despotic power** | *low* | feudal | bureaucratic |
| | *high* | imperial | authoritarian |

**Fig. 5.1** Two dimensions of state power

The *feudal* state is the weakest, for it has both low despotic and low infrastructural power. The medieval European state approximated to this ideal-type, governing largely indirectly, through infrastructure freely and contractually provided and controlled by the principal and independent magnates, clerics and towns. The *imperial* state possesses its own governing agents, but has only limited capacity to penetrate and co-ordinate civil society without the assistance of other power groups. It corresponds to the term 'patrimonial state' used by writers like Weber (1968) and Bendix (1978). Ancient states like the Akkadian, Egyptian, Assyrian, Persian and Roman approximated to this type. I hesitated over the term *bureaucratic* state, because of its negative connotations. But a bureaucracy has a high organizational capacity, yet cannot set its own goals; and the bureaucratic state is controlled by others, civil society groups, but their decisions once taken are enforceable through the state's infrastructure. Contemporary capitalist democracies approximate to this type as does the future state hoped for by most radicals and socialists. *Authoritarian* is intended to suggest a more institutionalized form of despotism, in which competing power groupings cannot evade the infrastructural reach of the state, nor are they structurally separate from the state (as they are in the bureaucratic type). All significant social power must go through the authoritative command structure of the state. Thus it is high on both dimensions, having high despotic power over civil society groups and being able to enforce this infrastructurally. In their different ways, Nazi Germany and the Soviet Union tend towards this case. But they probably traded off some loss of infrastructural penetration for high despotic powers (thus neither attained as high a level of social mobilization during World War II as the 'despotically weak' but participatory Great Britain did). Nor is this to deny that such states contain competing interest groups which may possess different bases in 'civil society'. Rather, in an authoritarian state power is transmitted through its directives and so such groups compete for direct control of the state. It is different in the capitalist democracies where the power of the capitalist class, for example, permeates the whole of society, and states generally accept the rules and rationality of the surrounding capitalist economy.

These are ideal-types. Yet my choice of real historical examples which roughly approximate to them reveals two major tendencies which are obvious enough yet worthy of explanation. Firstly, there has occurred a long-term historical growth in the infrastructural power of the state, apparently given tremendous boosts by industrial societies, but also perceptible within both pre-industrial and industrial societies considered separately. Second, however, within each historical epoch have occurred wide variations in despotic powers. There has been no general development tendency in despotic powers – non-despotic states existed in late fourth millennium B.C. Mesopotamia (the 'primitive democracy' of the early city-states), in first millennium B.C. Phoenicia, Greece and Rome, in medieval republics and city-states, and in the modern world alike. The history of despotism has been one of oscillation, not development. Why such wide divergencies on one dimension, but a developmental trend on the other?

### *The development of state infrastructural power*

The growth of the infrastructural power of the state is one in the logistics of political control. I will not here enumerate its main historical phases. Instead, I example some logistical techniques which have aided effective state penetration of social life, each of which has had a long historical development.

a) A division of labour between the state's main activities which it co-ordinated centrally. A microcosm of this is to be found on the battlefields of history where a co-ordinated administrative division between infantry, cavalry and artillery, usually organized by the state, would normally defeat forces in which these activities were mixed up – at least in 'high intensity' warfare.
b) Literacy, enabling stabilized messages to be transmitted through the state's territories by its agents, and enabling legal responsibilities to be codified and stored. Giddens (1981) emphasizes this 'storage' aspect of state power.

c) Coinage, and weights and measures, allowing commodities to be exchanged under an ultimate guarantee of value by the state.
d) Rapidity of communication of messages and of transport of people and resources through improved roads, ships, telegraphy, etc.

States able to use relatively highly-developed forms of these techniques have possessed greater capacity for infrastructural penetration. This is pretty obvious. So is the fact that history has seen a secular process of infrastructural improvements.

Yet none of these techniques is specific to the state. They are part of general social development, part of the growth of human beings' increasing capacities for collective social mobilization of resources. Societies in general, not just their states, have advanced their powers. Thus none of these techniques necessarily changes the relationship between a state and its civil society; and none is necessarily pioneered by either the state or civil society.

Thus state power (in either sense) does not derive from techniques or means of power that are peculiar to itself. The varied techniques of power are of three main types: military, economic and ideological. They are characteristic of all social relationships. The state uses them all, adding no fourth means peculiar to itself. This has made reductionist theories of the state more plausible because the state seems dependent on resources also found more generally in civil society. If they are all wrong, it is not because the state manipulates means of power denied to other groups. The state is not autonomous in *this* sense.

Indeed, the fact that the means used are essentially also the means used in all social relationships ensures that states rarely diverge far from their civil societies. Let us examine what happens when a state pioneers an increase in logistic powers. A characteristic, though slow-paced, example is literacy.

The first stages of literacy in Mesopotamia, and probably also in the other major independent cases of the emergence of civilization, occurred within the state. In this respect, the state was largely codifying and stabilizing two kinds of emergent norms, 'private' property rights and community rights and duties. The first pictograms and logograms enabled scribes at city-state temple-storehouses to improve their accountancy systems, and denote more permanently who possessed what and who owed what to the community. It solidified relations radiating across the surrounding territory and centred them more on itself. Writing then simplified into syllabic cuneiform script still essentially within the state bureaucracy, and performing the same dual functions. Writing was an important part of the growth of the first imperial states, that is of the Akkadian and subsequent Empires of the third and second millennia B.C. Literacy was restricted to the bureaucracy, stabilized its systems of justice and communications and so provided infrastructural support to a state despotism, though apparently in some kind of alliance with a property-owning economic class.

Yet the general utility of literacy was now recognized by civil society groups. By the time that the next simplifications, alphabetic script and parchment, became common (around the beginning of the first millennium

B.C.) state domination had ended. The main pioneers were now not despotic states but decentralized groups of peasant-traders, village priests, and trading peoples organized into loose federations of small city- or tribal-states (like the Arameans, the Phoenicians and the Greeks). From then on, the power of such groups, usually with non-despotic states, rivalled that of the despotic empires. What had started by bolstering despotism continued by undermining it when the techniques spread beyond state confines. The states could not keep control over their own logistical inventions. And this is generally the case of all such inventions, whatever period of history we consider. In our time we have instances such as 'statistics': originally things which appertain to the state, later a useful method of systematic information-gathering for any power organization, especially large capitalist corporations.

However, converse examples are not difficult to find either, where states appropriate infrastructural techniques pioneered by civil society groups. The course of industrialization has seen several such examples, culminating in the Soviet Union whose state communications, surveillance and accountancy systems are similar to those pioneered by capitalist enterprises (with their states as junior partners) in the West. In this example what started in civil society continued in state despotism. Infrastructural techniques diffuse outwards from the particular power organizations that invented them.

Two conclusions emerge. First, in the whole history of the development of the infrastructure of power there is virtually no technique which belongs necessarily to the state, or conversely to civil society. Second, there is some kind of oscillation between the role of the two in social development. I hope to show later that it is not merely oscillation, but a dialectic.

The obvious question is: if infrastructural powers are a general feature of society, in what circumstances are they appropriated by the state? How does the state acquire in certain situations, but not others, despotic powers? What are the origins of the autonomous power of the state? My answer is in three stages, touching upon the *necessity* of the state, its *multiplicity of functions*, and its *territorialized centrality*. The first two have often been identified in recent theory, the third is I think novel.

## Origins of state power

### *The necessity of the state*

The only stateless societies have been primitive. There are no complex, civilized societies without any centre of binding rule-making authority, however limited its scope. If we consider the weak feudal cases we find that even they tend to arise from a more state-centred history whose norms linger on to reinforce the new weak states. Feudal states tend to emerge either as a check to the further disintegration of a once-unified larger state (as in China and Japan) or as a post-conquest division of the spoils among the victorious, and obviously united, conquerors (see Lattimore 1957). Western European feudalism embodies both these histories, though in varying mixtures in different regions. The laws of the feudal states in Europe were reinforced by rules descending from Roman law (especially property law), Christian codes of conduct, and Germanic notions of loyalty and honour. This is a further glimpse of a process to which I will return later: a perpetual dialectic of movement between state and civil society.

Thus societies with states have had superior survival value to those without them. We have no examples of stateless societies long enduring past a primitive level of development, and many examples of state societies absorbing or eliminating stateless ones. Where stateless societies conquer ones with states, they either themselves develop a state or they induce social regress in the conquered society. There are good sociological reasons for this. Only three alternative bases for order exist, force, exchange and custom, and none of these are sufficient in the long-run. At some point new exigencies arise for which custom is inadequate; at some point to bargain about everything in exchange relations is inefficient and disintegrating; while force alone, as Parsons emphasized, will soon 'deflate'. In the long-run normally taken for granted, but enforceable, rules are necessary to bind together strangers or semi-strangers. It is not requisite that all these rules are set by a single monopolistic state. Indeed, though the feudal example is extreme, most states exist in a multi-state civilization which also provides certain normative rules of conduct. Nevertheless most societies seem to have required that some rules, particularly those relevant to the protection of life and property, be set monopolistically, and this has been the province of the state.

From this necessity, autonomous state power ultimately derives. The activities of the state personnel are necessary to society as a whole and/or to the various groups that benefit from the existing structure of rules which the state enforces. From this functionality derives the potentiality for exploitation, a lever for the achievement of private state interests. Whether the lever is used depends on other conditions, for – after all – we have not even established the existence of a permanent state cadre which might have identifiable interests. But necessity is the mother of state power.

*The multiplicity of state functions*

Despite the assertions of reductionists, most states have not in practice devoted themselves to the pursuit of a single function. 'Binding rule-making' is merely an umbrella term. The rules and functions have been extremely varied. As the two-dimensional models recognize, we may distinguish domestic and international, or economic, ideological and military functions. But there are many types of activity and each tends to be functional for differing 'constituencies' in society. I illustrate this with reference to what have been probably the four most persistent types of state activities.

a) *The maintenance of internal order.* This may benefit all, or all law-abiding, subjects of the state. It may also protect the majority from arbitrary usurpations by socially and economically powerful groups, other than those allied to the state. But probably the main benefit is to protect existing property relations from the mass of the property-less. This function probably best serves a dominant economic class constituency.

b) *Military defence/aggression*, directed against foreign foes. 'War parties' are rarely coterminous with either the whole society or with one particular class within it. Defence may be genuinely collective; aggression usually has more specific interests behind it. Those interests may be quite widely shared by all 'younger sons' without inheritance rights or all those expansively-minded; or they might comprise only a class fraction of an aristocracy, merchants or capitalists. In multi-state systems war usually involves alliances with other states, some of whom may share the same religion, ethnicity, or political philosophy as some domestic constituency. These are rarely reducible to economic class. Hence war and peace constituencies are usually somewhat idiosyncratic.
c) *The maintenance of communications infrastructures*: roads, rivers, message systems, coinages, weights and measures, marketing arrangements. Though few states have monopolized all of these, all states have provided some, because they have a territorial basis which is often most efficiently organized from a centre. The principal constituencies here are a 'general interest' and more particular trade-centred groups.
d) *Economic redistribution*: the authoritative distribution of scarce material resources between different ecological niches, age-groups, sexes, regions, classes, etc. There is a strongly collective element in this function, more so than in the case of the others. Nevertheless, many of the redistributions involve rather particular groups, especially the economically inactive whose subsistence is thus protected by the state. And economic redistribution also has an international dimension, for the state normally regulates trade relations and currency exchanges across its boundaries, sometimes unilaterally, sometimes in alliance with other states. This also gives the state a particular constituency among merchants and other international agents – who, however, are rarely in agreement about desirable trade policy.

These four tasks are necessary, either to society as a whole or to interest groups within it. They are undertaken most efficiently by the personnel of a central state who become indispensable. And they bring the state into functional relations with diverse, sometimes cross-cutting groups between whom there is room to manoeuvre. The room can be exploited. Any state involved in a multiplicity of power relations can play off interest groups against each other.

It is worth noting that one example of this 'divide-and-rule' strategy has been a staple of sociological analysis. This is the case of a 'transitional state', living amid profound economic transformations from one mode of production to another. No single dominant economic class exists, and the state may play off traditional power groups against emergent ones. Such situations were discussed by both the classic stratification theorists. Marx analysed and satirized Louis Bonaparte's attempts to play off the factions of industrial and finance capital, petite bourgeoisie, peasantry and proletariat to enhance his own independent power. This is the 'Bonapartist balancing act', so stressed by Poulantzas (1972) – though Marx (and Poulantzas) rather under-estimated Bonaparte's ability to succeed (see Perez-Diaz 1979). Weber was struck by the ability of the Prussian State to use a declining economic class, the agrarian landlord Junkers, to hold on to autocratic power in the vacuum created by the political timidity of the rising bourgeois and proletarian classes (see Lachman 1970: 92–142). All the various groups in both examples needed the state, but none could capture it. Another example is the development of absolutism in early modern Europe. Monarchs played off against each other (or were unable to choose between) feudal and bourgeois, land and urban, groups. In particular, military functions and functions performed in relation to dominant economic classes were different. States used war as a means of attempting to reduce their dependence on classes (as Skocpol 1979 and Trimberger 1978 both argue).

These are familiar examples of the state balancing between what are predominantly classes or class factions. But the balancing possibilities are much more numerous if the state is involved in a multiplicity of relations with groups which may on some issues be narrower than classes and on others wider. Because most states are pursuing multiple functions, they can perform multiple manoeuvres. The 'Bonapartist balancing act' is a skill acquired by most states. This manoeuvring space is the birthplace of state power.

And this is about as far as the insights contained within current two-dimensional theory can be expanded. It is progress, but not enough. It does not really capture the *distinctiveness* of the state as a social organization. After all, necessity plus multiplicity of function, and the balancing-act, are also the power-source and stock-in-trade of any ruthless committee chairperson. Is the state only a chair writ large? No, as we will now see.

### *The territorial centrality of the state*

The definition of the state concentrates upon its institutional, territorial, centralized nature. This is the third, and most important, precondition of state power. As noted, the state does not possess a distinctive *means* of power independent of, and analogous to, economic, military and ideological power. The means used by states are only a combination of these, which are also the means of power used in all social relationships. However, the power of the state is irreducible in quite a different *socio-spatial* and *organizational* sense. Only the state is inherently centralized over a delimited territory over which it has authoritative power. Unlike economic, ideological or military groups in civil society, the state elite's resources radiate authoritatively outwards from a centre but stop at defined territorial boundaries. The state is, indeed, a *place* – both a central place and a unified territorial reach. As the principal forms of state autonomous power will flow from this distinctive attribute of the state, it is important that I first prove that the state does so differ socio-spatially and organizationally from the major power groupings of civil society.

*Economic* power groupings – classes, corporations, merchant houses, manors, plantations, the *oikos*, etc. – normally exist in decentred, competitive or conflictual relations with one another. True, the internal arrangements of some of them (e.g. the modern corporation, or the household and manor of the great feudal lord) might be relatively centralized. But, first, they are oriented outwards to further opportunities for economic advantage which are not territorially confined nor subject to authoritative rules governing expansion (except by states). Economic power expansion is not authoritative, commanded – it is 'diffused', informally. Second, the scope of modern and some historic economic institutions is not territorial. They do not exercise general control of a specific territory, they control a specialised function and seek to extend it 'transnationally' wherever that function is demanded and exploitable. General Motors does not rule the territory around Detroit, it rules the assembly of automobiles and some aspects of the economic life-chances of its employees, stockholders and consumers. Third, in those cases where economic institutions have been authoritative, centralized and territorial (as in the feudal household/manor of historic nobilities) they have either been subject to a higher level of territorial, central control by the (imperial) state, or they have acquired political function (administering justice, raising military levies, etc.) from a weak (feudal) state and so become themselves 'mini-states'. Thus states cannot be the simple instrument of classes, for they have a different territorial scope.

Analogous points can be made about ideological power movements like religions. Ideologies (unless state-led) normally spread even more diffusely than economic relations. They move diffusely and 'interstitially' inside state territories, spreading through communication networks among segments of a state's population (like classes, age-cohorts, genders, urban/rural inhabitants, etc.); they often also move transnationally right through state boundaries. Ideologies may develop central, authoritative, Church-like institutions, but these are usually functionally, more than territorially, organized: they deal with the sacred rather than the secular, for example. There is a socio-spatial, as well as a spiritual, 'transcendence' about ideological movements, which is really the opposite of the territorial bounds of the state.

It is true, however, that military power overlaps considerably with the state, especially in modern states who usually monopolize the means of organized violence. Nevertheless, it is helpful to treat the two as distinct sources of power. I have not the space here to fully justify this (see Mann 1986). Let me instead make two simple points. First, not all warfare is most efficiently organized territorially-centrally – guerillas, military feudalism and warrior bands are all examples of relatively decentred military organizations effective at many historical periods. Second, the effective scope of military power does not cover a single, unitary territory. In fact, it has two rather different territorial radii of effective control.

Militaristic control of everyday behaviour requires such a high level of organized coercion, logistical back-up and surplus extraction that it is practical only within close communications to the armed forces in areas of high surplus availability. It does not spread evenly over entire state territories. It remains concentrated in pockets and along communications routes. It is relatively ineffective at penetrating peasant agriculture, for example.

The second radius enables, not everyday control, but the setting of broad limits of outward compliance over far greater areas. In this case, failure to comply with broad parameters such as the handing over of tribute, the performance of ritual acts of submission, occasional military support (or at least non-rebellion), could result in a punitive expedition, and so is avoided. This radius of military striking power has normally been far greater than that of state political control, as Owen Lattimore (1962) brilliantly argued. This is obviously so in the world today, given the capabilities of modern armaments. It is also true of the Superpowers in a more subtle sense: they can impose 'friendly' regimes and de-stabilize the unfriendly through client military elites and their own covert para-military organizations, but they cannot get those regimes to conform closely to their political dictates. A more traditional example would be Britain's punitive expedition to the Falklands, capable of defeating and so de-legitimizing the Argentine regime, and remaining capable of repeating the punishment, but quite incapable of providing a political future for the Islands. The logistics of 'concentrated coercion' – that is, of military power – differ from those of the territorial centralized state. Thus we should distinguish the two as power organizations. The militarist theory of the state is false, and one reason is that the state's organization is not coterminous with military organization.

The organizational autonomy of the state is only partial – indeed, in many particular cases it may be rather small. General Motors and the capitalist class in general, or the Catholic Church, or the feudal lords and knights, or the US military, are or were quite capable of keeping watch on states they have propped up. Yet they could not do the states' jobs themselves unless they changed their own socio-spatial and organizational structure. A state autonomous power ensues from this difference. Even if a particular state is set up or intensified merely to institutionalize the relations between given social groups, this is done by concentrating resources and infrastructures in the hands of an institution that has different socio-spatial and organizational contours to those groups. Flexibility and speed of response entail concentration of decision-making and a tendency towards permanence of personnel. The decentred non-territorial interest-groups that set up the state in the first place are thus less able to control it. Territorial-centralization provides the state with a potentially independent basis of power mobilization being necessary to social development and uniquely in the possession of the state itself.

If we add together the necessity, multiplicity and territorial-centrality of the state, we can in principle explain its autonomous power. By these means the state elite possesses an independence from civil society which, though not absolute, is no less absolute in principle than the power of any other major group. Its power cannot be reduced to their power either directly or 'ultimately' or 'in the last instance'. The state is not merely a locus of class struggle, an instrument of class rule, the factor of social cohesion, the expression of core values, the centre of social allocation processes, the institutionalization of military force (as in the various reductionist theories) – it is a different socio-spatial organization. As a consequence we can treat states as *actors*, in the person of state elites, with a will to power and we can engage in the kind of 'rational action' theory of state interests advocated by Levi (1981).

*The mechanisms for acquiring autonomous state power*

Of course, this in itself does not confer a significant degree of actual power upon the state elite, for civil society groups even though slightly differently organized may yet be able to largely control it. But the principles do offer us a pair of hypotheses for explaining variations of power. (1) State infrastructural power derives from the social utility in any particular time and place of forms of territorial-centralization which cannot be provided by civil society forces themselves. (2) The extent of state despotic power derives from the inability of civil society forces to control those forms of territorial-centralization, once set up. Hence, there are two phases in the development of despotism: the growth of territorial-centralization, and the loss of control over it. First function, then exploitation – let us take them in order.

Because states have undertaken such a variety of social activities, there are also numerous ways in which at different times they have acquired a disproportionate part of society's capacity for infrastructural co-ordination. Let me pick out three relatively uncontentious examples: the utility of a redistributive economy, of a co-ordinated military command for conquest or defence, and of a centrally co-ordinated 'late development' response to one's rivals. These are all common conditions favouring the territorial-centralization of social resources.

The redistributive state seems to have been particularly appropriate, as anthropologists and archeologists argue, in the early history of societies before the exchange of commodities was possible. Different ecological niches delivered their surpluses to a central store-house which eventually became a permanent state. The case is often over-argued (e.g. by Service 1975), but it has often been archeologically useful (see Renfrew 1972).

The military route was, perhaps, the best-known to the nineteenth-century and early twentieth-century theorists like Spencer (1969 edition), Gumplowicz (1899) and Oppenheimer (1975 edition). Though they exaggerated its role, there is no doubt that most of the well-known ancient Empires had the infrastructural powers of their states considerably boosted by their use of centralized, highly organized, disciplined, and well-equipped military forces for both defence and further conquest. Rome is the example best-known to us (see Hopkins 1978).

Thirdly, the response of late industrial developers in the nineteenth and twentieth centuries to the interference of their early-industrializing rivals is well-known: a cumulative development, through countries like France, Prussia, Japan and Russia of more and more centralized and territorially-confined mobilization of economic resources with state financing and state enterprises sheltering behind tariff walls (classically stated by Gerschenkron 1962). But it also has earlier parallels – for example, in the history of Assyria or the early Roman Republic, imitating earlier civilizations, but in a more centralized fashion.

Note that in all cases it is not economic or military necessity *per se* that increases the role of the state, for this might merely place it into the hands of classes or military groups in civil society. It is rather the more particular utility of economic or military *territorial centralization* in a given situation. There are other types of economy (e.g. market exchange) and of military organization (e.g. feudal cavalry or chariotry, castle defence) which encourage decentralization and so reduce state power. In all these above examples the principal power groupings of civil society *freely* conferred infrastructural powers upon their states. My explanation thus starts in a functionalist vein. But functions are then exploited and despotism results. The hypothesis is that civil society freely gives resources but then loses control and becomes oppressed by the state. How does this happen?

Let us consider first that old war-horse, the origins of the state. In some theories of state origins, the loss of control by 'civilians' is virtually automatic. For example, in the militarist tradition of theory, the leading warriors are seen as automatically converting temporal, legitimate authority in war-time to permanent, coercive power in peace-time. Yet as Clastres (1977) has pointed out, primitive societies take great precautions to ensure that their military leaders do not become permanent oppressors. Similarly, the redistributive state of the anthropologists seems to have contained a number of checks against chiefly usurpation which makes its further development problematic. In fact, it seems that permanent, coercive states did not generally evolve in later pre-history. Only in a few unusual cases (connected with the regional effects of alluvial agriculture) did 'pristine' states evolve endogenously, and they influenced all other cases. (I make this argument at greater length in Mann 1986.) The problem seems to be that for centralized functions to be converted into exploitation, organizational resources are necessary that only actually appeared with the emergence of civilized, stratified, state societies – which is a circular process.

However, the process is somewhat clearer with respect to the intensification of state power in already-established, stratified, civilized societies with states. It is clearest of all in relation to military conquest states. We know enough about early Rome and other, earlier cases to extend Spencer's notion of 'compulsory co-operation' (outlined in Mann 1977). Spencer saw that conquest may put new resources into the hands of the conquering centralized command such that it was able to attain a degree of autonomy from the groups who had set it in motion. But Spencer's argument can be widened into the sphere of agricultural production. In pre-industrial conditions increasing the productivity of labour usually involved increasing the intensity of effort. This was most easily obtained by coercion. A militarized economy could increase output and be of benefit to civil society at large, or at least to its dominant groups. Obviously, in most agricultural conditions, coercion could not be routinely applied. But where labour was concentrated – say, in irrigation agriculture, in plantations, mines and in construction works – it could. But this required the maintenance of centralized militarism, because a centralized regime was more efficient at using a minimum of military resources for maximum effect.

This would really require considerable elaboration. In my work I call it 'military Keynesianism' (see Mann 1986) because of the multiplier effects which are generated by military force. These effects boost the despotic power of the state *vis-à-vis* civil society because they make useful the maintenance of centralized compulsory co-operation, which civil society cannot at first provide itself. It is an example of how centralization increases general social resources – and thus no powerful civil society group wishes to dispense with the state – yet also increases the private power resources of the state elite. These can now be used despotically against civil society.

Provided the state's activities generate extra resources, then it has a particular logistical advantage. Territorial-centralization gives effective mobilizing potentialities, able to concentrate these resources against any particular civil society group, even though it may be inferior in overall resources. Civil society groups may actually endorse state power. If the state upholds given relations of production, then the dominant economic class will have an interest in efficient state centralization. If the state defends society from outside aggressors, or represses crime, then its centrality will be supported quite widely in society. Naturally, the degree of centralization useful to these civil society interests will vary according to the system of production or method of warfare in question. Centrality can also be seen in the sphere of ideology, as Eisenstadt (1969) argues. The state and the interests it serves have always sought to uphold its authority by a claim to 'universalism' over its territories, a detachment from all particularistic, specialised ties to kin, locality, class, Church, etc. Naturally in practice states tend to represent the interests of particular kinship groupings, localities, classes, etc., but if they appeared merely to do this they would lose all claim to distinctiveness and to legitimacy. States thus appropriate what Eisenstadt calls 'free-floating resources', not tied to any particular interest group, able to float throughout the territorially-defined society.

This might seem a formidable catalogue of state powers. And yet the autonomous power achievements of historical states before the twentieth century were generally limited and precarious. Here we encounter the fundamental logistical, infrastructural constraints operating against centralized regimes in extensive agrarian societies. We return to the greater effective range of punitive military action compared to effective political rule. Without going into detailed logistical calculations here, but drawing on the seminal work of Engel (1978) and van Creveld (1977), we can estimate that in Near Eastern imperial societies up to Alexander the Great the maximum unsupported march possible for an army was about 60–75 miles. Alexander and the Romans may have extended it to nearly 100 miles, and this remained the maximum until the eighteenth century in Europe when a massive rise in agricultural productivity provided the logistical basis for far wider operations. Before then further distances required more than one campaigning phase, or – far more common if some degree of political control was sought – it required elaborate negotiations with local allies regarding supplies. This is enhanced if routine political control is desired without the presence of the main army. So even the most pretentious of despotic rulers actually ruled through local notables. All extensive societies were in reality 'territorially federal'. Their imperial rule was always far feebler than traditional images of them allows for (this is now well-recognized by many writers e.g. Kautsky 1982; Gellner 1983: Chap. 2; Giddens 1981: 103–4).

So we have in this example contrary tendencies – militaristic centralization followed by fragmenting federalism. Combining them we get a dialectic. If compulsory co-operation is successful, it increases both the infrastructure and the despotic power of the state. But it also increases social infrastructure resources in general. The logistical constraints mean that the new infrastructures cannot be kept within the body politic of the state. Its agents continually 'disappear' into civil society, bearing the state's resources with them. This happens continually to such regimes. The booty of conquest, land grants to military lieutenants, the fruits of office, taxes, literacy, coinage all go through a two-phase cycle, being first the property of the state then private (in the sense of 'hidden') property. And though there are cases where the fragmentation phase induces social collapse, there are others where civil society can use the resources which the despotic state has institutionalized, without needing such a strong state. The Arameans, Phoenicians and Greeks appropriated, and further developed, the techniques pioneered by the despotic states of the Near East. Christian Europe appropriated the Roman heritage.

My examples are relatively militaristic only because the process is easiest to describe there. It was a general dialectic in agrarian societies. In other words, imperial and feudal regimes do not merely oscillate (as Weber, Kautsky and many others have argued), they are entwined in a dialectical process. A range of infrastructure techniques are pioneered by despotic states, then appropriated by civil societies (or *vice-versa*); then further opportunities for centralized co-ordination present themselves, and the process begins anew. Such trends are as visible in early modern societies as in the ancient ones from which I have drawn my examples.

Such a view rejects a simple antithesis, common to ideologies of our own time, between the state and civil society, between public and private property. It sees the two as continuously, temporally entwined. More specifically it sees large private property concentrations – and, therefore, the power of dominant classes – as normally boosted by the fragmentation of successful, despotic states, not as the product of civil society forces alone. So the power autonomy of both states and classes has essentially fluctuated, dialectically. There can be no general formula concerning some 'timeless' degree of autonomous state power (in the despotic sense).

But the contemporary situation is relatively unclear. Power infrastructures leaped forward with the Industrial Revolution. Industrial capitalism destroyed 'territorially federal' societies, replacing them with nation states across whose territories unitary control and surveillance structures could penetrate (as Giddens has been recently arguing, e.g. 1981). Logistical penetration of territory has increased exponentially over the last century and a half.

What happens if a state acquires control of all those institutions of control divided historically and elsewhere between states, capitalist enterprises,

Churches, charitable associations, etc.? Is that the end of the dialectic, because the state can now keep what it acquires? Obviously, in macro-historical terms the Soviet Union can control its provincial agents, and hence its provinces, in a way that was flatly impossible for any previous state. Moreover, though its degree of elective authoritarianism can be easily exaggerated (as in 'totalitarian' theories, for example), its centralization tendencies are novel in form as well as extent. Group struggles are not decentralized, as they are substantially in the capitalist democracies, nor do they fragment as they did in agrarian societies. Struggle is itself centralized: there is something pulling the major contending forces – the 'liberals', 'technocrats', 'military/heavy industry complex', etc. – towards the Praesidium. They cannot evade the state, as agrarian dissenters did; they cannot struggle outside the state, as capitalists and workers often do. Does this authoritarian state exist despotically 'above' society, coercing it with its own autonomous power resources? Or does its authoritarian despotism exist in milder terms, firstly as a place in which the most powerful social forces struggle and compromise, and secondly as a set of coercive apparatuses for enforcing the compromise on everyone else? This has long been debated among theorists of the Soviet Union. I do not pretend to know the answer.

The bureaucratic states of the West also present problems. They are much as they were in relative power terms before the exponential growth in logistical powers began. Whatever the increases in their infrastructural capacities, these have not curbed the decentred powers of the capitalist class, its major power rival. Today agencies like multi-national corporations and international banking institutions still impose similar parameters of capitalist rationality as their predecessors did over a century ago. State elites have not acquired greater power autonomy despite their infrastructural capacities. Again, however, I am touching upon some of the central unsolved theoretical issues concerning contemporary societies. And, again, I offer no solution. Indeed, it may require a longer-run historical perspective than that of our generation to solve them, and so to decide whether the Industrial Revolution did finish off the agrarian dialectic I described.

Thus the impact of state autonomy on despotic power has been ambiguous. In terms of traditional theory results might seem disappointing: the state has not consistently possessed great powers – or indeed any fixed level of power. But I have discussed interesting power processes of a different kind. In agrarian societies states were able to exploit their territorial-centrality, but generally only precariously and temporarily because despotic power also generated its own antithesis in civil society. In industrial societies the emergence of authoritarian states indicates much greater potential despotism, but this is still somewhat controversial and ambiguous. In the capitalist democracies there are few signs of autonomous state power – of a despotic type.

But perhaps, all along, and along with most traditional theory, we have been looking for state power in the wrong place. By further examining infrastructural power we can see that this is the case.

**Results: infrastructural power**

Any state which acquires or exploits social utility will be provided with infrastructural supports. These enable it to regulate, normatively and by force, a *given* set of social and territorial relations, and to erect boundaries against the outside. New boundaries momentarily reached by previous social interactions are stabilized, regulated, and heightened by the state's universalistic, monopolistic rules. In this sense the state gives territorial bounds to social relations whose dynamic lies outside of itself. The state *is* an arena, the condensation, the crystallization, the summation of social relations within its territories – a point often made by Poulantzas (1972). Yet, despite appearances, this does not support Poulantzas' reductionist view of the state, for this is an *active* role. The state may promote great social change by consolidating territoriality which would not have occurred without it. The importance of this role is in proportion to its infrastructural powers: the greater they are or become, the greater the territorializing of social life. Thus even if the state's every move toward despotism is successfully resisted by civil society groups, massive state-led infrastructural re-organization may result. Every dispute between the state elite and elements of civil society, and every dispute among the latter which is routinely regulated through the state's institutions, tends to focus the relations and the struggles of civil society on to the territorial plane of the state, consolidating social interaction over that terrain, creating territorialized mechanisms for repressing or compromising the struggle, and breaking both smaller local and also wider transnational social relationships.

Let me give an example (elaborated in much more empirical detail in Mann 1980). From the thirteenth century onward, two principal social processes favoured a greater degree of territorial centralization in Europe. First, warfare gradually favoured army command structures capable of routine, complex co-ordination of specialised infantry, cavalry and artillery. Gradually, the looser feudal levy of knights, retainers and a few mercenaries became obsolete. In turn this presupposed a routine 'extraction-coercion cycle' to deliver men, monies and supplies to the forces (see the brilliant essay by Finer 1975). Eventually, only territorially-centred states were able to provide such resources and the Grand Duchies, the Prince-Bishops and the Leagues of Towns lost power to the emerging 'national' states. Second, European expansion, especially economic expansion taking an increasingly capitalistic form, required (1) increased military protection abroad, (2) more complex legal regulation of property and market transactions, and (3) domestic property forms (like rights to common lands). Capitalistic property owners sought out territorial states for help in these matters. Thus European states gradually acquired far greater infrastructural powers: regular taxation, a monopoly over military mobilization, permanent bureaucratic administration, a monopoly of law-making and enforcement. In the long-run, despite attempts at absolutism, states failed to acquire despotic powers through this because it also enhanced the infrastructural capacities of civil society groups, especially of capitalist property-holders. This was most marked in Western Europe and as the balance of geo-political power tilted Westwards – and especially to Britain – the despotically weak state proved the general model for the modern era. States governed with, and usually in the interests of, the capitalist class.

But the process and the alliance facilitated the rise of a quite different type of state power, infrastructural in nature. When capitalism emerged as dominant, it took the form of a series of territorial segments – many systems of production and exchange, each to a large (though not total) extent bounded by a state and its overseas sphere of influence. The nation-state system of our own era was not a product of capitalism (or, indeed, of feudalism) considered as pure modes of production. It is in that sense 'autonomous'. But it resulted from the way expansive, emergent, capitalist relations were given regulative boundaries by pre-existing states. The states were the initially weak (in both despotism and in infrastructure) states of feudal Europe. In the twelfth century even the strongest absorbed less than 2% of GNP (if we could measure it), they called out highly decentralized military levies of at most 10 to 20,000 men sometimes only for 30 days in the campaigning system, they couldn't tax in any regular way, they regulated only a small proportion of total social disputes – they were, in fact, marginal to the social lives of most Europeans. And yet these puny states became of decisive importance in structuring the world we live in today. The need for territorial centralization led to the restructuring of first European, then world society. The balance of nuclear terror lies between the successor states of these puny Europeans.

In the international economic system today, nation-states appear as collective economic actors. Across the pages of most works of political economy today stride actors like 'The United States', 'Japan', or 'The United Kingdom'. This does not necessarily mean that there is a common 'national interest', merely that on the international plane there are a series of collectively organized power actors, nation-states. There is no doubting the economic role of the nation-state: the existence of a domestic market segregated to a degree from the international market, the value of the state's currency, the level of its tariffs and import quotas, its support for its indigenous capital and labour, indeed, its whole political economy is permeated with the notion that 'civil society' is its territorial domain. The territoriality of the state has created social forces with a life of their own.

In this example, increasing territoriality has not increased despotic power. Western states were despotically weak in the twelfth century, and they remain so today. Yet the increase in infrastructural penetration has increased dramatically territorial boundedness. This seems a general characteristic of social development: increases in state infrastructural powers also increase the territorial boundedness of social interaction. We may also postulate the same tendency for despotic power, though it is far weaker. A despotic state without strong infrastructural supports will only claim territoriality. Like Rome and China it may build walls, as much to keep its subjects in as to keep 'barbarians' out. But its success is limited and precarious. So, again we might elaborate a historical dialectic. Increases in state infrastructural power will territorialize social relations. If the state then loses control of its resources they diffuse into civil society, decentering and de-territorializing it. Whether this is, indeed, beginning to happen in the contemporary capitalist world, with the rise of multi-national corporations outliving the decline of two successively hegemonic states, Great Britain and the United States, is one of the most hotly-debated issues in contemporary political economy. Here I must leave it as an open issue.

In this essay I have argued that the state is essentially an arena, a place – just as reductionist theories have argued – and yet this is precisely the origin and mechanism of its autonomous powers. The state, unlike the principal power actors of civil society, is territorially bounded and centralized. Societies need some of their activities to be regulated over a centralized territory. So do dominant economic classes, Churches and other ideological power movements, and military elites. They, therefore, entrust power resources to state elites which they are incapable of fully recovering, precisely because their own socio-spatial basis of organization is not centralized and territorial. Such state power resources, and the autonomy to which they lead, may not amount to much. If, however, the state's use of the conferred resources generates further power resources – as was, indeed, intended by the civil society groups themselves – these will normally flow through the state's hands, and thus lead to a significant degree of power autonomy. Therefore, *autonomous state power is the product of the usefulness of enhanced territorial-centralization to social life in general*. This has varied considerably through the history of societies, and so consequently has the power of states.

I distinguished two types of state power, despotic and infrastructural. The former, the power of the state elite over civil society classes and elites, is what has normally been meant by state power in the literature. I gave examples of how territorial-centralization of economic, ideological and military resources has enhanced the despotic powers of states. But states have rarely been able to hold on to such power for long. Despotic achievements have usually been precarious in historic states because they have lacked effective logistical infrastructures for penetrating and co-ordinating social life. Thus when states did increase their 'private' resources, these were soon carried off into civil society by their own agents. Hence resulted the oscillation between imperial/patrimonial and feudal regimes first analysed by Max Weber.

By concentrating on infrastructural power, however, we can see that the oscillation was, in fact, a dialectic of social development. A variety of power infrastructures have been pioneered by despotic states. As they 'disappear' into civil society, general social powers increase. In Volume I of my work (Mann 1986), I suggest that a core part of social development in agrarian societies has been a dialectic between centralized, authoritative power structures, exemplified best by 'Militaristic Empires', and decentralized, diffused power structures exemplified by 'Multi-Power Actor Civilizations'. Thus the developmental role of the powerful state has essentially fluctuated – sometimes promoting it, sometimes retarding it.

But I also emphasized a second result of state infrastructural powers. Where these have increased, so has the territoriality of social life itself. This has usually gone unnoticed within sociology because of the unchallenged status of sociology's masterconcept: 'society'. Most sociologists – indeed, most people anywhere who use this term – mean by 'society' the territory of a state. Thus 'American society', 'British society', 'Roman society', etc. The same is true of synonyms like 'social formation' and (to a lesser extent) 'social system'. Yet the relevance of state boundaries to what we mean by societies is always partial and has varied enormously. Medievalists do not generally characterize 'society' in their time-period as state-defined; much more likely is a broader, transnational designation like 'Christendom' or 'European society'. Yet this change between medieval and modern times is one of the most decisive aspects of the great modernizing transformations; just as the current relationship between nation states and 'the world system' is crucial to our understanding of late twentieth-century society. How territorialized and centralized are societies? This is the most significant theoretical issue on which we find states exercising a massive force over social life, not the more traditional terrain of dispute, the despotic power of state elites over classes or other elites. States are central to our understanding of what a society is. Where states are strong, societies are relatively territorialized and centralized. That is the most general statement we can make about the autonomous power of the state.

## References


Bendix, R., *Kings or People* (Berkeley, University of California Press, 1978).

Clastres, P., *Society against the State* (Oxford, Blackwell, 1977).

Creveld, M. van, *Supplying War: logistics from Wallenstein to Patton* (Cambridge, Cambridge University Press, 1977).

Eisenstadt, S. M., *The Political Systems of Empires* (New York, The Free Press, 1969).

Engel, D. W., *Alexander the Great and the logistics of the Macedonian Army* (Berkeley, University of California Press, 1978).

Finer, S., State and nation-building in Europe: the role of the military, *in* Ch. Tilly (ed.), *The Formation of National States in Western Europe* (Princeton, Princeton University Press, 1975).

Gellner, E., *Nations and Nationalism* (Oxford, Blackwell, 1983).

Gerschenkron, A., *Economic Backwardness in Historical Perspective* (Cambridge, Mass., Belknap Press, 1962).

Giddens, A., *A Contemporary Critique of Historical Materialism* (London, Macmillan, 1981).

Gumplowicz, L., *The Outlines of Sociology* (Philadelphia, American Academy of Political and Social Science, 1899).

Hintze, O., *The Historical Essays of Otto Hintze* (ed. F. Gilbert) (New York, Oxford University Press, 1975).

Hopkins, K., *Conquerors and Slaves* (Cambridge, Cambridge University Press, 1978).

Kautsky, J. H., *The Politics of Aristocratic Empires* (Chapel Hill, University of North Carolina Press, 1982).

Lachman, L., *The Legacy of Max Weber* (London, Heinemann, 1970).

Lattimore, O., Feudalism in history: a review essay, *Past and Present* (1957) No. 12, 47–57.

Lattimore, O., *Studies in Frontier History* (London, Oxford University Press, 1962).

Levi, M., The predatory theory of rule, *Politics and Society*, X (1981), 431–65.

MacIver, R. M., *The Modern State* (Oxford, Clarendon Press, 1926).

Mann, M., States, ancient and modern, *Archives européennes de sociologie*, XVIII (1977), 262–298.

Mann, M., State and society, 1130–1815: an analysis of English state finances, *in* M. Zeitlin (ed.), *Political Power and Social Theory*, Vol. I (Connecticut, J.A.I. Press, 1980).

Mann, H., *The Sources of Social Power. Vol. I: a history of power from the beginning to A.D. 1760* (Cambridge, Cambridge University Press, 1986).

Mann, M., *The Sources of Social Power. Vol. II: the rise of classes and nation-states, 1760–1914* (Cambridge, Cambridge University Press, 1993).

Oppenheimer, F., *The State* (1975 edition: New York, Free Life Editions).

Perez-Diaz, V., *State, Bureaucracy and Civil Society: a critical discussion of the political theory of Karl Marx* (London, Macmillan, 1979).

Poulantzas, N., *Pouvoir politique et classes sociales* (Paris, Maspero, 1972).

Renfrew, C., *The Emergence of Civilization: the Cyclades and the Aegean in the third millennium B.C.* (London, Methuen, 1972).

Rüstow, A., *Freedom and Domination: a historical critique of civilization* (Princeton, Princeton University Press, 1982).

Service, E., *Origins of the State and Civilization* (New York, Norton, 1975).

Skocpol, T., *States and Social Revolutions* (Cambridge, Cambridge University Press, 1979).

Spencer, H., *Principles of Sociology* (one-volume abridgement) (1969 edition: London, Macmillan).

Therborn, G., *What does the Ruling Class do when it Rules* (London, New Left Books, 1978).

Tilly, Ch., *The Formation of National States in Western Europe* (Princeton, Princeton University Press, 1975).

Tilly, Ch., *As Sociology Meets History* (New York, Academic Press, 1981).

Trimberger, E., *Revolution From Above: military bureaucrats and development in Japan, Turkey, Egypt and Peru* (New Brunswick, N.J., Transaction Books, 1978).

Weber, M., *Economy and Society* (New York, Bedminster Press, 1968).

Wolin, S., *Politics and Vision: continuity and innovation in western political thought* (Boston, Little, Brown, 1961).

---

# 6 Sankaran Krishna, 'Cartographic Anxiety: Mapping the Body Politic in India'

Reprinted in full from: *Alternatives: Social Transformation and Humane Governance* 19(4), 507–21 (1994)

---

The parliamentary elections of May 1991 in India were disrupted by the assassination of the former prime minister, Rajiv Gandhi. In the aftermath of this event, one of the ubiquitous election posters of the Congress (I) party showed a picture of Rajiv Gandhi's identity card.[1] On the identity card, the space allocated for designation of religion was filled by the word *Indian*. The not-so-subtle implication was that the slain leader had, in his lifetime, transcended divisive societal identities such as Hindu or Muslim and defined himself primarily as a secular citizen of a nation-state.[2]

The effacement of alternative identities inherent in this political stratagem, and their replacement by one based on the modern, sovereign nation-state – that of *citizen* – is interesting. At one level, it can be seen as an attempt to rewrite India in terms of a univocal narrative of modern nationalism – a nationalism that is supposedly secular and hostile to all other forms of identity. In view of nationalism – or the "retrospective illusion" of nationalism, as it has been described[3] – alternative ideas of the self, be they religious, regional, linguistic, or ethnic, are rendered spurious, reactionary, and vestigial. Moreover, modernity comes to be defined in the disciplining of ambiguity and its intolerance for multiple or layered notions of identity or sovereignty: citizenship is invariably a matter of either-or.[4]

Yet, at another level, the equation of *religion* with *Indian* highlights the fact that terms like nationalism, sovereignty, and citizenship are themselves implicated in practices that are, in their own way, the rituals of a modern faith. Contemporary nationalism may then, according to this line of thought, be regarded as one in a long line of discourses that have sought to construct social reality and identity for people in a given space.[5] The production of national identity is a contested process everywhere and the struggle to produce citizens out of recalcitrant peoples accounts for much of what passes for history in modern times.[6] In India, as in many another settings, the process has been accompanied by an enormous degree of violence, both physical and epistemic. This essay looks at one small facet of this ongoing process of nation-building in India: that of cartography.

By *cartography* I mean more than the technical and scientific mapping of the country. I use the term to refer to representational practices that in various ways have attempted to inscribe something called India and endow that entity with a content, a history, a meaning, and trajectory. Under such a definition, cartography becomes nothing less than the social and political production of nationality itself. Within this, I am specifically interested in the visual depiction of competing representations of India in public culture, such as the way the country is depicted in political posters, government maps, newspaper and newsmagazine articles, election campaign literature, political rhetoric, and other such artifacts that seek to create and reproduce the symbolic universe of the Indian nation.

The argument of this article can be summarized as follows: in a postcolonial society such as India, the anxiety surrounding questions of national identity and survival is particularly acute. While this anxiety is writ large over the political culture of the country, its cartographic manifestations are particularly interesting and revealing. Specifically, the ubiquity of cartographic metaphors, the production of *inside* and *outside* along the borders of the country, reveal both the epistemic and physical violence that accompanies the enterprise of nation-building. At the same time, people who live along borders are wont to regard this latest discursive universe of nationality and territoriality as, at a minimum, one more minefield to be navigated safely, or – better – one to be profited from. The encounters between the state and the people along frontiers are suggestive of the contested and imbued production of sovereign identity. Ultimately, cartographic anxiety is a facet of a larger postcolonial anxiety: of a society suspended forever in the space between the "former colony" and "not-yet-nation." This suspended state can be seen in the discursive production of India as a bounded, sovereign entity and the deployment of this in everyday politics and in the country's violent borders. Quotidian life along the frontiers and its micropolitics renders transparent the arbitrariness and the violence of the discourse of nationhood. Cartographic anxiety, from this perspective, becomes a prominent signifier of the postcolony.

## Reading maps as texts

> Maps are too important to be left to cartographers alone.
>
> *–J. B. Harley*[7]

Despite his firm belief in the timeless existence of a spiritual and civilizational entity called India, Jawaharlal Nehru nevertheless felt compelled to begin his appropriately titled *Discovery of India*[8] with a solid and physicalistic description of the country's "natural" frontiers. Nehru's imaginative geography depicted impassable mountain ranges, vast deserts, and deep oceans that produced a natural "cradle" for what became India. And in Nehru's *Autobiography*,[9] anxiety regarding the physical boundaries of the nation is inscribed early. The narrative script that runs through Nehru's definitive work in imagining India clearly traces the country's downfall to porous frontiers and, more importantly, to an unfortunate timing by which a disunited and fragmenting India encountered the cresting and united civilization of the British. The encounter produced not only colonial rule but, with it, Nehru argued, the sources of India's eventual redemption: modernity, science, the rational spirit, and, most importantly, national unity.[10]

The worst fears regarding physical boundaries were realized in 1947 with the partition of India at independence into India and Pakistan. Nehru's recollections of that event are worth quoting:

> All our communications were upset and broken. Telegraphs, telephones, postal services and almost everything as a matter of fact was disrupted. Our services were broken up. Our army was broken up. Our irrigation systems were broken and so many other things happened. . . . *But above all, what was broken up which was of the highest importance was something very vital and that was the body of India.* That produced tremendous consequences, not only those that you saw, but those that you cannot imagine, in the minds and souls of millions of human beings.[11]

As this creation-by-amputation attempted to achieve nationhood, the corporeal element of its existence has remained in the foreground. It is perhaps unsurprising that this child of partition, India, has cartographic anxiety inscribed into its very genetic code.

In the years since, history can be (and often is) read as a series of encounters with this anxiety: if China in 1962 represented a demoralizing defeat, Bangladesh in 1971 was recorded as an important victory. Since independence (or more accurately, since partition) the anxiety has been showcased perfectly in the space of desire called Kashmir. More importantly, "accurate" representations of the body politic in maps and insignia are watched with an intensity that is perhaps unequalled elsewhere.

In India today, there is an evident obsession with "alien" infiltration, with shadowy "foreign hands" out to destabilize the country, with the need for a blue-water navy to secure the peninsula's coastlines; with plans to construct a fence along the borders with Pakistan and Bangladesh; and with various other phenomena perceived as threats to the "unity and integrity" of the nation.[12] The prevalence of cartographic themes in political discourse can be seen in the following spectacle. During the Republic Day celebrations in January 1992, the leader of the Bharatiya Janata Party (BJP), Murli Manohar Joshi, embarked on what was described as an Ekta Yatra (literally, Unity Pilgrimage: note the religious metaphor), planned to begin at the southern tip of India and to end in a grand flag-hoisting ceremony in Srinagar, the capital of Kashmir, in the north. When it appeared that the yatra would be delayed, if not aborted, by avalanches and bad weather, the nation was exhorted by the BJP's leaders to draw an outline map of India in the soil nearest them and plant the Indian tricolor on the spot representing Srinagar.[13]

Cartographic themes were prominent in the parliamentary elections of 1989, in which there was an unprecedented use of the mass media to gain popular support. The ruling Congress (I), as in other years at least since 1977, played on fears of impending national disintegration and portrayed itself as the sole party capable of averting balkanization. Congress (I)'s series of political commercials mostly centered around the physical map of India. The ads, which were published widely in leading newspapers and magazines and which included regional-language versions, hammered home this theme of the nation in danger.[14] The invocation of the map of India and the notions of anxiety and danger were made obvious – perhaps too obvious – as reflected in one of the more dramatic posters that queried: "Will this be the last time you see India in this shape?"

As the physical map of India gains ubiquity as an iconic representation of the body politic, it becomes the terrain for competing efforts to define, and possess, the self. In response, opposition parties put out their own versions of India, also in the form of a map. Listing a litany of their complaints against the ruling party, the opposition maps were dominated by a narrative that contrasted the normal and the pathological. The utility of such a trope in discussions of the body politic has a long genealogy[15] and was represented in this instance by literally inscribing the problems of the nation across the maps. The BJP's maps focused on the territorial "losses" of the Congress government, an argument underlain by an implicit reference to the Hindu fundamentalist ideas of Akhand Bharat (Undivided India).[16] The theme showed the body politic being diminished, on the one hand, by loss of territory, and on the other, by disease and impurity.

The theme of a perceived indispensability of "secure" or "inviolate" borders for national development, in fact for nationality itself, is not limited to politics. In Indian society, it is a recurring theme. A typical comment regarding borders runs:

> According to political scientists one of the prerequisites for the ordered growth of a modern nation state is settled boundaries. Once a country has well defined borders, the planned development of the various sectors of the economy becomes easier and predictable. Also, as a member of the comity of nations it will have more credibility, if not confidence, in its relations with other nations when it knows that its internationally accepted borders cannot draw it into the mire of territorial chauvinism. . . . After almost 40 years as an independent country, India still has undefined borders with two of its neighbours.[17]

The theme is repeated in news reports and editorials. Physical preservation of the borders has become synonymous with the state of the union.

What these invocations regarding the sanctity of the body politic usually push to the margins is the violence that produces the border. A classic instance of just such a production of borders is ongoing in the conflict between India and Pakistan over the Siachen Glacier. Interestingly, this part of the Indo-Pakistan border was left unmapped at Partition because the cartographer considered definition to be unnecessary: the terrain was so inhospitable and the details so sketchy that it was not anticipated the area would become a matter of contention.[18] Since April 1984, the two sides have lost well over two thousand soldiers, 97 percent of them killed by the "weather and the terrain." Only 3 percent have been lost to enemy fire. The glacier begins at 12,000 feet above sea level, in the Saltoro Range of the Himalayas, and ascends to 22,000 feet. (By way of comparison, Mount Everest, the highest point on earth, is a little over 29,000 feet high.) Most of India's soldiers manning the station are in the base camp; Pakistani soldiers are in the valley below. The Indians take turns serving at the forward posts, situated between 18,000 and 22,000 feet on the glacier. One in two soldiers posted to Siachen will die, and at any moment in time, over two thousand men are stationed at Siachen. On average, soldiers who return alive have lost from 7 to 12 kilograms in weight during their three-month stint. They undergo one of the most harrowing physical and psychological experiences conceivable. The following lengthy excerpt from a report conveys a sense of the surreal violence at this frontier.

> Last fortnight, we were witness to the funeral of the latest victim, Chandra Bhan, 28, a sepoy of the Rajput Regiment stationed at the forward post of *Pahalwan* (20,000 feet) which often comes under concentrated Pakistani artillery fire. Bhan, however, did not die because of enemy action. The cause of death was pulmonary embolism – a blood clot in the lungs caused by the rarefied air. . . . Bhan had actually died 18 days earlier. His body was laboriously carried down the steep snow-covered slope to *Zulu* (19,500 feet), the only post in the area with a helipad. . . . Eventually, pilots . . . braved gusty tail winds of over 60 kmph to land the fragile Cheetah helicopter on the narrow, snow-covered helipad. Bhan's body proved too big for the Cheetah. His legs had to be broken to ferry it to *Base Camp*. . . . Bhan's widow – expecting his third child – will get only an urn of her husband's ashes and

> his ribbons, including a grey and white one awarded to all Siachen veterans. She is fortunate. Many others won't even get that. The bodies of their menfolk – who were trapped in crevasses, buried under avalanches, or lost in blizzards – have never even been recovered.[19]

The Indo-Pakistan border, here as elsewhere, is literally being drawn in blood. Such violence is routinely finessed in news reports about life on the frontiers – reports that rewrite it into a moral economy of sacrifice for the good of the nation. Jean Elshtain has noted, "To preserve the larger civic body, which must be 'as one', particular bodies must be sacrificed."[20] Operating out of this exculpatory narrative, a retired Indian diplomat averred: "The public will have to be educated into understanding that the bloody war . . . and the miscellaneous frontier clashes which have led to the loss of life would be amply recompensed if India can be assured of secure and inviolable borders in the decades to come."[21]

Recently, the anxiety over incursions from Pakistan has reached a new high. Subversives are allegedly trained in Pakistan to foment unrest in India's provinces of Punjab and Kashmir. To circumvent the erosion of its borders with Pakistan and Bangladesh, India has created a scheme for a "security belt" along the whole of its northwestern and northeastern frontiers. Moreover, in May 1993, the Union cabinet approved a plan to issue identity cards to all border area residents in thirty-three districts in nine states. A leading newspaper, in a schizoid reading of this plan, supported it at the beginning of an editorial but ended pessimistically:

> On the face of it the Centre's decision to issue identity cards to people living in 33 border districts bordering Pakistan and Bangladesh will help remove a thorn in the side of India's body politic. On the western border the cards will be a deterrent to the infiltrator armed with guns and secessionist thoughts. In the east the card is expected to help stem the tide of millions of Bangladeshi immigrants flowing into India. . . . Cards are likely to be just another unsuccessful, largely symbolic exercise in futility like the planned border fence and the odd police combing operation. Without more imaginative administrative thinking and greater political will the identity card will be little more than an expensive successor to the ration card.[22]

In a way, this contradictory reading is a result of the narrow discursive space within which most news coverage of security matters in India operates. The concluding invocations of "political will" and "imaginative administration" reveal both a chafing at the limits of this space and lack of vision or vocabulary to transcend it. In this, as in many other instances, the hegemony of realpolitik literally narrates the nation.

The preoccupation with borders and with "alien" infiltration sounds especially disingenuous when India is faced with numerous secessionist threats from within. Many of these ethnic, linguistic, and religious movements can hardly be wished away as arising solely out of "external" meddling in "domestic" affairs. The operation of the inside/outside antinomy serves not so much to prevent "foreign infiltration" as it does to discipline and produce the "domestic(ated) self."

**Everyday life on the borderlines**

What the map cuts up, the story cuts across.
–*De Certeau*[23]

Despite the efforts of states and theorists of politics to present the post-Westphalian world of territorially sovereign nation-states as a timeless essence, this world order is historically contingent, violently produced, and contested. Along the borders of India, Pakistan, and Bangladesh are people whose lives are abstracted by the discourses of citizenship, sovereignty, and territoriality. Their recalcitrances and compromises are not to be read exclusively in a register of heroism or romantic evasions, or even as instances of everyday resistance. Rather, quotidian realities can reveal how cartography produces borders, the arbitrariness involved in the creation of normality, and the fluid definitions of space and place that prevail in the midst of efforts to hegemonize territory.

Consider the following vignette from everyday life along the Indian border with Bangladesh, as was reported in a Calcutta newspaper.

> Hoseb Ali, a resident of Nabinnagar village in Nadia (a district in West Bengal, India), sat in his courtyard, lit a *bidi* and gently tossed the matchstick away. The matchstick, still smouldering, landed in Bangladesh. "Uncle, come over, I have something to tell you," he shouted.[24]

Hoseb Ali – as the report went on to reveal – was calling his maternal uncle, Emdadul, to discuss the upcoming village-level elections being held in the state of West Bengal. They were neighbors, but it so happened that the international boundary between India and Bangladesh cut across their courtyard rendering them citizens of different countries. The border at this point was marked by a makeshift fence – a nuisance more than a frontier. At various other points, the border was a three-foot-tall bamboo fence, and elsewhere there wasn't even that.

Emdadul, irritated by the efforts of the Border Security Force (BSF – an Indian paramilitary organization) and the Bangla Desh Rifles (BDR) to keep the border a reality at least for the duration of the local village elections, asked querulously: "How can you segregate us? We have grown up together and belong to the same family. We will cross what you people call the 'border' and visit each other. Can the BSF or the BDR keep a constant watch on us?" The sector commander of the BSF, Balbir Singh Shahal, was similarly irritated. He asked: "How can we stop the infiltration? We do not understand Bengali. These people speak the same language, wear similar clothes and look no different. It is impossible to differentiate between a Bangladeshi and an Indian. Also, many live in houses adjacent to each other." In his own words, the arbitrary and violent production of a "border" by the Border Security Force becomes transparent. In the face of a reality that does not allow him to distinguish a "Bengali Indian" from a "Bengali Bangladeshi," the commander is forced to rely on the production of an alternative border – that of the nation-state and of citizenship. Commander Shahal angrily concludes that "Indians should be issued identity cards immediately." Given the impossibility of producing difference out of religious, regional, linguistic, and physical characteristics, he plumps for nationality. Yet, in the subcontinent (as elsewhere), the differentiation of nations supposedly rests upon some combination of precisely these "essentialized" characteristics.

In the meantime, people on both sides were operating within a moral economy that had comfortably internalized the border into their everyday lives. This was evident in the elections in Petrapole, another village on the border. All three candidates vying for the last seat in the *gram panchayat* (village-level administration) were well-known smugglers whose main trade was in illegal immigrants. Chitta Sardar was the Congress (I) candidate, Khagen Mistry represented the BJP, and Atiar Sheikh was from the Communist Party of India (Marxist), or CPI(M). In a very matter-of-fact way, Sardar laid out his main difference with Sheikh: "Just because his family owns the land right on the border we have to pay him 'ghat duty' [a transit fee] to transport Bangladeshis," he complained. Mistry noted that up to one thousand Bangladeshis a day could be transported across the border, at a cost of Rs30 per person. Mistry's problem was that while the Bangladeshis did all the work, Sheikh gave them only 10 percent of the proceeds and kept the rest for himself. Winning the local election was thus mainly a matter of controlling the smuggling operation.[25]

The organization supposedly charged with maintaining the border, the BSF, is also thoroughly implicated in this economy. An officer of the local bank that operates accounts for BSF personnel revealed that their repatriation of earnings to home villages far exceeded their salaries. In other words, border transgression is good business for the BSF. Meanwhile, in Mudafat, a village that has seen many Bangladeshi Hindus cross over and settle in this predominantly lower-caste area, resident Amal Sarkar noted that there was no tension among the residents. In fact, their panchayat had been adjudged the best for the 1990/91 year in its district for implementation of poverty alleviation schemes.[26] Mudafat effortlessly gives lie to the discourse of danger that invariably accompanies reports on Bangla "infiltration" into India.

Even the armed forces of the two countries have become somewhat domesticated in this context. In the neighboring state of Tripura (which also shares a border with Bangladesh), at Akhaura, the BSF checkpost stands alongside that of the BDR. Every morning (and evening) the flags of both countries go up (and come down) at a single command issued by personnel on either side of the border. Trade in staples such as fish, eggs, and clothing routinely moves uninterruptedly across the border, while marriages between parties on either side are quite frequent. As the journalist reporting on the region notes: "The bridegroom's party, accompanied by band music, crosses the border without hindrance and goes back with the bride and presentations. Sometimes it's the other way about."[27]

Running through all the above vignettes of daily life along the borders – or in the places that, on maps, are bisected by a line – are the recurring themes of discipline and abstraction. They serve to acquaint us with a part of the world we have pulverized in our minds into a space dominated by the concept, *the border*. Overcoded with the discourses of citizenship, geopolitics, and sovereignty, life here is *abstract* in the most violent sense of that word.[28] As

Michel De Certeau notes in the context of the transition from the detailed and sensuous medieval tours and itineraries to the modern map, that ensemble of abstract places,

> the map, a totalizing stage on which elements of diverse origin are brought together to form the tableau of a "state" of geographical knowledge, pushes away into its prehistory or into its posterity, as if into the wings, the operations of which it is the result or the necessary condition. It remains alone on the stage. The tour describers have disappeared.[29]

The Indian state attempts, with its maps and its various surveillances, the production and dissemination of this geopolitical second nature against the unconsciously recalcitrant practices of people along the frontiers. The status of these peoples gets rewritten in the defence of the country. An example of such a rewriting, with an interesting reversal in the valorization of human and material "resources," is seen in the following exhortation toward border management (sic):

> For effective border management it is essential that the people staying in border areas be firmly integrated with the rest of their countrymen and function like the eyes and ears of the paramilitary forces/army entrusted with guarding the frontiers. Only then will it be possible to effectively monitor our borders without incurring infrastructural expenditure, which only diverts the precious resources of the country.[30]

Rendered as synecdochical organs on a larger body politic ("the eyes and ears") people in the border areas are thus literally reduced to abstractions less than human.

## The illusion of postcoloniality

> Every established order tends to produce ... the naturalization of its own arbitrariness.
>
> –*Bourdieu*[31]

A recent collection of essays[32] on issues of postcoloniality points to the degree to which orientalist essentializations underlie both the colonial regime and the independent states of South Asia. The derivative discourses of nationality have, even as they rail against orientalism, relied upon it for their own ammunition and for their desired visions of the future. A crucial question that emerges is this: is there anything *post* about the postcolony at all? If we examine the degree of anxiety revealed by the state over matters of cartographic representation; the inordinate attention devoted to notions of security and purity; the disciplinary practices that define *Indian* and *non-Indian, patriot* and *traitor, insider* and *outsider, mainstream* and *marginal*; and the physical and epistemic violence that produces the border – if we examine these, the answer to this question is negative.

Cartographic anxiety may be described, then, as one symptom, among many, of a postcolonial condition. This preoccupation with national space and with borders reveals an obsession to approximate a historical original that never existed, except as the telos of the narrative of modernity: a pure, unambiguous community called the homeland. In this sense, postcoloniality may be defined as a condition marked by the perpetual effort of colonized societies to catch up with the putative pasts and presents of colonizing societies who anyway do not accept that they are in a race. Hence, we are back in the realm of the ironic: the definitive marker of the postcolonial society is that of one trapped in time – former colony but pre-nation. And yet, the way to modern nationhood can be only through the complete colonization of the self. Thus, to decolonize the self may mean denationalizing the narratives that embody space and time.

## Notes

1 The letter *I* in the party name stands for the late Indira Gandhi.
2 In the political universe of contemporary India, the Congress (I) has portrayed itself as the champion of secularism as opposed to the explicitly Hindu orientation of the Bharatiya Janata Party (BJP). The BJP has seen a sharp improvement in its electoral fortunes, going from a relatively insignificant share of the popular vote in the elections of 1980 and 1984 to, following the general election of 1991, being the largest opposition party in Parliament. While the religious, indeed fundamentalist, character of the BJP is in little doubt, the Congress (I)'s real commitment to secularism has recently been an issue of much contention.
3 See Etienne Balibar, "The Nation Form: History and Ideology," in Etienne Balibar and Immanuel Wallerstein, *Race, Nation, Class: Ambiguous Identities*, Chris Turner, trans. (London: Verso, 1991), pp. 86–106.
4 It is the dichotomous nature of this choice that, among other reasons, led me to title this piece, *Cartographic Anxiety*. The direct referent is Richard Bernstein's "Cartesian Anxiety," about which he writes: "*Either* there is some support for our being, a fixed foundation for our knowledge, *or* we cannot escape the forces of darkness that envelop us with madness, with intellectual and moral chaos." See Richard Bernstein, *Beyond Objectivism and Relativism: Science, Hermeneutics and Praxis* (Philadelphia: Univ. of Pennsylvania Press, 1985), p. 18. Emphasis in the original. Foucault aphorized this famously as the "blackmail of the Enlightenment" and it is a particularly appropriate notion in the context of discourses on citizenship. In the modern world system, nationality exhausts the discursive space of identity and impoverishes our imagination for alternative, multiple, or layered understandings of selves. Hence, the sterile choices presented are invariably of the order of inside/outside, patriot/traitor, nationalist/communalist, and so forth.
5 Studies that depict the emergence of the modern world of sovereign nation-states and its accompanying ideologies of nationalism, citizenship, sovereignty, and security have long commented on the synonymously religious or eschatological character of these "newest" faiths. Certainly Marx, Durkheim, and Weber, to mention only three, were cognizant of this aspect of the discourse of statism and *citoyen*. For a superb discussion, see Derek Sayer, *Capitalism and Modernity: An Excursus on Marx and Weber* (New York: Routledge, 1991). In the specific context of India, the following quote from Ashis Nandy makes my point with brevity: "Certainly in India, the ideas of nation-building, scientific growth, security, modernization and development have become parts of a left-handed technology with a clear touch of religiosity – a modern demonology, a *tantra* with a built-in code of violence. . . . To many Indians today, secularism comes as a part of a larger package consisting of many standardized ideological products and social processes

– development, mega-science and national security being some of the most prominent among them. This package often plays the same role vis-a-vis the people of the society – sanctioning or justifying violence against the weak and the dissenting – that the church, the *ulema*, the *sangha*, or the Brahmans played in earlier times." See Ashis Nandy, "The Politics of Secularism and the Recovery of Religious Tolerance," in R. B. J. Walker and Saul Mendlovitz, eds, *Contending Sovereignties: Redefining Political Community* (Boulder: Lynne Rienner, 1990), p. 134.

6 A recent work that presents a genealogy of the contested production of the terrain called Siam (Thailand) can be found in Thongchai Winichakul, *Siam Mapped: A History of the Geo-Body of the Nation* (Honolulu: Univ. of Hawaii Press, 1994) and an exemplary account of this process in France can be found in Eugen Weber, *Peasants into Frenchmen: The Modernization of Rural France, 1870–1914* (Stanford, Calif.: Stanford Univ. Press, 1976). For a discussion of the production of national boundaries via a dialectic between the local and the national see Peter Sahlins, *Boundaries: The Making of France and Spain in the Pyrenees* (Berkeley: Univ. of California Press, 1989).

7 J. B. Harley, "Deconstructing the Map," in Trevor J. Barnes and James S. Duncan, *Writing Worlds: Discourse, Text and Metaphor in the Representation of Landscape* (London: Routledge, 1992).

8 Jawaharlal Nehru, *The Discovery of India* (Delhi: Oxford Univ. Press, 1946).

9 See Jawaharlal Nehru, *An Autobiography* (London: John Lane, 1936).

10 The ambivalent discourse of anticolonial nationalism is discussed in, among others, Ashis Nandy, *The Intimate Enemy: Loss and Recovery of the Self Under Colonialism* (Delhi: Oxford Univ. Press, 1983) and Partha Chatterjee, *Nationalist Thought and the Colonial World: A Derivative Discourse* (Delhi: Oxford Univ. Press, 1986). For a discussion focusing on Nehru and the production of national identity, see Sankaran Krishna, "Inscribing the Nation: Nehru and the Politics of Identity in India," in Stephen Rosow, Naeem Inayatullah, and Mark Rupert, eds, *The Global Economy as Political Space* (Boulder: Lynne Rienner, 1994).

11 Jawaharlal Nehru, *Independence and After: A collection of speeches 1946–1949* (Delhi: Government of India, 1956), p. 247. Emphasis added.

12 As an illustration, consider the following collage of newspaper and magazine headlines: "We will not cede an inch: PM", "To seal a porous border"; "Bangla infiltration worries Center"; "Identity-Cards for Border Area Residents"; "Special ID cards soon on 33 border districts"; "Borderline Case"; "Laying Down the Line"; " 'Live Borders' of the North-East"; "The Border Security Belt"; "Charting India"; "The Border Security Force: Battling at the Sandy Frontier." These headlines were taken at random from recent issues of the *Statesman* and the *Telegraph* (both Calcutta-based English-language newspapers) and *Frontline*, a newsmagazine published in Madras.

13 The yatra, and its quixotic (in the original sense of that term) finale, was well covered by Indian newspapers. See especially the *Statesman* (Calcutta: January 25, 26, and 27, 1992).

14 For the whole series of advertisements centering on various maps of India, see Dilip Sarwate, *Political Marketing: The Indian Experience* (New Delhi: Tata-McGraw Hill, 1990), pp. 157–163. The ads drew the wrath of many for their elitism, pretentiousness, and highly melodramatic tone. Needless to say, Indian cartoonists had a field day and parodied the ads mercilessly. Sarwate's book also includes some choice selections of these ripostes.

15 For a discussion of the notions of the normal and the pathological in discussions

centering on the body politic, see David Campbell, *Writing Security: United States Foreign Policy and the Politics of Identity* (Minneapolis: Univ. of Minnesota Press, 1992), pp. 85–101.

16 In the BJP's desired cosmology, India's frontiers extend all the way to Afghanistan in the West and Burma in the East, and would include Nepal, Bhutan, Tibet, and other neighboring countries. Maps of the so-called Akhand Bharat served as the frontispiece for their publication, *The Organiser*, and are regularly used in BJP pamphlets, posters, and party literature. Akhand Bharat is the invariable benchmark from which the BJP criticizes the Congress (I) and (in an earlier time) was the reason for the assassination of Gandhi. For an insightful analysis of the Hindu right's origins and its hypermasculinity as a complete internalization of the values and pathology of the colonizer, see Ashis Nandy, *At the Edge of Psychology: Essays in Politics and Culture* (Delhi: Oxford Univ. Press, 1980).

17 See Appan Menon, "Time for a Political Solution" in *Frontline* (Madras, May 16–29, 1987): 14.

18 See Ravi Rikhye, *The Militarization of Mother India* (Delhi: Chanakya, 1990), pp. 24–27.

19 This, and much of the information in this section, is taken from "Siachen: The Forgotten War," a report by W. P. S. Sidhu and Pramod Pushkarna, in *India Today* (May 31, 1992): 58–71. Siachen, incidentally, means Rose Garden.

20 Jean Bethke Elshtain, "Sovereignty, Identity, Sacrifice," in *Millennium: Journal of International Studies* 20, no. 3 (1991): 397.

21 N. B. Menon, "Laying Down the Line," *Frontline* (Madras, August 8–21, 1987): 40.

22 See "Borderline Case," editorial in the *Telegraph* (Calcutta: May 27, 1993).

23 Michel De Certeau, *The Practice of Everyday Life*, Steven Rendall, trans. (Berkeley: Univ. of California Press, 1988).

24 From Aloke Banerjee's report in the *Statesman* (Calcutta: May 29, 1993), "Where Two Nations Slip into Each Other." The remainder of this paragraph and the next two paragraphs of text rely heavily on this report.

25 See report by Diptosh Majumdar, "3 Smugglers Fight for Panchayat Seat," in the *Telegraph* (Calcutta: May 27, 1993).

26 Pijush Kundu, "Decadence, privation and politicking," in the *Statesman* (Calcutta: May 19, 1993).

27 Anil Bhattacharjee, "Indo-Bangla Border," in *Frontline* (Madras: May 3–16, 1986).

28 Consider Lefebvre's expansion on this theme: "*There is a violence intrinsic to abstraction* . . . abstraction's *modus operandi* is devastation, destruction. . . . The violence involved does not stem from some force intervening aside from rationality, outside or beyond it. Rather, it manifests itself from the moment any action introduces the rational into the real, from the outside, by means of tools which strike, slice and cut – and keep doing so until the purpose of their aggression is achieved." Emphasis in original. See Henri Lefebvre, *The Production of Space*, Donald Nicholson, trans. (Cambridge, Mass.: Blackwell, 1991), p. 289.

29 De Certeau, note 23, p. 121.

30 A. V. Liddle, "The Role of Locals in Border Management," in D. V. L. N. Ramakrishna Rao and R. C. Sharma, eds, *India's Borders: Ecology and Security Perspectives* (New Delhi: Scholars Publishing Forum, 1990), p. 204.

31 Pierre Bourdieu, *Outline of a Theory of Practice*, Richard Nice, trans. (Cambridge: Cambridge Univ. Press, 1977).

32 Carol Breckenridge and Peter van der Veer, eds, *Orientalism and the Postcolonial Predicament: Perspectives from South Asia* (Philadelphia: Univ. of Pennsylvania Press, 1993).

# SECTION THREE

## *GEOPOLITICS*

### Editor's introduction

The term 'geopolitics' has been both the major public face of political geography (often used synonymously with it) and its major historical burden. Originally coined at the turn of the the twentieth century, the word became closely associated with a variety of *realpolitik* in which places were turned into 'strategic commodities' in a global struggle for mastery between contending Great Powers. In the United States and Britain after the Second World War intellectuals tended to shy away from the term because of its dual connotations of use by the Nazi regime in Germany to rationalize (if not guide) their territorial expansionism and its connection with a geographical determinism in which locations at a world scale were ascribed a determining role in the rise and fall of dominant states. The ideas of Friedrich Ratzel and Halford Mackinder were central to this tradition. They lived on in textbooks and to a degree found their way into the everyday language of politicians and others involved in the study of 'foreign affairs' (both Richard Nixon and Ronald Reagan, for example, were fond of expressing themselves in geopolitical language).

The term's close relationship to the formal models of Mackinder and others, however, obscured the extent to which all foreign-policy making is guided by geographical understandings including the ranking of places by strategic importance, the specification of policies in geographical terms (as in the policy of 'containment' pursued by the United States during the Cold War against the former Soviet Union) and the use of geographical metaphors (such as the 'domino effect' linking distant events back 'home'). These policies, of course, are not constructed in a political-economic vacuum (Agnew and Corbridge 1995). They are part and parcel of the practice of statecraft in historical settings in which different political-economic considerations enter into play. Today, for example, trading and other economic concerns seem to be more central to the foreign policies of the most powerful states than was the case only recently when military concerns were key (e.g. Rosecrance 1986). But there is also continuity, in the envisioning of the world in its entirety as a theatre of global conflict and in the designation of different world regions as more or less 'modern' and hence, as more or less 'serious' barriers to be overcome in the pursuit of 'primacy' by the Great Powers.

The main theoretical divisions among students of geopolitics reflect these differences over the inheritance from the past (can anything be learned about today's world politics from Mackinder?) and the relative importance of discursive representations as against political-economic processes in the practices of geopolitics. In both political-economic and postmodern usage the term 'geopolitics' refers to the ways in which geographical designations and labels enter into state practices rather than in the older usage which implies a strong causal relationship between global physical geography and state capacity to influence world events. There is a 'school' of geopolitics, in which Gray (1988) is a major figure, that seeks to update and apply Mackinder's 'heartland' model of global geopolitics to the contemporary world. From this point of view, the world-power situation is still best thought of as a struggle between the interests of the sea-powers (such as the United States?) and the land-powers (such as Russia). Even the 'apparent' end of the Cold War has not dulled enthusiasm for the basic geopolitical formula.

More typically, however, the older approach lives on more interestingly in the attempts of those such as **Alan K. Henrikson** (**Reading 7**) to structure the history of foreign relations (in this case American ones) in a centre–periphery model of world geography. In this construction, it is the *perception* of a state's centrality or peripherality within world affairs that is seen as driving the way in which foreign policy is framed. As the geographical frame shifts, so does the content of the policies. Identifying and analysing the frames that have come and gone over the years in the global positioning of the United States allows Henrikson also to summarize and evaluate relative to different historical epochs the various geopolitical schemes (such as those of Mackinder and Spykman) that were previously cast in a more determinist vein. It could be said that Henrikson adapts these schemes to inform his own spatial-analytic perspective on geopolitics but without succumbing to a direct determinism (other spatial-analytic perspectives would include Cohen 1973; O'Sullivan 1986; O'Loughlin 1986; and Nijman 1992).

Be that as it may, one criticism of this approach could be its neglect of the changing world-economic context in which the geographical framing of foreign policies takes place (Taylor 1993). It is increasingly common to hear not only that state boundaries are increasingly 'permeable' but that interstate competition is now more geoeconomic than geopolitical. By this is meant that state economic capacities are now important in their own right (in deciding standards of living and employment levels, for example) rather simply as a means to the end of enhanced state power at a global scale but that one state's gain still indicates another's loss. It is the same old zero-sum game but now about economic and not military competition. At the same time he calls into question the efficacy of older geopolitical formulae, **Stuart Corbridge** (**Reading 8**) also suggests that the change in global conditions is more profound than a substitution of economic for military competition (between the so-called Great Powers; there are still many other military conflicts around the world). Corbridge argues that the global context has changed to the extent that states are no longer the singular actors of world politics. All manner of international regulatory institutions and private business organizations are now important actors in world politics (also see, for example, Strange 1986; Helleiner 1993; Corbridge *et al.* 1994). The contemporary 'geopolitical economy' has three power centres – the United States, Japan and the European Community – but these are themselves fragmented among a series of city hierarchies linked by flows of capital and by financial and trade connections to distant sites around the world. In a 'new world' of quasi-states and uncertain sovereignties there are both opportunities for a more egalitarian world and the creation of new instabilities within and between states.

This historicist perspective (arguing for the role of changed historical conditions in creating new geographies of interstate relations) is not the only kind of political-economic perspective available. Even more popular are perspectives which emphasize the permanent importance of the 'geopolitics of capitalism': seeing the contemporary era as only the latest in a long series of reorganizations of business at different scales to resolve the long-term tendency of the rate of profit to decline without a constant search for a 'spatial fix' (Harvey 1982). Others, such as world-systems theorists and advocates of cyclical models of history, stress the emergence of new hegemonic (state) powers out of the ruins of previous epochs because of, respectively, technological advantages (Wallerstein 1993) or fewer military obligations relative to national economic capacity (Kennedy 1986).

As Corbridge notes early in his article, the end of the Cold War has coincided not only with an upsurge in writing about geopolitics (uncertain times seem always to invite this!) but also with a questioning of the geographical assumptions upon which the practices of foreign policy are predicated. A self-consciously 'critical geopolitics' has developed in which the ways in which politicians represent places and their strategic significance (to themselves and to publics) take centre-stage (see, e.g., Ó Tuathail and Agnew 1992; Ó Tuathail 1996). Informed by the writings of such philosophers/savants as Derrida, Foucault and Baudrillard, emphasis is placed on deconstructing the discursive strategies used to make foreign policy 'situations', 'crises' and 'wars' intelligible with reference both to a global 'big picture' and to past events seen as analogies to present ones (e.g. Munich/appeasement, Vietnam/quagmire). **Gearóid Ó Tuathail** (**Reading 9**) offers a detailed analysis of the Gulf Crisis (and War) of 1990–91 in terms of two main themes: (1) the placing of the Gulf 'situation' in the big picture of the end of the Cold War and desire of American political leaders to 're-territorialize' the US as the sole Superpower, and (2) the war as an example of the 'dematerialization' of place through the use of new high-tech weapons and the immediate visibility of the war on television screens around the world as a type of 'war-game' in which the place it was happening was largely incidental to the demonstration of high-tech capability. From this point of view, the American representations of the crisis and the war, therefore, are *the* key to understanding how they were played out. Yet, as Luke (1991) points out, Kuwait and Iraq were not innocent bystanders in all of this. Kuwait had liquid assets stashed abroad that allowed it to pay the United States for its liberation; Iraq was prepared to fight an old kind of war using Soviet war-fighting doctrines. The US government was not all there was to the Gulf Crisis even though its representations were easily the most powerful ones. Although the spectacular television coverage of the bombing of Baghdad did give a 'media event' quality to the war (Stam 1992; Wark 1994), the use of high-tech weapons in the war as a whole was in fact quite limited (Mattelart 1994, 120). There were also some notorious failures, such as the poor performance of the Patriot air defence system and the inability of US military planners to locate Iraq's Scud missiles. Undoubtedly an 'imperial encounter' (Doty 1996), the Gulf Crisis speaks to the general character of 'North–South' relations in the late twentieth century, but not in a single voice.

Ó Tuathail reminds us, however, that geopolitical practice remains closely attached to the conduct of wars despite changes in the nature of the world economy. As guide and inspiration, the 'geopolitical imagination' reduces places to points on a map that then cease to have intrinsic qualities and only count in a calculus driven by global considerations. Of course, this global vision is not new but has a long genealogy going back to the European encounter with the rest of the world from the sixteenth century onwards. The content of the vision does change, however. The challenge to the student of geopolitics is to detect what is changing at a particular moment and to show whether it can be understood in *purely* representational terms (Agnew and Corbridge 1995, Chapter 3).

## References

Agnew, J. A. and Corbridge, S. 1995: *Mastering space: hegemony, territory and international political economy*. London: Routledge.

Cohen, S. B. 1973: *Geography and politics in a world divided*. Second edition. New York: Oxford University Press.

Corbridge, S., Martin, R. L., and Thrift, N. (eds) 1994: *Money, power and space*. Oxford: Blackwell.

Doty, R. L. 1996: *Imperial encounters: the politics of representation in North–South relations*. Minneapolis MN: University of Minnesota Press.

Gray, C. 1988: *The geopolitics of superpower*. Lexington KY: University Press of Kentucky.

Harvey, D. 1982: *The limits to capital*. Chicago: University of Chicago Press.

Helleiner, E. 1993: *States and the reemergence of global finance*. Ithaca NY:

Cornell University Press.

Kennedy, P. 1986: *The rise and fall of the great powers: economic change and military conflict from 1500 to 2000.* New York: Random House.

Luke, T. 1991: The disciplines of security studies and the codes of containment. *Alternatives*, 16: 315–44.

Mattelart, A. 1994: *Mapping world communication: war, progress, culture.* Minneapolis MN: University of Minnesota Press.

Nijman, J. 1992: The limits of superpower: the United States and the Soviet Union since World War II. *Annals of the Association of American Geographers*, 82: 681–95.

O'Loughlin, J. 1986: Spatial models of international conflict: extending current theories of war behavior. *Annals of the Association of American Geographers*, 76: 63–80.

O'Sullivan, P. 1986: *Geopolitics.* New York: St. Martin's Press.

Ó Tuathail, G. 1996: *Critical geopolitics.* Minneapolis MN: University of Minnesota Press.

Ó Tuathail, G. and Agnew, J. 1992: Geopolitics and discourse: practical geopolitical reasoning in American foreign policy. *Political Geography*, 11: 190–204.

Rosecrance, R. 1986: *The rise of the trading state: commerce and conquest in the modern world.* New York: Basic Books.

Stam, R. 1992: Mobilizing fictions: the Gulf war, the media and the recruitment of the spectator. *Public Culture*, 4: 101–26.

Strange, S. 1986: *Casino capitalism.* Oxford: Blackwell.

Taylor, P. J. (ed.) 1993: *The political geography of the twentieth century.* London: Belhaven Press.

Wallerstein, I. 1993: *Geopolitics and geoculture.* Cambridge: Cambridge University Press.

Wark, M. 1994: *Virtual geography: living with global media events.* Bloomington IN: Indiana University Press.

# 7 Alan K. Henrikson, 'America's Changing Place in the World: from "Periphery" to "Centre"?'

Reprinted in full from: Jean Gottmann (ed.), *Centre and Periphery: Spatial Variation in Politics*, Chapter 4. London: Sage (1980)

Americans today are not well "located" in international affairs. Having grown accustomed to world leadership since World War II, the modern nonterritorial equivalent of universal empire, they sense they are no longer "the central world power."[1] Many in the United States fear they are being relegated to a periphery – not the periphery of the old colonial "West," but the social and political periphery of a postindustrial "North." Outnumbered, outvoted, and sometimes even outproduced, they feel themselves in a state of incipient historical "decline" or "marginality."

The causes of this apparent descent are not readily understood. Americans are unsure whether their relative loss of status is due simply to the fact that other great powers have risen to challenge their primacy and centrality, and in doing so have "displaced" them, or whether it is due to an upheaval in the basis of the international system itself, in the underlying hierarchical–locational structure of international relations.[2]

President Carter, speaking at Notre Dame University on May 22, 1977, addressed this concern of his countrymen explicitly. "By the measure of history," he said, "our nation's 200 years are brief; and our rise to world eminence is briefer still. It dates from 1945, when Europe and the old international order both lay in ruins. Before then, America was largely on the periphery of world affairs. Since then, we have inescapably been at the center."

But perhaps no longer. "Historical trends" – the relaxation of the unifying threat of conflict with the Soviet Union, the moral crisis over the Vietnam War, the economic strains of the 1970s and the apparent inability of industrial democracy to provide sustained well-being, and, above all, the passing of colonialism with the resultant "new sense of national identity" in nearly 100 new countries – have weakened the foundations of the system in which the United States once predominated. It is, President Carter declared, "a new world," which calls for "a new American foreign policy."

Americans should no longer expect, Mr. Carter warned, that the other 150 countries will "follow the dictates of the powerful." The era of "almost exclusive alliance" with the Atlantic states is past, and "a wider framework of international cooperation" is needed. "We will cooperate more closely," he affirmed, "with the newly influential countries in Latin America, Africa and Asia. We need their friendship and cooperation in a common effort *as the structure of world power changes*" (emphasis added; Carter, 1977).

President Carter's use of the geometrical terms "center" and "periphery" in discussing this changing structure, and America's place in it, draws attention to a dimension of our theme that is nowadays often overlooked or minimized: the spatial or geographical. To see the history of American foreign relations in this way is, however, natural. The various changes that have occurred in America's "place in the world" probably add up to a greater actual and imagined movement in space than that experienced by any other country, at least in so brief a time.

How does the "center–periphery" schema help us to understand the shifts in America's world position? The answer, it seems to me, lies in the relationship of polarity that is inherent in the center–periphery model. Let me briefly explain.

A center–periphery system is characterized by a certain physical, social, and psychical distance between an interrelated core and margin (Eisenstadt, 1977: 72).[3] If a center–periphery structure is to endure, core and margin must be kept distinct. If essential elements of the center migrate to the periphery, or if substantial parts of the periphery are absorbed at the center, the relationship between them may break down. Similarly, if the center "sinks" or the periphery "rises," a breakdown in the relationship can occur. If either of these kinds of movement, horizontal or vertical, becomes massive enough – that is, if a periphery heavily engulfs a center or decisively surpasses it in some respect – then there can occur an actual reversal of roles: a switch in polarity.

Although the historical processes that underlie such crossovers are usually incremental and long term, the role reversals themselves may appear to be sudden. The explanation of this felt abruptness may in some cases be mainly psychological. Like the instantaneous image shifts in the experiments of Gestalt psychologists, center/periphery reversals in geopolitical fields may also seem to occur rapidly, especially when accompanied by dramatic events.

The structure of a fluid situation may require a new idea or theme in order to become crystallized. In international affairs this is especially the case, since the participants are so varied, their relationships so numerous, and their environments so vast. Often the best representation of an evolving international system is a map, a visual analogue to political and other facts in their geographic setting.[4] Some use will be made of maps, both imaginary spatial images and actual cartographic representations, in the present essay. Maps can have iconic value. They represent hierarchies and locations symbolically. As evidence and as emblem, they mould as well as mirror historical circumstances.

At a time of crisis, the mind may oscillate between two "views," or impressionistic interpretations, of the course of world events. If supported by a theory of international relations, that is, if a *perceptual* change is accompanied by a *conceptual* change, a sensed shift in polarity may become mentally fixed. A new "paradigm," in Kuhn's (1970) sense of that term, controls the future direction of thought.

The classic case of a lasting center/periphery reversal, of peculiar analogic appropriateness to our present theme, is the sixteenth-century shift from a Ptolemaic, earth-centered concept of the universe to the Copernican, suncentered world view. Once the new logic of heliocentrism was understood, the old "normal science" of geocentrism became not simply outmoded but nonsensical.[5] Similarly, if a new scientific theory is offered and accepted as a rationalization of a geopolitical role change, it may make the change intellectually (if not practically) irreversible.

The international history of the United States contains a number of such perceptual-conceptual shifts, most but not all of them related to shifts in the world position of Europe. The greatest of these was that to which President Carter referred – the great globewide center/periphery reversal of the 1940s.

A shift from periphery to center, one would think, assuming the irreversibility of such historical movements, could occur only once. In fact, America and Americans have arrived at, or acquired, a position of "centrality" a number of times. The apparent illogic of this is explained by the realization that America's centrality has been gained in different – increasingly larger and more complex – contexts.

Originally, as President Carter stated, America was "largely on the periphery." By this he presumably meant the periphery of world affairs as dominated by Europe. The idea is rooted in America's colonial past. For nearly 200 years the lives of British settlers in North America were regulated by the mercantilist policies of distant London.

The notion of American peripherality also derives, however, from ancient geographical tradition. For Eratosthenes, and most of his successors down through the Middle Ages, the earth had but three parts – Europe, Asia, and Africa – rimmed by a narrow, circumfluent ocean. The discovery by Columbus of the "Indies" (as he believed them to be) did not immediately destroy this inherited picture of earth. A new continent would be a "fourth part" of the world – a conceptual impossibility. Moreover, with six-sevenths of the *orbis* land, there was hardly any room for a new part! Until Martin Waldseemueller's famous 1507 map, which gave America its label, the new land was conceived, and cartographically represented, as the easternmost extension of Asia (James with James, 1972; Thrower, 1972; Boorstin, 1976). When the full expanse of the globe was finally appreciated, the American continents were often placed on the left side of joined-hemisphere maps – a solution that, while it kept Eurasia at the center of its orbit, made America central as well (Figure 7.1). The anomalous fourth part of the Old World became the core of a New World – a Great Satellite.

The first major step in the actual fulfillment of this schematic "Western Hemisphere" centrality was, of course, the American Revolution. The Declaration of Independence, a document of not only political but geographical import, proclaimed the need for Americans "to dissolve the political bands, which have connected them with another, and to assume among the powers of the earth, the separate and equal station to which the Laws of Nature and of Nature's God entitle them." Thomas Paine in *Common Sense* (1776) provided a Copernican-Newtonian justification for America's geopolitical independence. "In no instance," he argued, "hath nature made the satellite larger than its primary planet, and as England and America, with respect to each other, reverses the common order of nature, it is evident they

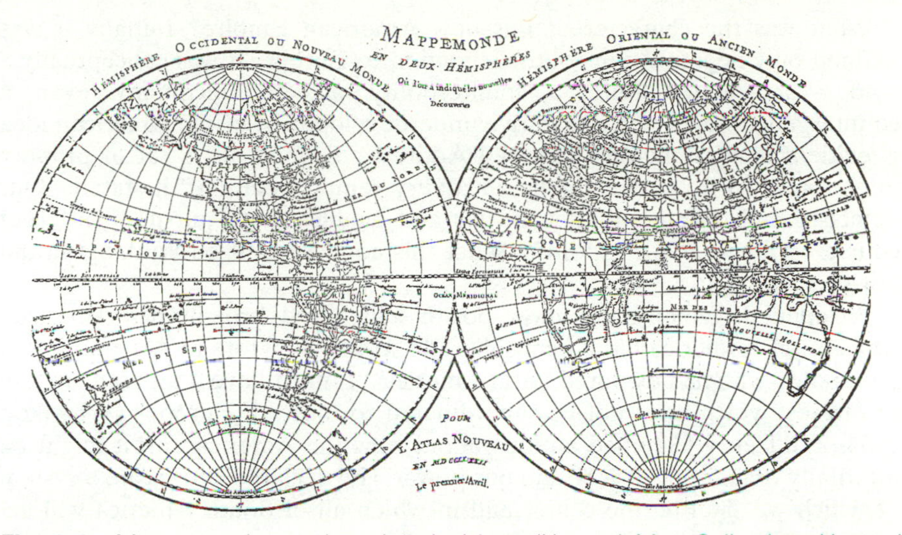


**Fig. 7.1** Mappemonde en deux hémisphères (Harvard Map Collection, Harvard College Library)

belong to different systems: England to Europe, America to itself" (Kramnick, 1976: 91).[6]

A quaint yet characteristic expression of this primitive America-centrism was the late eighteenth- and early nineteenth-century cartographic practice of placing the prime meridian of longitude within the American orbit. Some of the first maps of the United States show the central meridian running through Philadelphia. The "Meridian of Philadelphia" gradually gave place to a Washington, D.C., zero meridian. On L'Enfant's plan of Washington, the 0°0′ line passes through "Congress House." Other contemporary *projets* put the line through the "President's House" (Pratt, 1942).[7]

This use of locally defined zero meridians was, of course, partly just cartographic convenience – the tradition, familiar in Europe, of putting the 0°0′ line through the largest, most important city on the map. In the Philadelphia and Washington cases, however, it was a reflection of something more: an expression of an intention to lay down an entirely new geographical frame of reference. The Meridian of Greenwich remained a symbol of monarchism, a vestige of colonial subordination. "Are we *truly* independent, or do we appear so," a patriot asked in 1819, when on leaving his own country every American is "under the necessity of casting his 'mind's eye' across the Atlantic, and asking of England his relative position"? This was calculated to "wean the affections. . . . What American seaman has not experienced this moral effect?" (Stanton, 1975: 7).

A republican pole – or an alternative imperial pole – was needed. This position was assumed by Washington, the District of Columbia. Some of its builders ambitiously conceived of it as the "Rome of the New World." (Given the fact that most of South America lies longitudinally to the east of the United States, a Washington zero meridian is actually not a bad choice for the vertical axis of the New World.)

What was the character of this new American Empire? Initially it was defined only negatively, in contradistinction to Europe. It was conceptually a void – a counterculture, a counterpolity, a countereconomy, even a countergeography. The comparative underdevelopment of the American idea is evident in Washington's Farewell Address. "Europe has a set of primary interests, which to us have none, or a very remote relation," it stated. "Our detached and distant situation invites and enables us to pursue a different course" (Gilbert, 1961: 144–147). What this course might be Washington did not specify.

President Jefferson had a more positive, idealized, broadly pan-American conception. "America has a hemisphere to itself," he wrote to Alexander von Humboldt, who had explored Latin America. "It must have a separate system of interest which must not be subordinated to those of Europe" (Whitaker, 1954: 29). Henry Clay had an idea about how the American orbit might be internally organized. "It is within our power," he declared, "to create a system of which we shall be the center and in which all of South America will act with us" (Rippy, 1958: 4–5).

The Monroe Doctrine may be taken as the point at which, in American and in certain European eyes, the American system began to have parity with the European system. It seemed to be gaining gravitational equivalence. Such was the increase in America's overall mass that Jefferson, asked by Monroe whether to accept the British Foreign Secretary's proposal of a joint declaration on Latin American independence, imagined that, by acceding to Canning's proposition, the United States might "detach her from the bands, bring her mighty weight into the scale of free government, and emancipate a continent at one stroke" (Bartlett, 1970: 174–175).

John Quincy Adams, defending his administration's participation in an inter-American congress at Panama in 1826, argued that the trebling of the population, wealth, and territory of the United States since Washington's Farewell Address made possible a reversal of the founding father's formula. Rather than reemphasizing that Europe had a set of primary interests with which the United States must not interfere, Adams, in a message to the House of Representatives on March 15, 1826, stressed just the opposite: The time had arrived when "*America* has a set of primary interests which have none or a remote relation to Europe" (LaFeber, 1965: 136). America's independence, or polar position, was thus complete.

Westward expansion consolidated this image of hemispheric independence and unity. By quick, giant steps – the Louisiana bargain with France (1803), the Transcontinental Treaty with Spain (1819), the Oregon Treaty with Great Britain (1846), and the Treaty of Guadalupe Hidalgo with Mexico (1848) – the United States extended its domain across North America. In so doing, it defeated any European thought of maintaining a North American "balance of power" and any Latin American thought of offering direct rivalry.

The United States now "faced" both the Atlantic and the Pacific – a fundamental change in geographical outlook and focus. To some Americans, such as Senator Thomas Hart Benton, the country's intermediate situation between Europe and Asia presented it with an opportunity to become the prime route to the Orient, thus fulfilling the original dream of Columbus. "The European merchant, as well as the American," Benton prophesied in 1849, "will fly across our continent on a straight line to China. The rich commerce of Asia will flow through our centre" (Smith, 1950: 28–29). His home town of St. Louis, he hoped, would become the emporium of this global East–West trade.

This idea of mid-America as the next great center of international exchange was supported by a deterministic geographical theory derived from Humboldt's delineation of global equal-temperature zones: that the zone of (northern) temperate climates is also the path of world progress. Benton's associate, the speculator-soldier-politician William Gilpin, envisioned an "Isothermal Zodiac" – a great undulating common-temperature belt, about 30° wide, girdling the northern hemisphere between roughly the twenty-fifth and fifty-fifth parallels and, happily, including all of the territory of the United States. Through it ran an "axis of intensity," at about the fortieth parallel (St. Louis: 39° N), which had a mean annual temperature of 52°. "Within this isothermal belt, and restricted to it," wrote Gilpin, 'the column of the human family, with whom abides the sacred and inspired fire of civilization, accompanying the sun, has marched from east to west, since the birth of time." Upon it had been constructed "the great primary cities" – the Chinese, Indian, Persian, Grecian, Roman, Spanish, and British – and now "the republican empire of the people of North America."

Topography and hydrography, too, favored the world centrality of North America. The Great Basin of the Mississippi was to Gilpin "the amphitheatre of the world." In contrast to the interiors of the other continents, that of North America presented toward heaven "an expanded bowl," catching and fusing whatever entered within its rim. Other continents presented "a bowl reversed," scattering everything into radiant distraction. "In geography the antithesis of the old world," he judged, "in society we are and will be the reverse" (Boorstin, 1965: 233).[8]

Given such grandiose views, it is not surprising that the United States began around 1850 to appear at the center of world maps (Figure 7.2). Previously, on most American as well as European global maps, the United States had been off to the left, in the West. Following the US victory over Mexico, the discovery of gold in California, the triumph of the Union in the Civil War, and the purchase of Alaska, the country seemed to many American map makers to deserve a more central position.[9]

The principal reason for this shift of global frame was, one suspects, not the increase in power, wealth, and solidity of the young Uncle Sam, but the change in his geographical stance – that is, the fact that he now stood astride a continent, one foot on the Atlantic and the other on the Pacific.[10] It only "made sense" for the self-proclaimed broker between Occident and Orient to be in the middle of the picture.

Despite this spread-eagle imagery, the United States was, in a political and military sense, isolated and marginal in world affairs throughout most of the nineteenth century. Its position and perspective were still essentially "continental." Captain Alfred Thayer Mahan (1890: 42) described

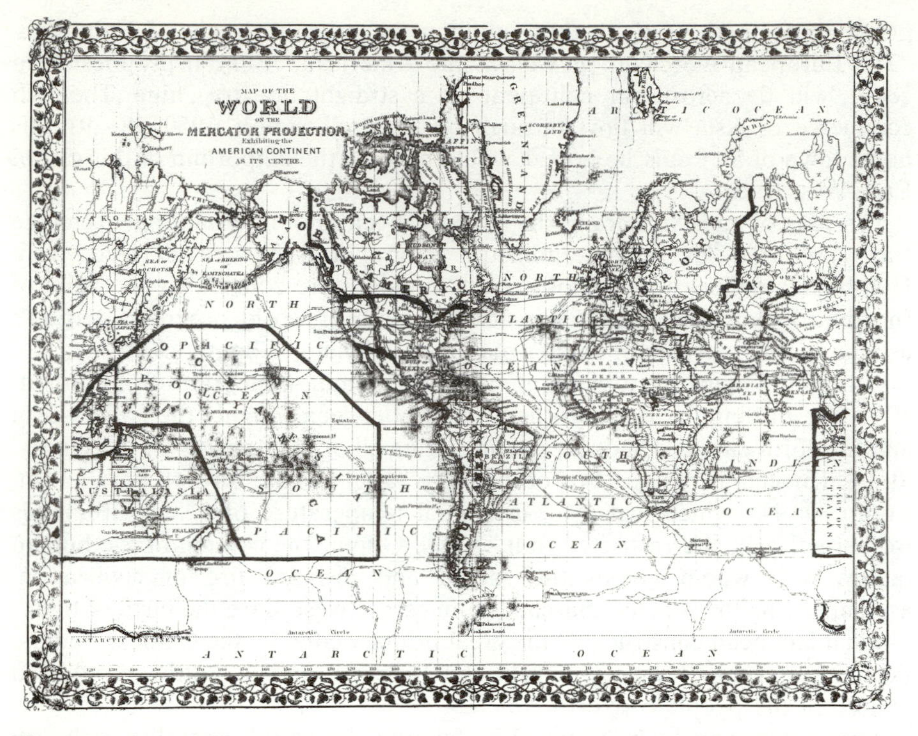


**Fig. 7.2** American world map with the United States at the center (from S. Augustus Mitchell, *New General Atlas*)

this self-contained outlook well in his book, *The Influence of Sea Power upon History*:

> Except Alaska, the United States has no outlying possession – no foot of ground inaccessible by land. Its contour is such as to present few points specially weak from their saliency, and all important parts of the frontiers can be readily attained – cheaply by water, rapidly by rail. The weakest frontier, the Pacific, is far removed from the most dangerous of possible enemies. The internal resources are boundless as compared with present needs; we can live off ourselves indefinitely in 'our little corner,' to use the expression of a French officer to the author.

The Spanish–American War broke this defensive, continentalist shell. For the first time in its history, the United States was connected geopolitically with the wider world. In the conquered Philippines, the one-time distant periphery gained its own distant periphery. Partly because of this possession, the American Republic was increasingly described as a "world power" (Coolidge, 1912; May, 1961). In truth, it was only a Caribbean power. Nonetheless, its Philippine commitment marked a major location shift.[11]

From the prospect of Europe, America was still remote – in some ways *more* remote than it had been. Halford Mackinder (Pearce, 1962: 67, 261, 262) in his famous 1904 lecture on "The Geographical Pivot of History," declared: "The United States has recently become an *eastern* power, affecting the European balance not directly, but through Russia, and she will construct the

Panama Canal to make her Mississippi and Atlantic resources available in the Pacific. From this point of view the real divide between east and west is to be found in the Atlantic ocean" (emphasis added). Mackinder's illustrative maps placed the American "satellite," as he described it, on the "outer or insular crescent" – on the right, or eastern, side of the map as well as on the left, or western, side.[12] If anything, America, pivoting elliptically around the Eurasian "Heartland," was thus more "peripheral" than it had been.

This impression was illusory. The very idea of marginality suggested mobility, and the offsetting of continental power by maritime power. Dewey's victory at Manila Bay and the round-the-world cruise of the Great White Fleet announced America's presence on the high seas.

In closely related developments, the diplomatic role of the United States also soon expanded. The participation of the United States in the Hague Peace Conferences, its singular part in ending the Russo–Japanese War, and its mediatory diplomacy during the first Moroccan crisis all bespoke a new and wider American centrality in world affairs. No continent was completely irrelevant to the United States any longer.

Theodore Roosevelt, probably the first American president to have a real sense of a *world* balance of power, saw heroic possibilities for the nation. In 1910 he told Baron van Eckardstein, former German ambassador in London, that, should Great Britain fail in her traditional role, "the United States would be obliged to step in at least temporarily, in order to restore the balance of power in Europe, never mind against which country or group of countries our efforts may have to be directed." "In fact," he judged, "we ourselves are becoming, owing to our strength and geographical situation, more and more the balance of power of the whole world" ( Beale, 1956: 447).

Woodrow Wilson forsook "world balancing" for "world leadership." If Roosevelt's typical image had been that of a global scale, Wilson's was that of a global platform. He simply thought on a different plane – that of ideology. On this level, his horizons were unlimited. In his second inaugural address (March 15, 1917), he declared that "we realize that the greatest things that remain to be done must be done with the whole world for stage and in cooperation with the wide and universal forces of mankind, and we are making our spirits ready for those things. We are provincials no longer" (Inaugural Addresses, 1961: 203–206). He said this even before the United States had entered World War I. Afterward, America's global destiny was even clearer to him: "There can be no question of our ceasing to be a world power. The only question is whether we can refuse the moral leadership that is offered us, whether we shall accept or reject the confidence of the world" (Weinberg, 1935: 470).

Wilson failed in his aim partly because he tried too hard to be cosmopolitan, to transcend the barriers of physical and cultural distance. In attempting to bring the United States into a world organization, he lost his sense of America's "place," and of his own base of political support. This was a "geopolitical" misjudgment far greater than his particular errors in, for example, defining the Italian–Austrian boundary or assigning "Class C" mandates. It is conceivable that had he possessed a better appreciation of the psychological gulfs between continents and, accordingly, proposed a more "regionalized" League of Nations, his experiment might have turned out more successfully.

World War I and Wilson's universalist leadership did, however, greatly expand Americans' world knowledge and interest. This was most impressively manifested in the economic realm, where an important reversal in polarity occurred. The United States, throughout its previous history a net borrower, became a net lender. "Since the beginning of the World War," Bowman noted (1928: iii), "the United States has increased its foreign investments fourfold, doubled its foreign commerce, and become the creditor of sixteen European nations." The geographical scope of its stake was second to none. "If our territorial holdings are not so widely distributed as those of Great Britain, our total economic power and commercial relations are no less extensive."[13]

New York emerged as a world financial center, rivaling London. The extent of the world's dependence on the American capital market did not become fully apparent, however, until after Black Thursday on Wall Street. As Braudel has reflected (*Time*, 1977), "In the world of exchange, there's always a central zone, an intermediary zone, and a peripheral zone. In 1929, the so-called Dark Year, the center of the world, which was London, passed to New York, peacefully."

World War II completed the shift of the United States from periphery to center on other levels. A group of distinguished European refugee and American intellectuals, stunned by the fall of France, issued a manifesto (Agar, 1940), *The City of Man: A Declaration on World Democracy*, in which they lamented that Europe's "solar period" was over. The new sun for mankind would have to be America – whose light, they argued, must shine more steadily and universally.[14]

This meant nothing less than a transatlantic "Copernican Revolution," a reversal in the direction of spiritual (if not yet cultural) gravitation and radiation. The dawn of the new US-dominated age was proclaimed by publisher Henry R. Luce in his influential tract, *The American Century* (1941: 39). Objecting to former President Herbert Hoover's observation that America was "fast becoming the sanctuary of the ideals of civilization," he retorted that it now "becomes our time to be the powerhouse from which the ideals spread throughout the world."

There was a reversal in America's world geostrategic situation as well. Before Pearl Harbor, the manor military doctrine of the United States was "hemisphere defense." After Pearl Harbor, it became "global offense" (Figure 7.3).[15] Either encircle or be encircled, President Franklin D. Roosevelt warned in a fireside chat of February 23, 1942. "We must all understand and face the hard fact that our job now is to fight at distances which extend all the way around the globe" (Rosenman, 1950). Otherwise, the United States would become detached from its allies and end up fighting the Axis alone.

From the American perspective and, to a lesser extent, from the perspective of other nations, World War II was, in an unprecedentedly literal sense, a global war. The relationship of the United States to the major theaters of battle was such that a new picture of the world – a new global strategic map

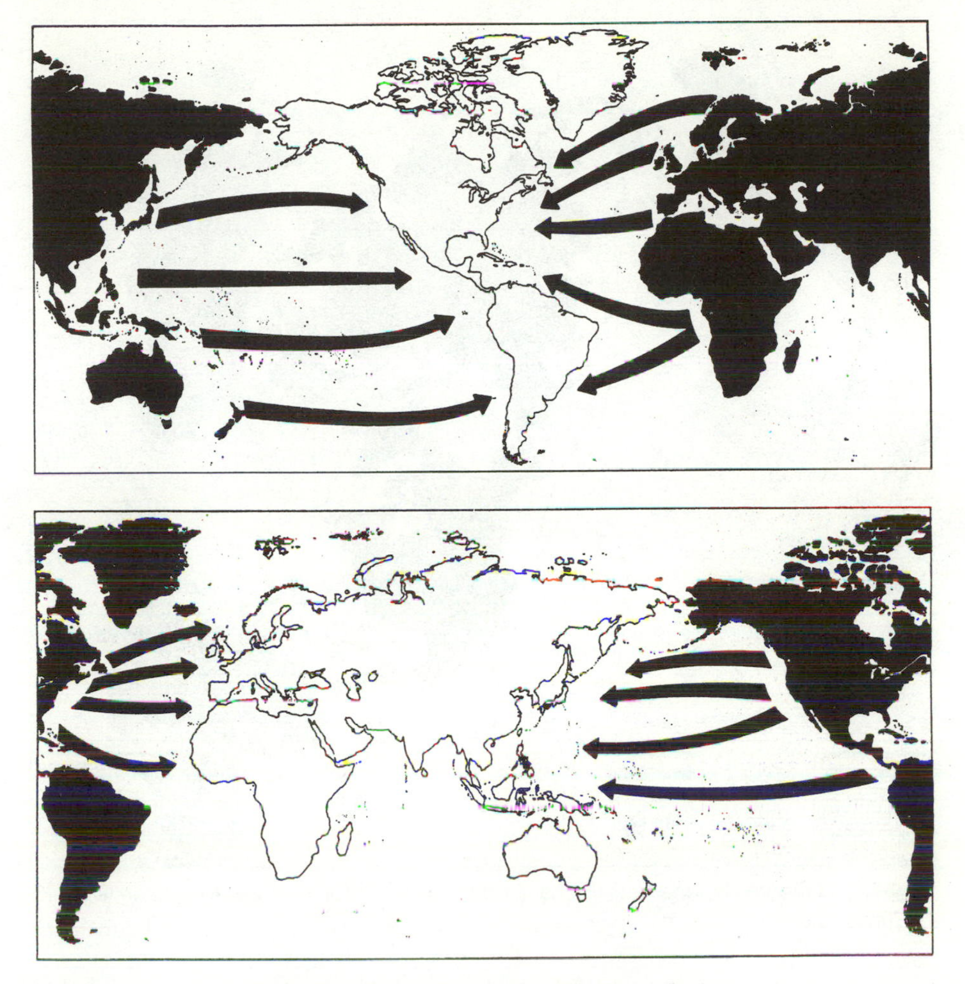

The future of the Western hemisphere

**Fig. 7.3** From Western hemisphere defense to global offense (map by J McA. Smiley from *The Geography of the Peace* by Nicholas John Spykman, reproduced by permission of Harcourt Brace Jovanovich, Inc.)

– was needed. Cylindrical map projections, such as the conventional equator-based Mercator, failed to show the continuity, unity, and organization of the "worldwide arena," as Roosevelt called it. Hence, other map projections came into fashion, notably the North Pole-centered azimuthal projection. The masterpiece of this type was probably Richard Edes Harrison's "One World, One War" (Figure 7.4), appearing originally in *Fortune* magazine and widely reproduced for military training and other purposes.[16]

The position of the United States on these polar maps was usually a central one – at or just below the geometrical center (usually 90° N) and on the vertical axis (usually 90° W–90° E). This was, perceptually speaking, the prime location. As Arnheim (1977) points out: "Any picture space is dominated by

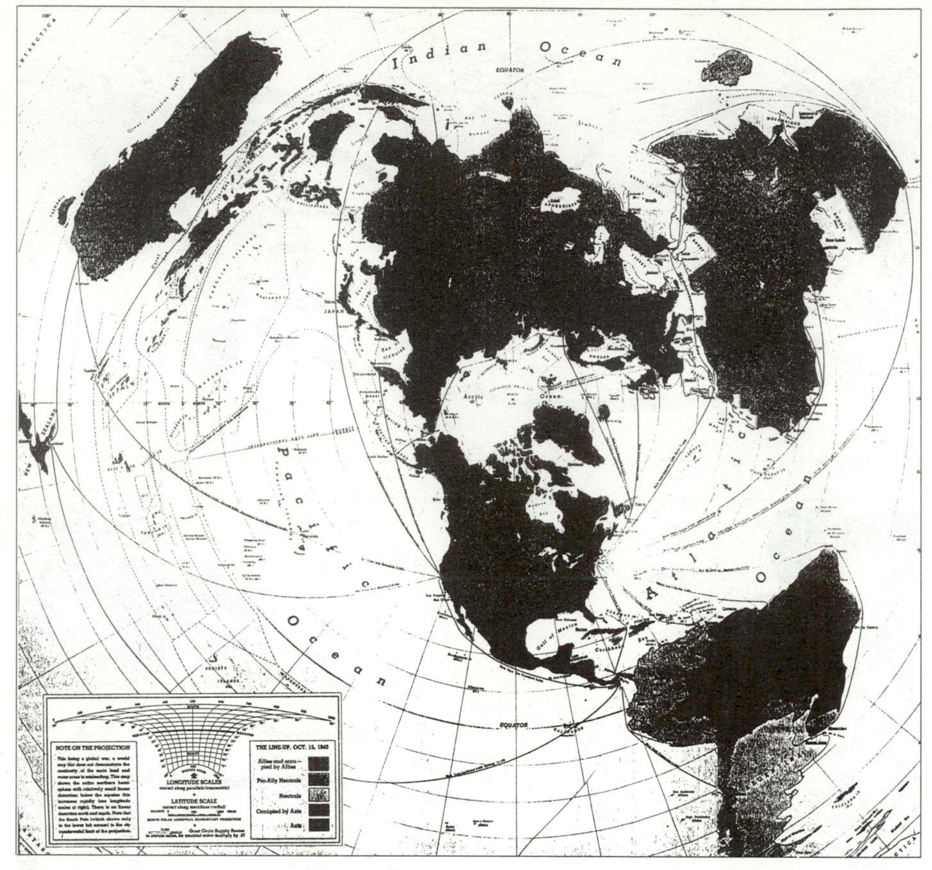


**Fig. 7.4** One world, one war (Richard Edes Harrison for *Fortune Magazine* © 1942 by Time, Inc.)

the vertical central spine, which is experienced as pointing upward – probably in conformity to the forward direction of the viewer. When I look at a picture, the upward direction in the picture corresponds to where my nose points and where I go if I move. The lateral directions are symmetrical to each other and constitute a (less dynamic) base."[17]

The new transpolar focus did not come naturally to Americans. Although their maps had conventionally been North-"oriented," the course of their history had largely run eastward and westward. Except for a few Arctic enthusiasts, such as the explorer-scientist Vilhjalmur Stefansson, the Northland was frozen, inert, and Canadian. Alaska, in most minds, was still "Seward's Folly." In order to make the North the scene of potential activity – to make it "dynamic" – cartographers drew arrows to suggest cross-polar movement.[18]

An important consequence of viewing these wartime North Pole-centered maps was a new awareness of the proximity of North America to Eurasia, and vice versa. The Arctic Ocean became, psychologically, an inland sea – a circumpolar "Mediterranean." Russia, previously thought of as being on the opposite side of the earth, suddenly appeared overhead. Russia had been situated "behind" the countries of western Europe and eastern Asia, now it was located directly "in front of" them. No longer, as Lippmann (1943: 145) ominously warned, would American–Russian relations be controlled by "the historic fact that each is for the other a potential friend in the rear of its potential enemies." What factors, then, *would* control relations between the two powers that Tocqueville had identified as marked out by the will of heaven to sway the destinies of half the globe?

The old "Heartland" theory, according to which the United States was a mere "satellite" in orbit around the planetary "World Island", was clearly no longer tenable. The United States was now too independent, too strong, and too far-reaching in its influence to be considered only a peripheral power. Columbia University geographer George Renner (1944: 44, 47), fascinated by the technology of polar aviation, proposed an adjustment: that the Heartland be expanded and shifted upward to include the interior parts of all the land masses ringing the Artic Mediterranean. Mackinder (1943: 598, 602) himself adjusted his concept by encompassing the Heartland and "Midland Ocean" (North Atlantic–Arctic) with a new "great feature of global geography: a girdle, as it were, hung around the north polar regions" (Figure 7.5). Spykman (1944: 40–41) further detracted from the Heartland's centrality by focusing on the critical "Rimland" (Mackinder's "Inner or Marginal Crescent"). To this vital intermediate zone between land power and sea power the United States, like most other great powers, had direct access. Cressey (1945: 245–246) completed the displacement of the old world core. "If there is anywhere a world citadel or Heartland," he proposed, "it may well lie in North America rather than in Eurasia." He justified his conclusion: "The American continent has adequate size, compact shape, internal accessibility, a central location, good boundaries, access to two oceans, favorable topography, rich minerals, excellent climate, and a dynamic spirit of its people."[19]

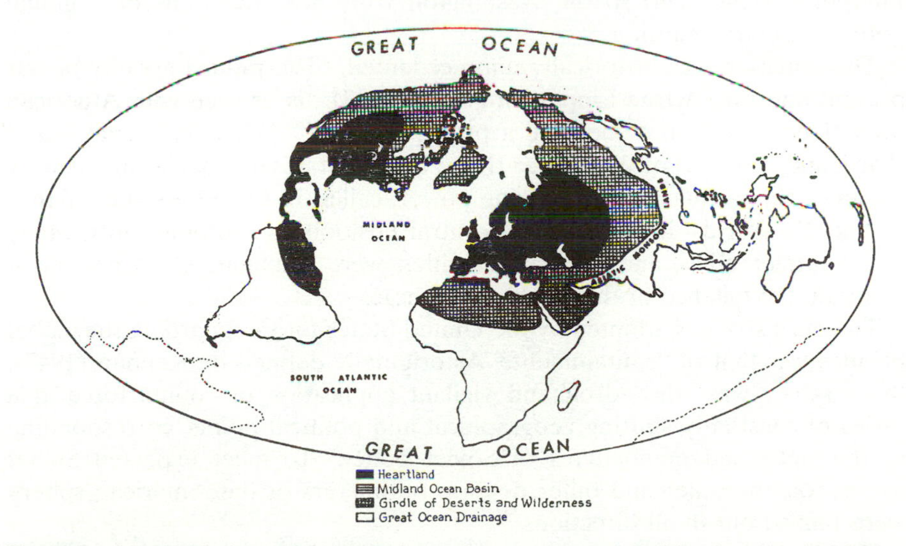


**Fig. 7.5** Mackinder's world (1943, from *Geography and Politics in a World Divided* by Saul Bernard Cohen © 1963 by Random House, Inc.)

At war's end, the United States thus emerged as the geostrategic center of the world, at least the geographic center of global strategic decisions. Political lines followed military lines. Washington, D.C., was finally the "Rome" to which all roads – not just those of the New World – seemed to lead. Americans began to think of themselves, as Woodrow Wilson had urged, not merely as US nationals but as citizens of the world. Others, too, began to look upon the United States as world capital – an image strengthened by the selection of New York as the site of the United Nations Organization (Eichelberger, 1977).

Europeans, particularly, had a sense of the reconstellation of world forces. They felt themselves in a position similar to that of the Greeks in 146 B.C.: America's power center had coalesced and Europe's had collapsed. Luce's *The American Century* had its complement in Fischer's *The Passing of the European Age* (1948) and similarly titled works (Weber, 1948; Holborn, 1963). "The capital fact," André Malraux reflected,

> is the death of Europe. When I was twenty years old [in 1921] the United States was approximately in the position of Japan today in terms of world importance. Europe was at the heart of things and the great superpower was the British empire. But now all dominating forces in today's world are foreign to Europe. The great power in the world is the United States and, to the side, there is the Soviet Union. Europe has virtually disappeared as a factor and it took astonishingly little time for this change to come about. Two centuries ago the United States was not even a nation; now it is a colossus [Sulzberger, 1973: 4–5].

The American colossus, which hitherto had towered only over the Western Hemisphere, now had a foot in Europe and another in Asia. By subtle degrees, its overseas wartime missions and occupation responsibilities were transformed into alliance commitments – the Rio Pact, NATO, ANZUS Pact, Philippine Treaty, and so on. Washington truly held the reins of a global politico-military coalition.[20]

This situation was historically unprecedented. "The radical novelty of our present position," wrote Lippmann (1952: 31–33), "is, as seen with American eyes, that we have now become a principal power." Whereas during World War I, and even part of World War II, the United States had been "an auxiliary power, a supporting and reinforcing power, called in to redress the balance of the Old World," it now had the central position and responsibility. Many of its former allies, and even its enemies, were "auxiliaries" in helping it maintain the balance of the world as a whole.

The main strategy adopted by the United States for safeguarding the global balance was that of "containment." As originally defined by Kennan (1947), this strategy was "the adroit and vigilant application of counterforce at a series of constantly shifting geographical and political points, corresponding to the shifts and manoeuvres of Soviet policy." To meet expected Soviet aggression, the inner and outer defense perimeters of the American sphere were fanned out in all directions.[21]

The flaw in the containment strategy, as Lippmann (1947: 21) and other critics pointed out, was that it relied on the "feeble or disorderly nations, tribes and factions around the perimeter of the Soviet Union" rather than on the compact nucleus of the Atlantic Community – the center. Moreover, the strategy seemed negative, reactive rather than affirmative. The adversary seemed to have the initiative. To follow such a policy could only debilitate.[22]

There was a further, psychological danger in the containment policy: by focusing on the outer margin of the Soviet bloc rather than on the Western sphere, American officials might shift the "center" to Moscow, leaving the West "centerless." On a contemporary map of the American perimeter of defense around the communist core, the larger American sphere can easily become recessive and the smaller but more compact Soviet–Chinese sphere dominant (Figure 7.6).[23] This perceptual switch has its counterpart in the ideological realm: the "Free World" becomes a mere anti-"Slave World," much as "America" had once been a mere anti-"Europe."

With the 1955 Bandung Conference, the central position of the United States in world politics began to erode in another way. New political centers in the "South" – Djakarta, New Delhi, Cairo, Lagos, Accra, Brasilia – emerged. A "Third World" rose to challenge the Old and the New Worlds. This

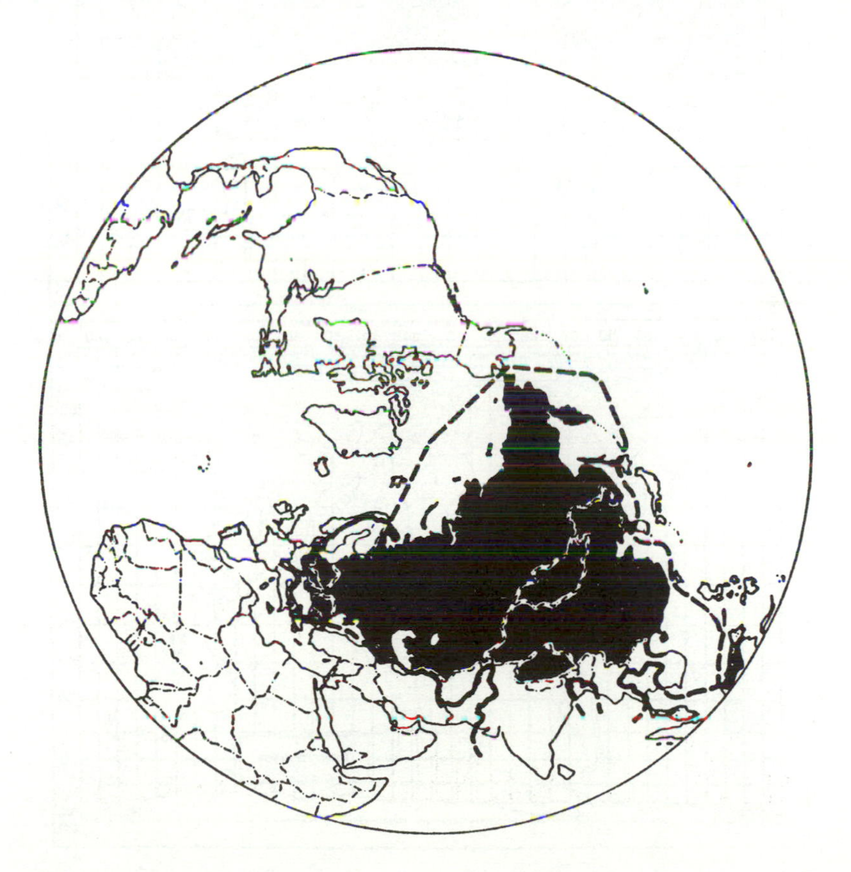

**Fig. 7.6** The American perimeter of defense (1955, from *Principles of Political Geography* by Hans W. Weigert et al. © 1957 by Appleton-Century-Crofts, Inc.)

southward shift in a hitherto largely "northern" international system was supported by global demographic trends.

So fundamental were these changes that a new *Weltbild* was needed. The German historian Arno Peters produced one cartographically. Noting that on a conventional Mercator-projection map the central horizontal axis is well above the equator, a placement that has the effect of exaggerating the size of North America and Eurasia and dwarfing Latin America and Africa, he recommended an equal-area projection, centered on the equator, in which this "Northern Hemisphere" bias is rectified (Figure 7.7).[24] This shift in cartographic emphasis, while directed mainly against "Europe-centrism," also

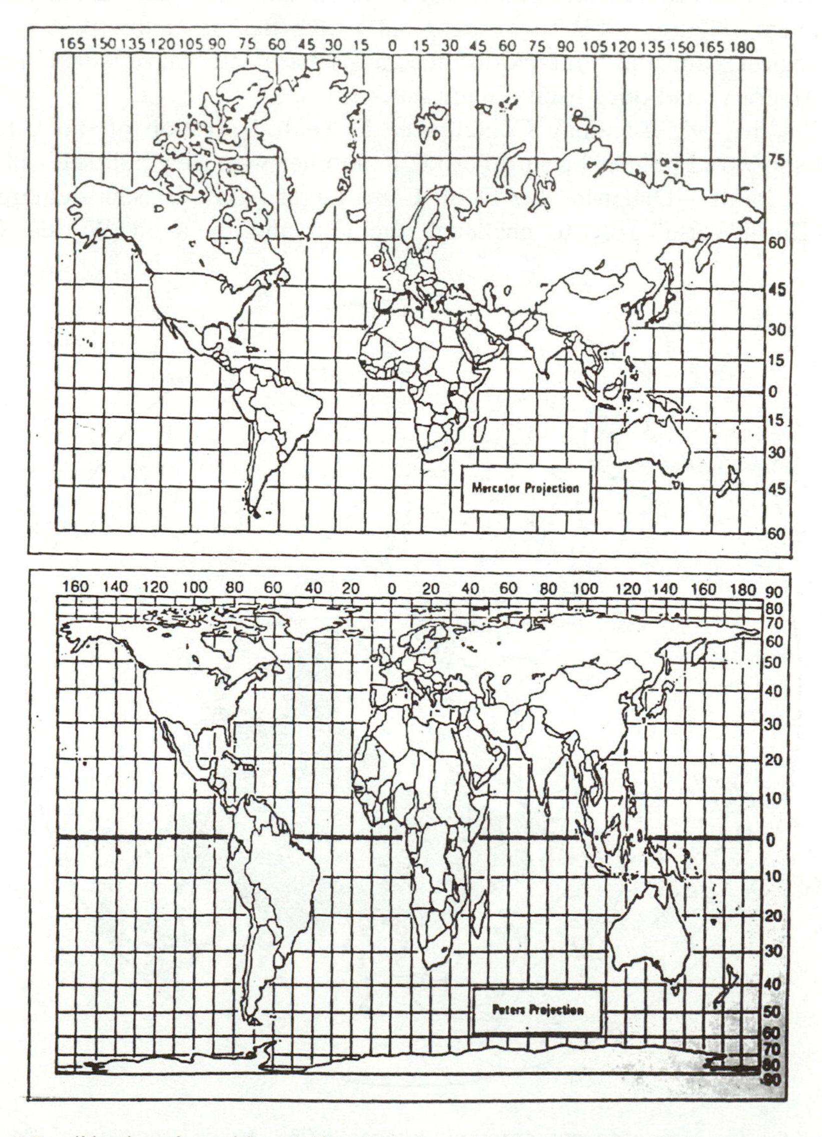


**Fig. 7.7** 'Northern' world and 'Southern' world (*Los Angeles Times* map by Don Clement © 1973, reprinted by permission)

counteracts "America-centrism." The United States is shunted into an upper corner, a "northern" periphery.

There was as yet no gainsaying, however, the continued world centrality of the United States in economic and military affairs. This power, during the 1950s often latent, became overt in the 1956 Suez crisis and, later, the 1962 Cuban missile crisis. Britain and France could not sustain their invasion of Egypt without the concurrence of Washington, upon whom they were dependent for oil, credits, and even confidence. During the Cuban crisis, the Soviet Union could not in the end complete its installation of ballistic missiles near the United States. The reasons for these failures were the same: the acknowledged role of the United States as the primary upholder of order in the world.

In the 1960s it was common to refer to the United Slates as "the only truly global power." The perverse proof of this was the war in Vietnam – America's "first imperial war," as Liska (1967) characterizes it.[25] This conflict was waged over a distance of 8000 miles, with seemingly inexhaustible reserves, ostensible moral conviction, and apparent political impunity. No military objective seemed impossible. Wohlstetter (1968), rejecting the assumption that a nation's power and knowledge must fall off steeply with distance from home, argued that, to the contrary, the advancing technology of transport and communication made even the most faraway country accessible – and, by implication, conquerable.

The American antiwar reaction, which grew in the aftermath of the 1968 Tet offensive, led to a full-blown reexamination of America's historic place in the world. One result was a rediscovery of what Ball (1968: 351–352) has termed "the difference between center and periphery." Americans had been navigating, Ball suggested, "by a distorted chart – like something drawn by a medieval cartographer, in which Vietnam appears as a major continent lying just off our shores and threatening our national existence."[26] He urged that it be put into true scale and back into its proper Southeast Asian setting again.

The new foreign policy search, as Ball (1968: 353, 356–357) put it, needed to be for "a solid base" on which some durable system of power could be erected. For him, as for like-minded members of the subsequently organized Trilateral Commission, this was the "nuclear," or central, relationship with the other nations of the Atlantic Community, plus Japan. In particular, he favored a unified Western Europe, which could assume with the United States "a political partnership that fully reflects the common interests of a common civilization."

Other theorists considered such an alliance too narrow, too insufficient. The United States, Henry Kissinger indelicately pointed out, had global interests, whereas Europe had only regional interests. The best way for American diplomacy to resolve the central issues, especially the strategic arms race, and even to handle such peripheral matters as Vietnam, was to deal directly with the Soviet Union and the People's Republic of China. Hence the diplomacy of Détente and the unprecedented presidential trips to Peking and Moscow.

The resulting fluidity of international relations was superficially rationalized by the concept of a "pentagonal" world in which all great powers

– allies or adversaries – functioned more or less alike. President Nixon explained: "I think it will be a safer world and a better world if we have a strong, healthy United States, Europe, Soviet Union, China, and Japan, each balancing the other, not playing one against the other, an even balance" (*Time*, 1972: 14). Implicitly, this was an admission of the descent of the United States from superiority to parity, from being *the* center to being *a* center.

The 1973 October War and Arab oil embargo further detracted from an American sense of world centrality. The Nixon Administration's attempt to organize a "multilateral" response to Arab economic blackmail failed. Europe, which imported more than 80 percent of its petroleum supplies from the Middle East, could not afford the likely penalty of such a broad confrontation. A map of world oil movements (Figure 7.8), showing Europe's much greater vulnerability, makes the logic of Europeans' refusal to follow American leadership plain.[27] Given this divergence of interest, it began to seem once again, as it had for Mackinder in 1904, that "the real divide" between East and West lay in the Atlantic. The emerging rift was overcome, however, as America's imports of Middle Eastern oil also rose to huge amounts. Soon the United States was dealing "bilaterally" with the Arab oil producers like everyone else. Israeli diplomatist Abba Eban has even taunted the United States with becoming a "satellite" of Abu Dhabi and Dubai!

The disproportion between Middle Eastern and American oil production was so great as to suggest a new polarization in world affairs. Whereas the countries of the Middle East had an estimated 53 percent of known petroleum reserves, the United States had only about 5 percent. Represented visually by means of the contiguous-area cartogram, as in a series of Exxon Corporation

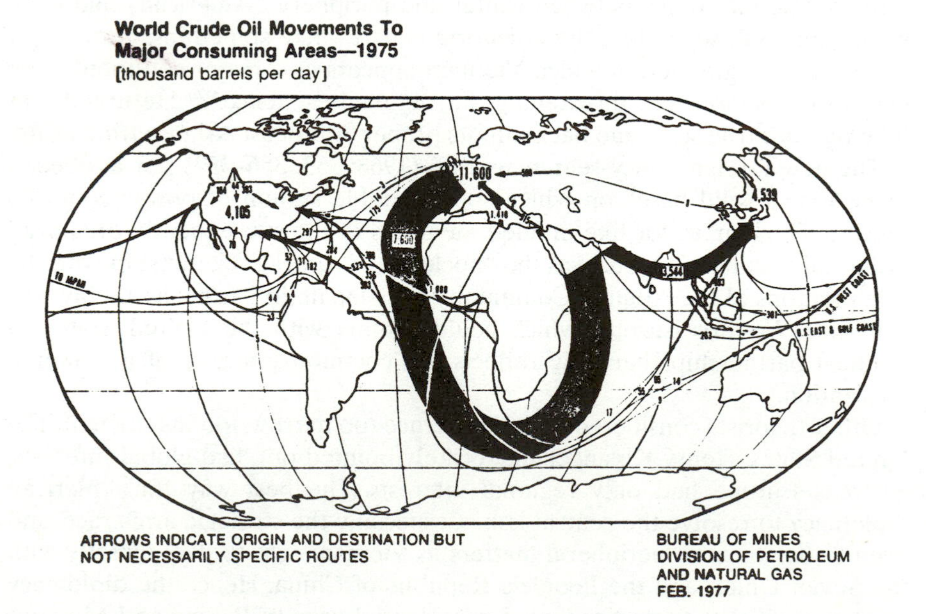


**Fig 7.8** Relative American independence and European dependence (Bureau of Mines, U.S. Department of the Interior, February 1977)

advertisements, these statistics made a striking portrait of dependency. The coal situation, however, is quite different. The United States is believed to have the largest deposits in the world (Figure 7.9).[28] America's long-term energy "picture" thus would appear to be balanced, and its independence basically secured.

Nonetheless, the crisis of 1973 seemed to have a profound impact upon the hierarchical-locational structure of international relations and upon the position of the United States within it. Coupled with the policy of Détente and a nonideological, tilting conduct of alliances, it has introduced what may be termed a "new relativity" into international affairs. There are few absolute or fixed positions any more. What is center and what is periphery is no longer clearly apparent. Moreover, it is no longer even clear what dimension of international relations is most important. The result for many Americans, as mentioned at the outset, is a sense of geopolitical flux.

How can this uncertainty of Americans about their place in the world be understood? Several suggestive theories, or explanatory paradigms, of the fluidity of the international affairs have been offered. One is the "Diffusion of Power" thesis. According to it, the mastery of modern technology by more and more nations has reduced, and will finally eliminate the advantages enjoyed by the industrial nations – including the United States. The gradients between economic centers and peripheries will thus be obliterated (Rostow, 1968). A second, more unusual, offering is the "Politectonic Zone" idea. Based on a loose analogy with the plate-tectonics theory in geology, this explanation regards political blocs as if they were freely floating on a liquid core. The accent is on nations' geological resources, geographical edges, and long-term chronological drifts. Most importantly for our present discussion, the position of any one bloc is relative to the positions of the others. None can logically be conceived as either central or peripheral (Cline, 1975).[29] A third, more cosmic, paradigm is the "Spaceship Earth" doctrine – the notion that, in the words of one of its leading proponents, "planet earth, on its journey through infinity, has acquired the intimacy, the fellowship, and the vulnerability of a spaceship" (Ward, 1966: v; Fuller, 1970). Patently, in such a world system, the physical, social, and psychic distances necessary to the maintenance of center–periphery hierarchies among peoples and places disappear. The

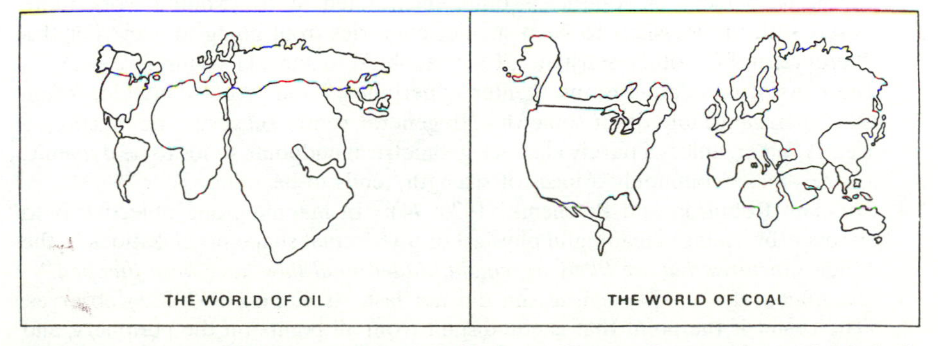


**Fig. 7.9** Oil insufficiency and coal sufficiency (Exxon Corporation and Mark S. Monmonier, Syracuse University)

United States, as the most distance-overcoming and altruistic of all nations, is usually expected, by Spaceship Earth theorists, to be the first to adopt this new selfless ethic.

Whether Americans can or will do so may depend upon their having a steady perspective and a firm sense of their own centrality. Is this objectively possible? The world position of the United States, though no longer supreme, is still a uniquely advantaged one. As Hassner (1976: 74–75) points out, "America alone is a decisive actor in every type of balance and issue." Lacking absolute power, it may be able to use its *relative* strengths to offset its *relative* weaknesses. To do this entails "the art of selectively separating or linking issues, regions and dimensions." Such a "differentiated" role, if played skillfully and with a singleness of vision, may enable the United States to hold the world center it has won in the short period of two centuries. If not, the American world position, and the international order still dependent upon it, may fragment and blur, like a cubist painting. From such a kaleidoscopic process, however, new, powerful focal centers may emerge – much as the United States itself emerged from a disintegrating world order.

## Notes

1 The expression "central world power" appears to have gained currency following the change in U.S. policy or, in locational terms, reorientation toward mainland China – a more historic, traditional, and possibly self-assured "center of the world." See, for example, the argument of Naval War College professor Thomas H. Etzold, who, doubting that the United States is today "the center of international politics," maintained that it is "wrong to imagine that the United States is central, or even very important, in the larger concerns of the Communist leaders of modern China" (Boston Sunday Globe, 1977). Compare the remark of George W. Ball, who, noting that Henry Kissinger visited Peking nine times, that two American presidents had been there, and that no leading Chinese official had as yet come to Washington, asked: "Are we, or are we not, vassals of the Middle Kingdom?" (Ball, 1977).

2 The confusion about the reason for the shift in the American world position – great-power competition or systemic instability – is partially indicated by a Harris Survey (1977) finding that, despite a sense of continuing rivalry with the Soviet Union, 60 percent of Americans think the United States "should work more closely with the Russians to keep smaller countries from going to war." For this reference, and for other assistance, I am indebted to John H. Maurer.

3 The pairs of correlative terms, "center"–"periphery" and "core"–"margin," often used interchangeably, differ somewhat. In general, center suggests a point, core an area. A center, unless a purely abstract, geometrical midpoint, tends to be dynamic; a core, though commonly a locus of strength, tends to be static.

4 Compare Robinson and Petchenik (1976: 74): "In mapping, one objective is to discover (by seeing) meaningful physical or intellectual shape organizations in the milieu *structures that are likely to remain hidden until they have been mapped*."

5 The notion of a unique *central* sun did not last. As Kuhn (1957: 233) observes, "The center is the point that is equidistant from all points on the periphery, and that condition is satisfied by every point in an infinite universe or by none."

6 A more ornate astronomical analogy was drawn by Georges-Louis Leclerc, Comte

de Buffon, who represented Britain as the "Sun," France as a "Comet." When the French comet, in its eccentric political movements, collided with the British sun, it rolled into orbit a new "Planet," "the American Empire" (Boyd, 1952: 443–445).

7 For examples of Philadelphia-based maps, see Lunny (1961: 38, 43) and Ristow (1977: 2). Examples of Washington-based maps are in Ristow (1972: 137, 146). In 1884 an International Meridian Conference, though held in Washington, fixed the prime meridian at Greenwich. For astronomical purposes, however, a Washington prime meridian (running through the Naval Observatory) remained legally in force until 1912.

8 For early cartographic representations of the Humboldt temperature-zonal system, see Robinson and Wallis (1967). See also "Gilpin's American Economic, Just, and Correct Map of the World," illustrating William Gilpin, *The Cosmopolitan Railway. Compacting and Fusing Together all the World's Continents* (1890). On the map a global linear "Isothermal Axis" is shown, roughly paralleled by the projected route of an intercontinental "Cosmopolitan Railway,' connecting North America and Siberia at the Bering Straits. Reproduced in Karnes (1970: 150).

9 Among the earliest U.S.-centred world maps were David H. Burr's "The World, on Mercator's Projection," published by J. Haven in Boston in 1850, and "Colton's New Illustrated map of the World on Mercator's Projection," published by J. Colton in New York in 1851. I am indebted for this information to John A. Wolter, Chief of the Geography and Map Division, Library of Congress. Figure 7.2 is from Mitchell (1878: 4–5). All editions of Mitchell's *Atlas* from 1860 include this map, as I am informed by Richard W. Stephenson, Head of the Reference and Bibliography Section of the Library of Congress Geography and Map Division. A further, more intrinsically cartographic, reason for centering world maps on the United States at mid-century was the need to exhibit, in a unified way, the Wilkes Track, the route of the U.S. Exploring Expedition into the Pacific-Antarctic region (1838–1842). On a Greenwich-centered global chart, the Pacific Ocean is divided.

10 This position is shown figuratively in a Currier and Ives print, "The Stride of a Century," commemorating the 1876 Centennial. Behind the continent-straddling Uncle Sam figure, a westward-moving railroad train carries the commerce of the Atlantic to the Pacific, where a ship awaits. Reproduced in Paterson et al. (1977).

11 The turn-of-the-century shift in American locational perspective is exemplified by the opening passage of Semple's (1903: 1) *American History and Its Geographic Conditions*: "The most important geographical fact in the past history of the United States has been their location on the Atlantic opposite Europe; and the most important geographical fact in lending a distinctive character to their future history will probably be their location on the Pacific opposite Asia." Semple's prophecy has some verification in current reality. The present American "global center," or the geometric focus of the smallest circle including all of the United States and its outlying areas, is in the vicinity of Kodiak Island off the southern coast of Alaska (U.S. Department of State, 1965: 1).

12 Other Europeans were more conscious of a direct American influence. Stead (1902: 19, 26) argued that within the "solar system" of English-speaking peoples the United States was becoming the "sun" and was able "to exert the pull" upon the United Kingdom. The increased Pacific orientation of Americans did not reduce the effect of proximity. As Stead noted: "Dublin is not half as far from New York as Manila is from San Francisco."

13 Compare Woodrow Wilson: "The United States has become the economic center of the world, the financial center. Our advice is constantly sought. Our economic engagements run everywhere, into every part of the globe." (Foley, 1923).

14 The principal author of the statement was the Italian emigré G. A. Borgese.
15 Figure 7.3 is from Spykman (1944) *The Geography of the Peace*, edited by Helen R. Nicholl, with an introduction by Frederick Sherwood Dunn, and maps by J. McA. Smiley.
16 Figure 7.4 is from Harrison (1944: 8–9), *Look at the World: The FORTUNE Atlas for World Strategy*, with text by the editors of *Fortune*. For a comprehensive examination, see Henrikson (1975).
17 See also Arnheim (1976). He regards a map "not as an assembly of shapes but as a configuration of forces," and stresses sensitivity to orientation. The images of countries on maps are transformed when rotated, as the United States would be if shown sideways or upside down. The "upright" position of the United States on polar maps minimizes (American) viewer disorientation.
18 For a map emphasizing polar routes, see Spykman (1944: 56).
19 Figure 7.5, adapted from Mackinder (1943), is from Cohen (1963: 53).
20 For a conventional cartographic representation of the American alliance system, with lines radiating from Washington, D.C. see U.S. Department of State (1977a).
21 For graphic illustrations of the various U.S. defense perimeters, see Pearcy (1964).
22 For a recent historical discussion, see Gaddis (1977).
23 Figure 7.6 is from Weigert (1957: 274).
24 Figure 7.7 is from Morris (1973: 72–73). The maps were prepared by a *Milwaukee Journal* artist from originals by Don Clement of the *Los Angeles Times*. The virtues of Peters's map are extolled in Government of the Federal Republic of Germany (1977). Professional cartographers, discounting the novelty and technical merits of the Peters map (it has faulty standard parallels and a "squeezed" look), have been sharply critical of the excessive claims made for it. Nonetheless, it has served its provocative purpose.
25 "In a unifocal international system," as Liska (1967: 27) classifies the U.S.-centered world, "the relationship of any one state to the imperial state is operationally more significant for its role and status than is its position in a regional hierarchy and balance or its declaratory stance on matters of global concern."
26 This exaggeration and displacement of the Vietnam image are probably causally related, to a degree, to the repeated experience of seeing the map of Vietnam (without an accompanying scale) on television. The image "filled" the screen, as did weather maps of the United States. Both countries occupied the same "space." The actual geographical relationship – distance and direction – of Vietnam and the United States was rarely, if ever, shown.
27 Figure 7.8 is from U.S. Department of State (1977b: 17). *The United States and World Energy: A Discussion Paper*, Publication 8904, General Foreign Policy Series 304 (November 1977), p. 17.
28 Percentage figures in the Exxon advertisement are from *Oil and Gas Journal* (December 25, 1972). Figure 7.9 is from Monmonier (1977: 19). On the official emblem of the Organization of Petroleum Exporting Countries (OPEC) – an oval shape showing all the members, from Ecuador to Indonesia, cartographically – the United States is not even on the map! (Allen, 1978: 23).
29 On a world map on the frontispiece, drawn to illustrate the Politectonic thesis, the "North America-Central America" zone appears to be floating upside down above Eurasia (the Heartland). For the notion of relativity in plate tectonics, see McKenzie (1977).

**References**


Agar, H. 1940: *The City of Man: A Declaration of World Democracy*. New York.

Allen, L. 1978: "Not so wild dream: OPEC's amazing rise to world economic power." *Harvard Magazine* 80, 5: 22–28.

Arnheim, R. 1976: "The perception of maps."*American Cartographer* 3, 1: 5–10.

Arnheim, R. 1977: Letter to the author. June 9.

Ball, G. 1968: *The Discipline of Power: Essentials of a Modern World Structure*. Boston.

Ball, G. 1977: "Against 'Cravenly Yielding' to Peking," *New York Times* (August 24).

Bartlett, J. 1970: *The Record of American Diplomacy: Documents and Readings in the History of American Foreign Relations*. New York.

Beale, H. K. 1956: *Theodore Roosevelt and the Rise of America to World Power*. Baltimore: Johns Hopkins Press.

Boorstin, D. J. 1965: *The Americans: The National Experience*. New York: Random House.

*Boston Sunday Globe* (1977) August 21.

Boorstin, D. J. 1976: *The Exploring Spirit: America and the World, Then and Now*. New York: Random House.

Bowman, I. 1928: *The New World: Problems in Political Geography*. Yonkers-on-Hudson, NY.

Boyd, J. [ed.] 1952: "Francis Hopkinson letter, January 4, 1784," in *The Papers of Thomas Jefferson*. Princeton, NJ: Princeton University Press.

Carter, J. 1977: "Address at Notre Dame University." *New York Times* (May 23).

Cline, R. S. 1975: *World Power Assessment: A Calculus of Strategic Drift*. Washington, D.C.: Georgetown Center for Strategic and International Studies.

Cohen, S. B. 1963: *Politics in a World Divided*. New York: Oxford University Press.

Coolidge, A. C. 1912: *The United States as a World Power*. New York: Macmillan.

Cressey, G. B. 1945: *The Basis of Soviet Strength*. New York: McGraw-Hill.

Eichelberger, C. M. 1977: *Organizing for Peace: A Personal History of the Founding of the United Nations*. New York: Harper & Row.

Eisenstadt, S. N. 1977: "Sociological theory and an analysis of the dynamics of civilizations and of revolutions." *Daedalus* 106, 4: 59–78.

Fischer, E. 1948: *The Passing of the European Age: A Study of the Transfer of Western Civilization and Its Renewal in Other Continents*. Cambridge, MA: Harvard University Press.

Foley, H. [ed.] 1923: *Woodrow Wilson's Case for the League of Nations*. Port Washington, NY: Kennikat.

Fuller, R. B. 1970: *Operating Manual for Spaceship Earth*. New York: Simon & Schuster.

Gaddis, J. L. 1977: "Containment: a reassessment." *Foreign Affairs* 55, 4: 873–887.

Gilbert, F. 1961: *To the Farewell Address: Ideas of Early American Foreign Policy*. Princeton, NJ: Princeton University Press.

Government of the Federal Republic of Germany (1977) "Peters' projection – to each country its due on the world map." *Bulletin* 25, 17: 126–128.

*Harris Survey* 1977: June 9.

Harrison, R. E. 1944: *Look at the World: The FORTUNE Atlas for World Strategy*. New York: Knopf.

Hassner, P. 1976: "Europe and the contradictions in American policy," pp. 60–86 in R. Rosecrance (ed.) *America as an Ordinary Country: U.S. Policy and the Future*. Ithaca, NY: Cornell University Press.

Henrikson, A. K. 1975: "The map as an 'idea': the role of cartographic imagery during the Second World War." *American Cartographer* 2, 1: 19–53.
Holborn, H. 1963: *The Political Collapse of Europe*. New York: Knopf.
Inaugural Addresses of the Presidents of the United States from George Washington 1789 to John F. Kennedy 1961 (1961) Washington, DC: Government Printing Office.
James, P. E. with E.W. James 1972: *All Possible Worlds: A History of Geographical Ideas*. Indianapolis: Bobbs-Merrill.
Karnes, T. L. 1970: *William Gilpin: Western Nationalist*. Austin: University of Texas Press.
Kennan, G. F. 1947: "The sources of Soviet conduct." *Foreign Affairs* 25, 4: 566–582.
Kramnick, I. [ed.] 1976: *Common Sense, by Thomas Paine*. Harmondsworth: Penguin.
Kuhn, T. 1957: *The Copernican Revolution: Planetary Astronomy in the Development of Western Thought*. Cambridge, MA: Harvard University Press.
Kuhn, T. 1970: *The Structure of Scientific Revolutions*. Chicago: University of Chicago Press.
LaFeber, W. [ed.] 1965: *John Quincy Adams and American Continental Empire: Letters, Papers and Speeches*. Chicago: Times Books.
Lippmann, W. 1943: *U.S. Foreign Policy: Shield of the Republic*. Boston.
Lippmann, W. 1947: *The Cold War: A Study in U.S. Foreign Policy*. New York: Harper & Row.
Lippmann, W. 1952: *Isolation and Alliances: An American Speaks to the British*. Boston: Little, Brown.
Liska, G. 1967: *Imperial America: The International Politics of Primacy*. Baltimore: Johns Hopkins Press.
Luce, H. R. 1941: *The American Century*. New York.
Lunny, R. M. 1961: *Maps of North America*. Newark, NJ.
Mackinder, H. J. 1943: "The round world and the winning of the peace." *Foreign Affairs*, 21, 4.
Mahan, A. T. 1890: *The Influence of Sea Power on History*, 1660–1873. Boston.
May, E. R. 1961: *Imperial Democracy: The Emergence of America as a Great Power*. New York: Harcourt Brace Jovanovich.
McKenzie, D. P. 1977: "Plate tectonics and its relationship to the evolution of ideas in the geological sciences." *Daedalus* 106, 3: 97–124.
Mitchell, S. A. 1878: *New General Atlas*. Philadelphia.
Monmonier, M. S. 1977: *Maps, Distortion and Meaning*. Resource Paper No. 75–4. Association of American Geographers.
Morris, J. A. 1973: "German's map of world improves on Mercator." *Military Review: Professional Journal of the US Army* 53, 11: 72-73.
Paterson, T. G., J. G. Clifford, and K. J. Hagan 1977: *American Foreign Policy: A History*. Lexington, MA: D. C. Heath.
Pearce, A. J. [ed.] 1962: *Democratic Ideals and Reality, with Additional Papers, by Halford J. Mackinder*. New York: Norton.
Pearcy, G. E. 1964: *Geopolitics and foreign relations*. Department of State Bulletin 50, 1288: 318–330.
Pratt, J. H. 1942: "American prime meridians." *Geographical Review* 32, 2: 233–244.
Renner, G.T. 1944: "Peace by the map." *Collier's* 113, 23.
Rippy, J. F. 1958: *Globe and Hemisphere: Latin America's Place in the Postwar Foreign Relations of the United States*. Chicago.
Ristow, W. W. 1972: *A la Carte: Selected Papers on Maps and Atlases*. Washington, D.C.: Library of Congress.
Ristow, W. W. 1977: *Maps for an Emerging Nation: Commercial Cartography in*

*Nineteenth-Century America*. Washington, D.C.: Library of Congress.

Robinson, A. H. and B.B. Petchenik 1976: *The Nature of Maps: Essays Toward Understanding Maps and Mapping*. Chicago: University of Chicago Press.

Robinson, A. H. and H. M. Wallis 1967: "Humboldt's map of isothermal lines: a milestone in thematic cartography." *Cartographic Journal* 4, 2: 119–123.

Rosenman, S. I. [ed.] 1950: *The Public Papers and Addresses of Franklin D. Roosevelt*. New York.

Rostow, W. W. 1968: *The Diffusion of Power: An Essay in Recent History*. New York.

Semple, E. C. 1903: *American History and Its Geographic Conditions*. Boston: Houghton Mifflin.

Smith, H. N. 1950: *Virgin Land: The American West as Symbol and Myth*. Cambridge, MA: Harvard University Press.

Spykman, N. J. 1944: *The Geography of the Peace*. New York: Harcourt Brace Jovanovich.

Stanton, W. 1975: *The Great United States Exploring Expedition of 1838–1842*. Berkeley: University of California Press.

Stead, W. T. 1902: *The Americanisation of the World, or the Trend of the Twentieth Century*. London.

Stefansson, V. 1947: "The North American Arctic," pp. 215–265 in H. W. Weigert and V. Stefansson (eds) *Compass of the World: A Symposium on Political Geography*. New York.

Sulzberger, C. L. 1973: *An Age of Mediocrity: Memoirs and Diaries, 1963–1972*. New York: Macmillan.

Thrower, J. W. 1972: *Maps and Man: An Examination of Cartography in Relation to Culture and Civilization*. Englewood Cliffs, NJ: Prentice-Hall.

*Time* 1977: "The master of the Mediterranean." 109, 21: 77–78.

*Time* 1972: "An interview with the President: the jury is out." 99, 1: 14–15.

U.S. Department of State 1965: *United States and Outlying Areas*. Geographic Bulletin No. 5. Washington, D.C.: Government Printing Office.

U.S. Department of State 1977a: *United States Collective Defense Arrangements*. Publication 8909, General Foreign Policy Series 303. Washington, DC: Government Printing Office.

U.S. Department of State 1977b: *The United States and World Energy: A Discussion Paper*. Publication 8904, General Foreign Policy Series 304. Washington, D.C.: Government Printing Office.

Ward, B. 1966: *Spaceship Earth*. New York: Columbia University Press.

Weber, A. 1948: *Farewell to European History, or the Conquest of Nihilism* (translated by R.F.C. Hull) New Haven, CT: Yale University Press.

Weigert, H. W. 1957: *Principles of Political Geography*. Englewood Cliffs, NJ: Prentice-Hall.

Weinberg, A. K. 1935: *Manifest Destiny: A Study of Nationalist Expansionism in American History*. Baltimore: Johns Hopkins Press.

Whitaker, A. P. 1954: *The Western Hemisphere Idea: Its Rise and Decline*. Ithaca, NY: Cornell University Press.

Wohlstetter, A. 1968: "Illusions of distance." *Foreign Affairs* 46, 2: 242–255.

# 8 Stuart Corbridge, 'Maximizing Entropy? New Geopolitical Orders and the Internationalization of Business'

Reprinted in full from: G. J. Demko and W. B. Wood (eds), *Reordering the World: Geopolitical Perspectives on the 21st Century*, Chapter 16. Boulder CO: Westview Press (1994)

The conduct and writing of geopolitics has changed greatly over the past twenty years. At one level, this reflects the end of the Cold War and the demise of the Soviet Union. The breakup of the state Socialist economies of Eastern Europe signaled further unease with the concept and practice of Socialism, and some commentators have prematurely proclaimed the end of history.[1] In geopolitical theory, the end of the Cold War has coincided with an upsurge in writings about geopolitics and the ability of the United States to manage the geopolitical world order.[2] At this level, political geographers are moving "inside" the surviving superpower to interrogate its governing myths and the words and practices by which it defines its external relations.[3]

Political geographers are also studying the growing transnationalization of economic and political relations. This approach to geopolitics moves away from an analysis of "the impact of *fixed* geographical conditions (heartlands/rimlands, lifelines, choke-points, critical strategic zones, domino effects) upon the activities of the 'Great Powers.' "[4] It focuses on industrial and financial capital flows that are redefining international relations and blurring national boundaries. Notwithstanding its military power, the United States is far from hegemonic in international economic and financial circles. In fact, the contemporary geopolitical economy has three centers of economic and political power – the United States, the European Community (EC), and Japan – that are defined by an emerging worldwide hierarchy of city-systems and by flows of capital, institutions, and agents. This new world of quasi-states and uncertain sovereignties presents opportunities for both stable international relationships and the creation of new instabilities.[5]

This chapter's focus is on the prospects of hegemonic stability and stable international relations in the face of rapidly internationalizing relations of economic production and exchange. Must this age of "late modernity" – a post-Cold War era of "deepening time-space compression"[6] – return us to a state of anarchy (or even entropy) in international political relations?

The second section of this chapter examines the internationalization and globalization of business (including the business of money and information) since 1945, briefly noting its changing geographies and raisons d'être. The third section discusses whether the recent internationalization of economic relations is linked to a period of hegemonic instability in which internationalist rules and regulations cannot easily be applied. Particular attention is paid to international trading relationships and the international debt crisis. An opposing hypothesis is considered in the fourth section: that the multilateralization of economic relationships and the creation of a decentralized global division of labor can support new and stable international relationships. This section also challenges the idea that nation-states must be at the heart of geopolitical analysis. Finally, in the fifth section, the contradictions between contemporary geopolitics and a changing global geopolitical economy are highlighted. Caution is urged for those who would proclaim the end of history and the end of geography and those who claim to already detect the outlines of a New World Order. Few such signs are clearly visible. Ours is still an uncertain world.

**Toward globalization: the world economy since 1944–1945**

The economic context for international political relations has changed dramatically over the past fifty years. This chapter provides a stylized and chronological account of the development of a global political economy since 1944–1945, paying particular attention to the differences between internationalization and globalization and to the geographical consequences and implications of these "global shifts."[7]

Although the world economy was never a collection of isolated regional and national economies, the first great era of economic internationalism coincided with Britain's rise to power as a global hegemony in the mid-to-late nineteenth century. Under the classical gold standard system that operated from 1870 to 1914, individual countries were required to run their economies in accordance with the economic imperatives of free trade and free capital movements.[8] At this time, too, large amounts of British capital were exported around the world, particularly to the United States, Canada, South Africa, and Argentina. The colonized countries were also expected to run their economies in line with the economic orthodoxies advanced in the colonial capitals, though a tendency to imperial zones of preference and creeping protectionism was already apparent by the turn of the century.

In the interwar years (1919–1939), this internationalized world economy fell apart. World War I had all but bankrupted Britain, and the United States was becoming the world's leading creditor country. Although the gold standard was reinvented in the 1920s, it could not survive past 1928, not least of all because Great Britain lacked the power to enforce the old rules of the game.[9] Following the crash of 1929 and the depression of 1929–1933, many countries – including Germany and Japan – sought to rebuild their economies by means of protectionism, competitive devaluations, and territorial expansionism. The world economy was now wracked by predatory economic and political mercantilism and the absence of international economic cooperation.

As World War II drew toward a close in 1944–1945, it became apparent that the United States would emerge as an unrivaled economic and financial superpower (although the Soviet Union would shortly match its military capabilities). It also became clear to the internationalists within the US government that the United States could not prosper after the war without also laying the foundations for economic growth and recovery in Europe (and possibly in Japan, although that became more obvious at the time of the Korean War). This, in turn, meant that the United States would have to establish and then police a system of international relations that encouraged economic integration and provided for stable financial conditions. This system was duly mapped out at the United Nations Monetary and Financial Conference, held at Bretton Woods in July 1944. At this conference, US negotiators closed the door on that version of US isolationism that called for the ruralization of Germany and the rebuilding of "fortress America";[10] instead, they adapted John Maynard Keynes's vision of a postwar economic order based upon free trade between independent countries, with currency convertibility tied to a fixed-peg exchange rate system. Propping up the Bretton Woods system would be the might of the US economy (perhaps half of the world economy in 1950), reflected in the US dollar. The US dollar would serve as the main unit of international account currency, and other currencies would price themselves in relation to this currency. The US government agreed to fix the value of the US dollar against gold and thus not to print excessive greenbacks.[11]

The Bretton Woods monetary system set the stage for the golden age of capitalism, which cushioned the advanced industrial world from 1945 to 1970.[12] With Bretton Woods establishing the new rules of the game in 1944 – and with them, the International Bank for Reconstruction and Development (IBRD, or World Bank) and the International Monetary Fund (IMF) – the US government moved to rebuild the economies of Western Europe and Japan by several means: IBRD transfers ($700 million between 1947 and 1952); the Marshall Plan for European Recovery (transfers of $13 billion between 1948 and 1952); increased spending on arms during both the Korean War and the Cold War (dwarfing official economic transfers); creating the world's first large-scale program of official development assistance (mainly to Latin America, Israel, and Southeast Asia); and encouraging investment overseas by private US firms.

In the 1950s, US firms – including Ford, General Motors, IBM, ITT, and Exxon – massively expanded their production facilities in Western Europe amid an estimated net private capital flow from the United States to Western Europe of $80 billion between 1950 and 1960. American corporations were attracted to countries across the Atlantic by the relatively low cost of labor in many European nations and by the access to burgeoning local markets (which had been rebuilt with US dollars and given further impetus by the founding of the European Economic Community [EC] in 1957). This first wave of postwar foreign investment by US-based multinational corporations was also made possible by new communications technologies, by new systems of corporate management and accounting, and by the emergence in Europe of currency markets that accepted dollar deposits beyond the regulatory reach of US authorities. As Richard O'Brien points out: "In the earliest days the demand for eurocurrency loans came as a result of an increase in activities in Europe by US multinationals. One of the first borrowers of eurodollars was IBM Europe. Although this was in the days of fixed exchange rates and thus limited foreign exchange risk, lending dollars to a company in Europe might have been considered problematic if that company had no clear stream of dollar earnings or access to dollars for repayment. IBM Europe, however, was seen as ultimately having access to dollars through its parent, and thus did not pose any 'geographical mismatch problem.' "[13]

The formation of the Euromarkets did not signal a truly global market in currencies, at least not in the 1950s and 1960s. The early eurocurrency markets were primarily offshore markets trading currencies that could clearly be linked back to one dominant country, the United States. This was also the case with the multinationalization of production. Although multinational corporations (MNCs) became more important in the 1960s and 1970s – with US corporations moving beyond Western Europe and with European and Japanese firms making their own presences felt worldwide – their production facilities were primarily meant to serve national markets, and the headquarters of most MNCs continued to have strong national affiliations. At this stage, few multinationals had become transnational or global corporations; rather, they were vehicles for the further internationalization of trade and private capital flows, particularly between Europe, Japan, and the United States. The blurring of discrete national economies was further encouraged in the mid-1960s by US government attempts to place restrictions on the domestic activities of US banks, which prompted them (led by the Bank of America and Citibank) to expand their offshore operations away from the eyes of US regulators.[14] In the 1960s, the rate of growth of net Eurocurrency transactions exceeded that of official reserves by a factor of five. Table 8.1 charts the increasing proportion of total assets of resident banking institutions, in different countries from 1960 to 1985, that can be accounted for by foreign-owned institutions.

In August 1971, the US president abandoned America's commitment to maintain a fixed exchange rate between the US dollar and gold. Henceforth, the United States was free to print dollars to the extent that non-US agents and countries were prepared to hold them as IOUs and as a means of international payment. The United States, for its part, sought to restore the competitiveness of its domestic economy, vis-à-vis a resurgent Germany and Japan, by driving down the value of the US dollar and by printing dollars to finance a growing balance-of-payments deficit.[15]

In the medium-term, this strategy triggered a rising rate of inflation on a global scale. In the early-to-mid-1970s, however, the monetization of the US deficit fueled only a further massive expansion of the privatized Eurocurrency markets. The economic powers of nation-states were slowly being ceded to the financial markets and their major players. The Eurocurrency markets were also boosted by the petrodollar deposits made in them by members of the Organization of Petroleum Exporting Countries (OPEC) following the third Arab–Israeli War of 1973 and the consequent threefold hike in oil prices. This war more or less coincided with the formal ending of the Bretton Woods system of fixed exchange rates. Currencies would now be allowed to float, and financial institutions making loans – not least of all from the Euromarkets to those newly industrialized and most industrialized countries (NICs and

**Table 8.1** Relative importance of assets of foreign-owned banking institutions operating in selected host countries, 1960–1985 (in percentages)[a]

| *Host country* | *December 1960* | *December 1970* | *December 1980* | *June 1985* |
|---|---|---|---|---|
| Belgium | 8.2[b] | 22.5 | 41.5 | 51.0 |
| Canada | – | – | – | 6.3 |
| France | 7.2 | 12.3 | 15.0 | 18.2[c] |
| Italy | n.a. | n.a. | 0.9 | 2.4 |
| Japan | n.a. | 1.3 | 3.4 | 3.6 |
| Luxembourg[d] | 8.0 | 57.8 | 85.4 | 85.4 |
| Netherlands[e] | n.a. | n.a. | 17.4[f] | 23.6 |
| Switzerland | n.a. | 10.3 | 11.1 | 12.2 |
| United Kingdom | 6.7 | 37.5 | 55.6 | 62.6 |
| United States[g] | n.a. | 5.8[h] | 8.7 | 12.0 |
| West Germany[i] | 0.5 | 1.4 | 1.9 | 2.4 |

[a]Percent of total assets refers to all banking institutions (domestic and foreign-owned) at end of month.
[b]December 1958.
[c]December 1984.
[d]Banks owned by Belgian residents are not considered foreign-owned banking institutions.
[e]Only universal branches of foreign-owned banking institutions.
[f]December 1983.
[g]Only agencies and branches of foreign-owned banking institutions.
[h]December 1976.
[i]Only branches of foreign-owned banking institutions.

Source: Bank for International Settlements, Recent Innovations in International Banking (1986, p. 152); after R. Bryant, *International Financial Intermediation* (Washington, D.C.: Brookings Institution, 1987), Table 3–4.

MICs) anxious to chase growth or to finance their balance-of-payments deficits – were forced to devise new financial instruments to protect themselves against the resulting foreign exchange risk.

The 1970s became the decade of the syndicated bank loan, with networks of commercial banks joining together under a lead bank to lend moneys to developing countries at floating rates of interest.[16] Once again, MNCs followed these flows of money into many poor countries, and for a short time, it became fashionable to speak about the "end of the Third World."[17] In the early 1960s, President Kennedy promised an "alliance for growth" in Latin America and ushered in the United Nations' "development decade." Both projects "failed," in part, because of a lack of official development assistance and private direct investment. In the 1970s, development would be different. Between 1970 and 1982, about nine money-center banks from the United States alone lent more than $80 billion to the non-oil-developing countries[18] (see Figure 8.1). At this time, too, a revitalized Mexican Border Industrialization Program (first introduced in 1965) brought rapid economic growth to northern Mexico, and parts of southern Brazil and Argentina seemed to emerge as ex-Third World regions. Astonishment was voiced at the phenomenal rates of growth in gross domestic product (GDP) and exports recorded in South Korea, Taiwan, Singapore, and Hong Kong. As the Ford

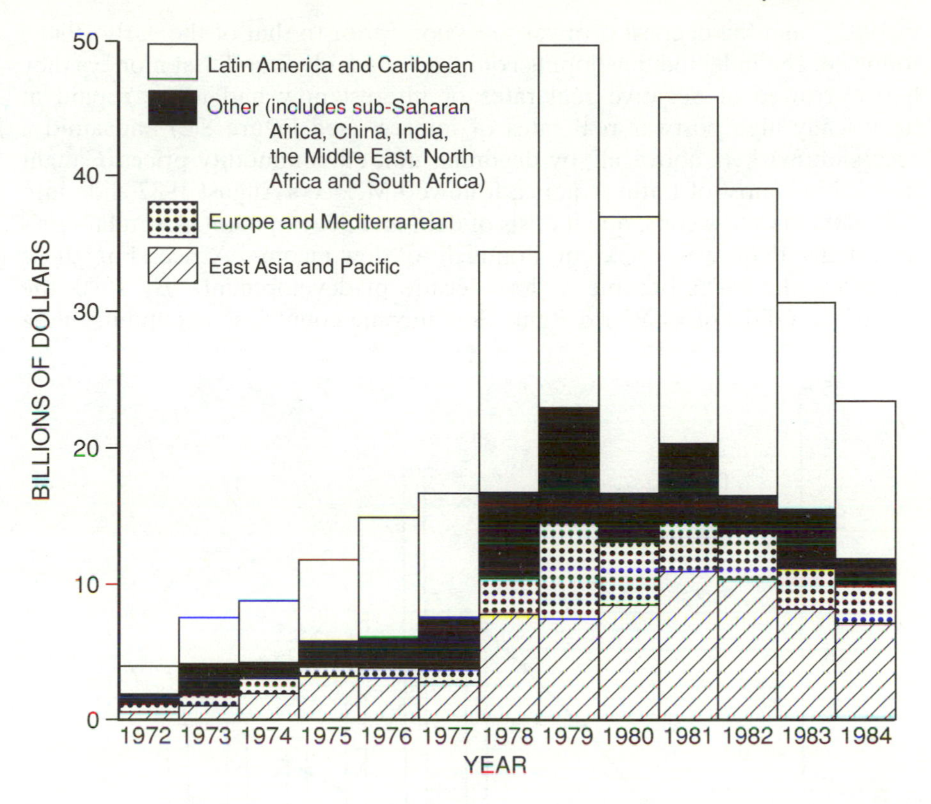


**Fig. 8.1** Syndicated Eurocurrency lending to developing countries, by region, 1972–1984 (S. Corbridge, *Debt and Development*, Oxford: Blackwell, 1992, p. 30, using OECD Financial Market Trends data)

Motor Corporation laid its plans for a truly global car, the Escort, the time did, indeed, seem ripe to abandon the Three Worlds idea that neatly divided a First World from a Second World and a Third World. In its *World Development Report* of 1980, the World Bank preferred to speak of "centrally planned economies," "capital-surplus oil exporters," "industrialized countries," "middle-income countries," and "low-income countries." The internationalization of business seemed set to herald the dawn of a world economy in which "the greatest challenge [would be] coming to terms with the inexorable spread of economic development over the next hundred years."[19]

The 1980s were not destined to bear out this vision. By 1978, it was apparent to President Carter that the US dollar could not be allowed to fall further and that inflation had reached an unforeseen and threatening level at 14 percent. (It was worse still in the United Kingdom and in many parts of the developing world.) In 1979, the chairman of the Federal Reserve Bank moved swiftly to tighten the money supply in the United States (and, indirectly, in much of the rest of the world). Prime interest rates in the United States climbed from an average of 9.5 percent in 1978 to an average of 15.1 percent in 1982,[20] and the deepest postwar recession (prior to that of the early 1990s) followed. The indebted developing countries suddenly found that moneys they had borrowed at negative real rates of interest now had to be repaid at historically high postwar real rates of interest (see Figure 8.2) and amid a recession marked, above all, by declining non-oil commodity prices. Caught in this bind, most of Latin America followed Mexico's August 1982 slide into debt defaults and a consequent crisis of debt and development. A similar crisis of official debt was apparent in much of low-income Africa. For these countries, the 1980s became a "lost decade of development." By 1990, the combined GDPs of 41 World Bank "low-income countries" (excluding India

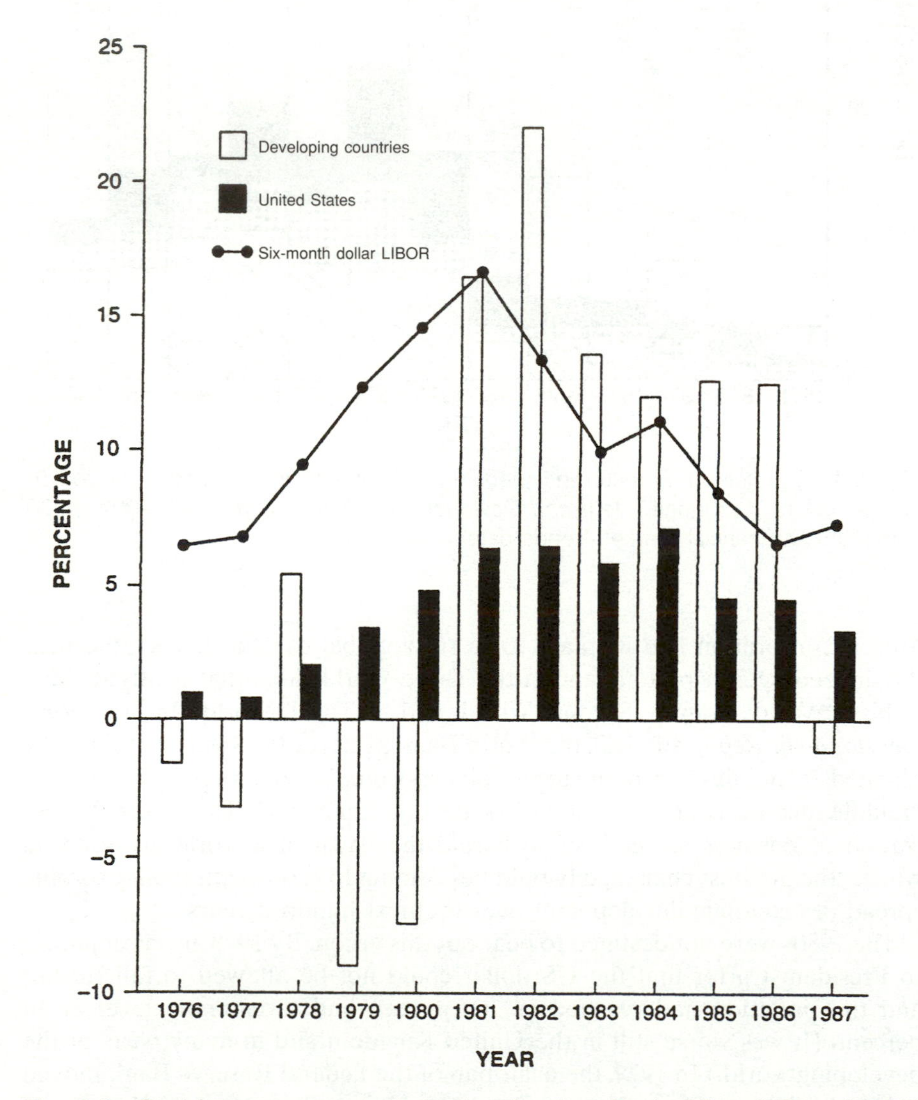


**Fig. 8.2** Real interest rates and the London Inter-Bank Offered Rate (LIBOR) (S. Corbridge, *Debt and Development*, Oxford: Blackwell, 1992, p. 39, using World Bank data)

and China) amounted to less than 1.5 percent of the total combined GDPs of 125 reporting countries. In low-income Africa, the "average" household was worse off in real terms in 1990 than it was in 1960.

In such dire circumstances, talk of globalization seems premature. The 1970s notion that the Third World is no more has been belied by the tragedies of maldevelopment that continue to haunt large parts of Africa, Asia, and Latin America. Although a rush of net inward investment did occur over parts of the Third World in the 1960s and 1970s, capital flows in the 1980s were outward. "To service its foreign debt, Latin America sent to the banks in the industrialized countries, in net terms, $159.1 billion from the end of 1981 to the end of 1986."[21]

The international debt crisis has clearly dented an earlier tendency to construct a global economy. At the same time, though, it has also perversely encouraged a global economy in which the Japanese economy and a globalization of new financial markets and information technologies figure much more prominently.

When Mexico defaulted on its debts in August 1982, it soon became apparent to US authorities that private US banks could fail if Mexico was not provided with the funds to roll over its debt repayments. In the short term, this meant that US banks had to make further, involuntary loans to Mexico (and other defaulting countries), under pressure from a newly empowered IMF. The indebted countries, in turn, would have to engage in structural adjustment programs. In the medium term, however, it was equally apparent that the indebted countries could only service their debts by first earning the necessary foreign exchange reserves.[22] Given the historical pattern of Latin America's trade, this implied that the US economy would have to be reinvigorated, even at the cost of increased borrowing from abroad. Thus began the second stage of the economic program known as Reaganomics. In 1982, following a period of tight fiscal and monetary policies in 1981 and 1982, Reagan moved to a loose fiscal policy at home, coupled with massively expanded foreign borrowings.[23] A historical buildup of US net assets abroad worth $141 billion by 1982 was turned into a net foreign debt of $112 billion by 1986.[24] By 1990, the United States owed close to 1 trillion US dollars to the rest of the world.

Pumped up by this massive foreign borrowing, the US domestic economy boomed from 1982 to 1989, even though the US current account worsened on the back of a strong US dollar and mounting deindustrialization at home. By 1988, however, it was apparent that the US territorial economy was in long-term decline, even though US firms performed as well as ever on the global stage[25] (see Table 8.2). It was further apparent that Japan had become a major economic power, along with a newly united Germany. Although the US economy is still larger than the economies of Germany and Japan combined, the rates of growth of the two latter countries suggest a changing balance of global economic power.[26] Japan signaled its newfound might by acting as the chief underwriter of US external debt in the 1980s while its corporations and citizens bought into American real estate and corporations. By the late 1980s, America's growing dependence on Japan and on financial decisions made in Tokyo were well known.[27]

**Table 8.2** World's top twenty-five corporations, 1992 (by market value)

| *Rank* | | | |
|---|---|---|---|
| *1992* | *1991* | *Name of corporation* | *Market value (in billions of US dollars)* |
| 1 | 2 | Royal Dutch/Shell Group (Netherlands/UK) | 77.82 |
| 2 | 1 | Nippon Telegraph & Telephone (Japan) | 77.52 |
| 3 | 2 | Exxon (US) | 75.30 |
| 4 | 6 | Philip Morris (US) | 71.29 |
| 5 | 4 | General Electric (US) | 66.00 |
| 6 | 14 | Wal-Mart Stores | 60.82 |
| 7 | 20 | Coca-Cola (US) | 58.41 |
| 9 | 18 | AT & T (US) | 55.85 |
| 10 | 7 | IBM (US) | 51.82 |
| 11 | 16 | Toyota Motor (Japan) | 43.97 |
| 12 | 26 | Glaxo Holdings (UK) | 42.64 |
| 13 | 19 | British Telecom (UK) | 40.45 |
| 14 | 9 | Mitsubishi Bank (Japan) | 39.84 |
| 15 | 17 | Bristol-Myers Squibb (US) | 37.60 |
| 16 | 11 | Sumitomo Bank (Japan) | 37.12 |
| 17 | 22 | du Pont (US) | 35.41 |
| 18 | 28 | Procter & Gamble (US) | 34.74 |
| 19 | 10 | Dai-Ichi Kangyo Bank (Japan) | 34.68 |
| 20 | 5 | Industrial Bank of Japan (Japan) | 34.04 |
| 21 | 8 | Fuji Bank (Japan) | 33.99 |
| 22 | 25 | Johnson & Johnson (US) | 32.28 |
| 23 | 12 | Sanwa Bank (Japan) | 30.41 |
| 24 | 39 | Unilever (Netherlands/UK) | 29.77 |
| 25 | 21 | Tokyo Electric Power (Japan) | 29.25 |

Source: *Business Week*, July 13, 1992.

The emergence of a tripolar world economy is not the only manifestation of the globalization in business affairs. The 1980s also witnessed a concerted attempt by various firms in the service sector to establish bases overseas and on a global stage. This was true not only of the traditional banking sector but also of finance, advertising, accountancy, and insurance companies. This internationalization of services, in turn, represented a still deeper globalization of financial markets and instruments. International trade commonly involves the physical movement of one commodity from one country to another, but the fungibility of money in a deregulated world economy means that even customs posts and national identities are often of little significance. The foreign exchange market is international in the sense that "all exchange rates involve different countries,"[28] but it is also globalized. All currencies are exchanged in one way or another. Indeed, there are few entry barriers to the market for it is a twenty-four-hour-a-day market with no special "home," played out on computer screens and telephones all over the world.

The significance of these changes in the global money market should not be underestimated. We have not yet reached "the end of history or geography," nor have we reached the "one world" prophesied by utopian thinkers. The differences between rich and poor places and peoples continue to widen.[29] Nevertheless, we do now live in a global economy in which production is widely internationalized and in which close to 40 percent of international trade is in the form of intrafirm commerce. We also live in a world in which "the markets" can defeat even the most concerted efforts by governments – and groups of governments – to defend particular exchange and interest rates. The withdrawal of the United Kingdom from the European Exchange Rate Mechanism in September 1992 is testimony to the insufficiency of country reserves in the face of massive currency trading on open markets. Today, more than ever, money is a type of information that is traded on computer screens and ignores national identities. The removal of exchange controls and financial deregulation in the early 1980s means that a global capital market now exists more clearly than ever before (see Table 8.3). Even powerful countries like the United States, Germany, and Japan must adapt their economic political ambitions to an internationalized world economy. In this new world, power is rapidly being devolved to the markets and to the IMF and the World Bank (in the case of many developing and ex-Socialist economies). True economic globalization may be in its infancy, but it is apparent nonetheless.

**Table 8.3** Number of companies listed on various stock exchanges at the end of 1990

| *Stock exchange/ location* | *Domestic stocks* | *Foreign stocks* | *Foreign stocks as percent of total* | *Total* |
|---|---|---|---|---|
| Amsterdam | 260 | 238 | 47.8% | 498 |
| London | 2,006 | 553 | 21.6 | 2,559 |
| Frankfurt | 389 | 354 | 47.6 | 743 |
| Paris | 443 | 226 | 33.8 | 669 |
| Zurich | 182 | 240 | 35.9 | 422 |
| Australia | 1,162 | 33 | 2.8 | 1,195 |
| Hong Kong | 284 | 15 | 5.0 | 299 |
| NASDAQ | 3,875 | 256 | 6.2 | 4,131 |
| NYSE | 1,678 | 96 | 5.4 | 1,774 |
| Tokyo | 1,627 | 125 | 7.1 | 1,752 |
| Toronto | 1,127 | 66 | 5.5 | 1,193 |

*Source: Quality of Markets Quarterly Review*, London Stock Exchange, January–March 1991; after R. O'Brien, *Global Financial Integration: The End of Geography* (London: Pinter/RIIA, 1992), Table 1.

## Uncertainty, instability, and rivalry in a posthegemonic world economy

If an internationalized world economy has emerged forcefully since 1970, what implications might this hold for contemporary geopolitics? The theory of hegemonic stability states that international relations will stabilize when those relationships are defined, regulated, and policed by an unrivaled hegemonic power. Such was the situation when Great Britain was the hegemonic power at the end of the nineteenth century and when free – if often unequal – trade flourished under its tutelage. Such was also the case in the 1950s and 1960s when the United States set the rules of the economic and financial worlds and policed them through its own domestic policies and through the World Bank, the IMF, and the General Agreement on Tariffs and Trade (GATT). At such times, the world economy grew rapidly (if again unequally), and major economic actors were assured a high degree of certainty when they entered into diverse economic transactions.[30] In the period of US hegemony, a complicating factor was the role and position of the Soviet Union as a rival military superpower. Even then, though, the Cold War and an associated balance in spheres of influence ensured that the world system was relatively stable.

According to the theory of hegemonic stability, international relations become fractious and unstable when the accepted hegemony comes under threat and is unable or unwilling to enforce rules that favor multilateral cooperation. This happened in the 1970s, and we continue to live with its consequences. The internationalization of economic relations in the 1950s and 1960s, together with a massive increase in global capital movements, ensured that the economy of the United States would enter a period of relative decline when compared with the EC and Japan and that the Bretton Woods system of fixed-peg exchange rates, limited capital flows, and balanced regional rates of growth would decay from within.[31] By the mid-1970s, the United States was pursuing an economic policy designed to boost the domestic economy in the short run, with little regard for the medium-term consequences for the world economy.[32] If, in the 1950s and early 1960s, the United States had acted as a benevolent despot, by the 1970s, it was prepared to live with OPEC oil-price rises in the expectation that such rises would hurt Germany and Japan more than the United States. America seemed prepared to stoke the fires of global inflation (by printing excess amounts of US dollars).

The EC and Japan retaliated against the US policies with competitive devaluations and tight money policies of their own. By 1980, new monetary policies were being adopted throughout Western Europe and the United States, and a world recession and international debt crisis followed. The United States, as a declining hegemonic power, was ceding some of its international authority in order to pursue its particular national ambitions.

Hegemonic instability will affect geopolitical-economic relationships of the 1990s in three particular areas: international trade, international capital flows and indebtedness, and the continuing mismatch between US military and economic capabilities.

As of October 1992, the Uruguay Round of the GATT talks continued to founder amid mutual recriminations between the United States and the EC. The range of issues that could not be resolved (international property rights and patents, agricultural policies and subsidies, First World protectionism, high-technology transfers, and so on) suggested a deeper disunity in the modern world economy. International trading relations have never been straightforward, but the GATT impasse suggested a degree of uncertainty and dislocation not apparent at the height of US hegemony. Many developing countries are now openly challenging the doctrine of free trade, even though many of them are being forced to export more primary commodities as part of their agreements with the IMF and the World Bank. The developing countries are also facing new economic rivals from the ex-Socialist world and creeping protectionism in the advanced industrial world. In 1985, the World Bank suggested that "an increase in protectionism [in industrial countries] big enough to produce a 10 percent deterioration in the terms of trade of Latin America would cost the region as much as the real interest cost of their entire debt."[33] For their part, some of the advanced industrial countries are striving to establish semiprotected trading "zones of influence", for example, the North American Free Trade Agreement.

As the world economy moves into its deepest recession since the 1930s, a new form of mercantilism is slowly taking hold, perhaps with a trading bloc in the Americas emerging in parallel to a yen zone in East Asia and Australasia.[34] Against such a backdrop and given the devastation of many of Africa's economies, Western Europe will consolidate its single market by the end of the 1990s while looking eastward to an expanded European Community that might even include former Soviet bloc countries and Turkey. The international trading economy may now resemble the world economy of the 1930s more than the benign trading environment of the 1960s. The present recession may itself reflect a greater volatility in international economic relations. In the 1930s, Keynes suggested that private business investment would only expand when the "animal instincts" of the world's entrepreneurs were undergirded by a general feeling of economic confidence. Confidence, in turn, was boosted to the extent that economic agents could predict a stable future. But in the mid-1980s, such expectations were regularly belied; the value of the pound against the US dollar between 1980 and 1986 fluctuated from close to 1 dollar to 1 pound to 2.4 dollars to 1 pound. Thus, while the world economy was internationalizing, the conditions for stable and expanded international trade were eroded because there was no hegemonic power to enforce order.

The international debt crisis further illustrates hegemonic instability. The origins of the developing countries' debt crisis lie firmly with the actions of OPEC in the 1970s and, more especially, with the unwillingness of the United States and the Bretton Woods institutions to provide official bilateral or multilateral development assistance to oil-importing developing countries that face balance-of-payment difficulties. By the mid-1970s, the United States could not pay for such a program of assistance. In the 1970s, it ceded a good deal of its economic power to the markets and encouraged the private banks to recycle OPEC's petrodollars through the private Eurocurrency markets. The risks associated with international money recycling fell on the indebted countries, which were then hit by the West's tight money policies of the early 1980s. Again, unilateral actions by a declining economic superpower hurt the indebted world very badly. The United States itself then fell into debt, and by the end of the 1980s, the world economy mapped out at Bretton Woods had been turned on its head: capital now flowed from the poor countries to the rich countries, and the world's most powerful economy was also the most indebted. Debt servicing was dictating a new geography of international trade. Fears grew in Europe and Japan that the United States would service its debt by driving down the US dollar (the currency in which its debt to Japan and other creditors would be made), while aggressively seeking out new markets for its tradable goods and services.

Finally, there is the matter of America's post-Cold War military might. The collapse of the Soviet Union might be read in two ways. It might seem that the world is now a safer place and that rapid steps can be made to rebuild civilian economies – the peace dividend. The United States might also seem to be in a position to lead a multilateral force against those countries, such as Iraq, that transgress international rules of behavior. On the other hand, the war in Iraq also demonstrated a certain weakness in the position of the United States. Although US military might was unrivaled, the mounting costs of modern warfare forced the United States to seek financial assistance from its allies.[35] Against a background of nuclear proliferation and humanitarian crises, America may again be provoked into military actions that it can ill afford. The economic difficulties of the United States – and the relative military weakness of Germany and Japan – will caution America against becoming involved even when it is invited. Either way, the connections between the geopolitical world of military strategy and the economic world of greater complexity and decentralization become increasingly clear and close.

### Reasons to be cheerful ... or hegemony revisited

Despite the preceding arguments, the consequences of a decline in US hegemony and an associated internationalization of the circuits of capital may not be all negative. Economic deterritorialization, decentralization, and multilateralism might create bold new opportunities for global political relationships, particularly the possible emergence of "dispersed power principles."[36]

The internationalization of business implies a measure of deterritorialization. The process is clearly evident in the United States where the territorial economy continues to suffer from relative decline even as US firms acquit themselves with distinction on the international stage. Deterritorialization can easily provoke a climate of economic protectionism that slides into a cultural xenophobia, such as the popular denunciation of the Japanese in America.[37]

The emergence of a transnational business class means that many economic and political power brokers in the major industrial countries are responsible for substantial assets in countries other than those in which they reside.[38] These multilateral, financial ties belie the notion that Germany, Japan, or the United States might ever go to war with one another to defend their domestic economies. In a very real sense, the "US economy" no longer exists; the same can be said for Japan and Germany, both of which have substantial capital stakes offshore.

Statistics on international trade continue to be collected and published, but their meaning is increasingly open to question. When a customer in Toronto, for example, buys a Chevrolet Spectrum, is it an American car? Ostensibly, it is, but in reality, it is little more than a Japanese Isuzu in disguise, built under license in the United States. When the car is shipped to Canada, can trade statistics accurately express the different geographical origins of value-added in this case? Similarly, how are cross-border, illicit financial transactions accounted for? The cityscapes of New York and Miami are powered, in large part, by massive flows of legal and illegal moneys, which bring together David Harvey's "fictitious capital" (or credit moneys) with an emerging network of "fictitious spaces."[39] The extraordinary scale of these transfers and of the illegal drug and arms trade to which they are sometimes linked illustrates the ever-greater fluidity of the modern world economy. Accountancy firms may be internationalizing, but international accountancy and accounting continue to lag behind the complexity and magnitude of current transactions. Today's geopolitical economy is dominated less by conventional economic centers, such as firms, plants, and industrial regions, than by the flows of money, information, business, and labor that link these centers. These vital economic flows make the defense of space and place, in a conventional geopolitical sense, more difficult than it once was. "Whose space?" and "whose place?" are pertinent questions. Where *do* today's "national" economies begin and end, and who commands or regulates them?

Deterritorialization is also linked to economic decentralization. When so-called geopoliticians and political scientists refer to the "end of history" and the prospects for a decentralized world polity based on democratic, market-based economies, they forget that the Third World, unlike the First World, is not free from immediate military threat. It might even be argued that the construction of today's New World Order is taking place on the backs of the Third World countries, turning an earlier East–West conflict into a North–South battle.[40]

At another level, the notion of decentralization does highlight a real redistribution and reconstruction of the postwar world economy. This is apparent in the emergence of the newly industrialized and newly industrializing countries in parts of Latin America and Eastern Europe. It is also apparent in the emergence of highly developed city-sectors, cities and city regions within even the low- and middle-income countries.

It would be easy to exaggerate the significance of these developments. Such changes matter very little to the lives of most men and women in India, for example, although they are symbols of a wider economic liberalization that does have far-reaching effects.[41] In terms of the contemporary geopolitical economy, the implication surely is that a hierarchy of nation-states is not the only means by which we might conceptualize, order, and classify the world around us. Such systems of classification are given in the annual reports of the World Bank, the IMF, and the UN (which might list countries in an economic hierarchy, ascending from Mozambique to Switzerland). But the modern world economy also resembles a gigantic central-place system, in which significant power and wealth is concentrated in just a few global cities

(London, New York, and Tokyo at a minimum – one for each of the major "eight-hour" time zones), with second-order cities like Paris, Frankfurt, Los Angeles, Sydney, and Hong Kong linking on to cities of the third, fourth, and fifth orders (which might include parts of Bombay and New Delhi, Sao Paulo, Lagos, and Manila).

As with the deterritorialized economy, these decentered spaces or places highlight the differences that beset the modern world economy while also emphasizing the similarities and linkages that bind today's geopolitical economy.[42] In theory, at least, the decentering of economic life should be associated with similar, if perhaps slower, decentering of political power. For the favored inhabitants of this favored central-place system, the defense of "national space" again becomes problematic. Leaders of indebted Latin America – who allowed capital earned in their countries to take flight to Miami and New York, even as they voiced anti-American rhetoric – proved unwilling to challenge the rules of an international financial system in which many had a large stake. The decentering of economic life (at least away from the old industrial heartlands) goes hand in hand with the further consolidation of an international business class and a parallel internationalization of "domestic" politics.

Finally, there is the matter of multilateralization. In the eyes of some commentators, the war against Iraq was evidence of the enhanced role played by the United Nations in international strategic affairs. The United States was able to prosecute a war against Iraq only after first gaining the support of its economic allies and the General Assembly of the UN. Others disagree, pointing out that the United States (like the EC and Japan) is still as likely to ignore international laws and institutions (e.g., in supporting the contras in Nicaragua) as it is to obey them. The UN has never been a powerful multilateral institution, especially in comparison with the growing powers of bodies working within the global political *economy*: from the World Bank and the IMF to the markets themselves.

For many years now, the World Bank has been the single largest source of official development assistance, but in the 1980s, its role and significance were greatly expanded in response to the global recession and the developing countries' debt crisis.[43] The World Bank began to intervene more directly in the economic management of indebted member countries but not in the economies of First World countries, which also had large balance-of-payment deficits. The bank – and the IMF – pressed economic reforms on poor countries designed to integrate them further into an open international economy. The World Bank also offered funds to countries facing the social unrest over IMF-mandated structural adjustment programs and provided some funds to compensate for unexpected changes in exchange rates and commodity prices.[44]

The IMF assumed a more visible role in international economic affairs, prompting many in Africa to berate the institution and its structural adjustment policies for "imposing misery and famine."[45] In the 1970s, the World Bank and the IMF had seen their power decline in tandem with a decline in their real resources, as compared to official country reserves and private Eurocurrency markets. In the 1980s, their coffers filled again as the United States and some other member countries looked to the two Bretton Woods institutions to repair the damage done by certain commercial banks. By 1990, some countries in low-income Africa (including Mozambique and Tanzania) relied upon official development assistance from multilateral institutions for more than 45 percent of their gross national products. The Bretton Woods institutions also began to play key roles in the reconstruction of Eastern Europe and the former Soviet Union. In all these, governmental decision-making has been compromised by dependence on multilateral financial agencies.

In the advanced industrial world, the capacity of what might be called "multilateralization" is still more limited. As in large parts of the developing world, the multilateralism that matters most is linked to economic and financial markets and the internationalization of business. Throughout the European Community, business interests now look with hope toward a European economic union that will deliver a single European currency by about the year 2000. Businesses are much less concerned about progress toward political unification. National cultural identities can be allowed to survive, but they should not interfere with economic transactions free of national customs posts. Businesses want a stable framework for international investment, which should be provided by a greater use of a common unit of currency such as the European Currency Unit (ECU). The power to regulate this currency will most likely fall upon Germany's Bundesbank. This might seem to suggest that the German national economy plays a leading role, but the Bundesbank, like the US Federal Reserve (and unlike the Bank of England), is free from direct political interference. Interest and exchange rate policies will be set by the Bundesbank and perhaps prompted by financial authorities in Japan and the United States and, much more importantly, by the financial markets themselves.

If developments in Europe are matched by similar developments elsewhere, the geopolitical economic (geopolinomic) world order of the early twenty-first century will be quite unlike the order of the mid-twentieth century. Geopolitics in the advanced industrial world, based on an internationalized community of capital, will have less regard for military might and far more regard for the disciplining of labor and resurgent nationalism. Geopolitics will not only be funded by a world economy, it will increasingly be obliged to serve it.

Cleavage in this New World Order will not be between East and West, nor between Europe, Japan, and the United States. In a deterritorializing, decentering, and multilateralizing world system, hegemony might reside not with a single country but with the markets themselves. Capital transactions will be even more abstract, with power exercised, unequally, through various nodal pressure points and networks in an expanding global central-place system. Within this decentralized system, the markets and various supranational institutions – linked by new lines of communication – might pressure the EC, the United States, and Japan to reach agreements on potentially divisive economic and political issues. Slowly, perhaps, but not unsurely, power might seep away from a set of fixed geographical regions and toward a less visible network of economic and political actors, institutions, and transactions.

## Conclusion

This chapter is a thumbnail sketch of some of the contradictory geopolitical implications that are associated with a growing internationalization of business. The path to a New World Order is unclear; signs point in different directions – to both an entropic disorder and a decentered orderliness. The future will remain uncertain because the internationalization of capital and information will continue to destabilize and even dissolve those fixed territorial units that conventionally have been at the heart of geopolitical theory and practice. We must now live with an uncertain world, in part because of a nascent decentering of production and exchange and an associated explosion in new information technologies. As the balance of power in our late-modern world shifts in country-location terms from the United States, Germany, and Japan, it is also changing in form and structure.

The advanced industrial world will slowly expand and will enjoy a period of relative, internal, geopolitical stability (at least in military terms). The south (or those Third World areas within the First World, like the Bronx and south-central Los Angeles) will not fare nearly so well. To the extent that these communities and even countries are only the "clients" of the markets and the multilateral institutions, their disempowerment will surely deny them the fruits of any new system of decentered hegemony that might emerge. For such peoples and places, physical violence will remain intrinsic. Only within the newly hegemonic center of decentered power will geopolitics be written in the language of economic costs and benefits. A geopolitical economy framework is a method of thinking about the modern world system; it will include not only economic warfare but also military warfare and extended use of physical force. Economic inequity continues to reinforce belligerence, but increasingly, it will do so away from the key nodes of a decentered network of hegemonic power.

## Notes

1 F. Fukuyama, "The End of History?" *The National Interest*, vol. 16 (1989), pp. 3–18; idem, *The End of History and the Last Man* (London: Hamish Hamilton, 1991).
2 P. Taylor, *Britain and the Cold War: 1945 as Geopolitical Transition* (London: Pinter, 1990).
3 S. Dalby, *Creating the Second World War: The Discourse of Politics* (London: Pinter, 1990).
4 J. Agnew and S. Corbridge, "The New Geopolitics: The Dynamics of Geopolitical Disorder," in *A World in Crisis? Geographical Perspectives*, 2d ed., R. Johnston and P. Taylor, eds (Oxford: Blackwell, 1989), p. 267.
5 R. Jackson, *Quasi-States: Sovereignty, International Relations and the Third World* (Cambridge: Cambridge University Press, 1990).
6 D. Harvey, *The Condition of Postmodernity* (Oxford: B1ackwell, 1989).

7 P. Dicken, *Global Shift: Industrial Change in a Turbulent World*, 2d ed. (London: Harper and Row, 1991).
8 M. de Cecco, *Money and Empire: The International Gold Standard, 1890–1914* (London: Blackwell, 1974).
9 G. Ingham, *Capitalism Divided? The City and Industry in British Social Development* (London: Macmillan, 1984).
10 H. Wachtel, *The Money Mandarins: The Making of a Supranational Economic Order* (New York: Pantheon Books, 1986).
11 R. Gilpin, *The Political Economy of International Relations* (Princeton: Princeton University Press, 1987).
12 S. Marglin and J. Schor, eds., *The Golden Age of Capitalism: Re-Interpreting the Post War Experience* (Oxford: Clarendon, 1990).
13 R. O'Brien, *Global Financial Integration: The End of Geography* (London: Pinter/RIIA, 1992), p. 30.
14 W. Greider, *Secrets of the Temple: How the Federal Reserve Runs the Country* (New York: Simon and Schuster, 1987).
15 R. Parboni, "The Dollar Weapon: From Nixon to Reagan," *New Left Review*, vol. 158 (1986), pp. 3–19.
16 K. Lissakers, *Banks, Borrowers and the Establishment: A Revisionist Account of the International Debt Crisis* (New York: Basic Books, 1991).
17 N. Harris, *The End of the Third World: Newly Industrializing Countries and the End of an Ideology* (Harmondsworth, England: Penguin, 1986).
18 S. Corbridge, *Debt and Development* (Oxford Blackwell, 1992).
19 M. Beenstock, *The World Economy in Transition* (London: Allen and Unwin, 1983), p. 236.
20 T. Congdon, *The Debt Threat* (Oxford: Blackwell, 1988).
21 S. Branford and B. Kucinski, *The Debt Squads: The US, the Banks and Latin America* (London: Zed, 1988), p. 1.
22 C. Diaz-Alejandro, "Latin American Debt: I Don't Think We Are in Kansas Anymore," *Brookings Papers on Economic Activity,* vol. 2 (Washington, DC: Zed, 1984), pp. 335–389.
23 S. Corbridge and J. Agnew, "The US Trade and Budget Deficits in Global Perspective: An Essay in Geopolitical-Economy," *Society and Space*, vol. 9 (1991), pp. 71–90.
24 B. Friedman, "Long-run Costs of US Fiscal Policy: The International Dimension," mimeo, 1987.
25 S. Cohen and J. Zysman, *Manufacturing Matters: The Myth of the Post-Industrial Economy* (New York: Basic Books, 1987).
26 L. Thurow, *Head to Head: The Coming Economic Battle Among Japan, Europe, and America* (New York: William Morrow, 1992).
27 M. Tolchin and S. Tolchin, *Buying into America* (New York: Times Books, 1988).
28 O'Brien, *Global Financial Integration*, p. 34.
29 G. Arrighi, "World Income Inequalities and the Future of Socialism," *New Left Review*, vol. 189 (1991), pp. 39–66.
30 J. Ruggie, "International Regimes, Transactions and Change: Embedded Liberalism in the Postwar Economic Order," *International Organization*, vol. 36 (1982), pp. 379–415.
31 E. Brett, *The World Economy Since the War: The Politics of Uneven Development* (London: Macmillan, 1985).
32 R. Cox, *Production, Power and World Order: Social Forces in the Making of History* (New York: Columbia University Press, 1987).

33 World Bank, *World Development Report: International Capital and Economic Development* (Oxford: Oxford University Press/World Bank, 1985), p. 38.
34 J. Frankel, "Is a Yen Bloc Forming in Pacific Asia?" in *Finance and the International Economy*, vol. 5, R. O'Brien, ed. (Oxford: Oxford University Press, 1991); see also M. Daly and M. Logan, *The Brittle Rim: France, Business and the Pacific Region* (Harmondsworth, England: Penguin, 1989).
35 G. Treverton, ed., *The Shape of the New Europe* (Washington, DC: Council on Foreign Relations, 1992).
36 Agnew and Corbridge, "The New Geopolitics," p. 284.
37 *Deterritorialization* refers to the increasingly rapid process of capital and technology flowing across boundaries with ease and rendering "national" space less important in economic location decisions.
38 D. Becker, "Development, Democracy and Dependency in Latin America: A Post-Imperialist View," *Third World Quarterly*, vol. 6 (1984), pp. 411–431.
39 S. Roberts, "Fictitious Capital, Fictitious Spaces? The Geography of Offshore Financial Flows," in *Money, Power and Space*, S. Corbridge, R. Martin, and N. Thrift, eds (Oxford: Blackwell, 1994).
40 N. Chomsky, *Deterring Democracy* (New York: Hill and Wang, 1992); S. George, *A Fate Worse than Debt* (Harmondsworth, England: Penguin, 1989).
41 S. Corbridge, "The Poverty of Planning or Planning for Poverty? An Eye to Economic Liberalization in India," *Progress in Human Geography*, vol. 15 (1991), pp. 467–476; World Bank, *India: An Industralizing Economy in Transition* (Washington, DC: World Bank, 1989).
42 A. King, *Global Cities: Post-Imperialism and the Internationalization of London* (London: Routledge, 1990).
43 William Wood, *From Marshall Plan to Debt Crisis: The Making of a Supranational Economic Order* (New York: Pantheon, 1986).
44 P. Mosley, J. Harrigan, and J. Toye, *Aid and Power: The World Bank and Policy-based Lending*, vol. 1 (London: Routledge, 1991).
45 C. Edwards, "The Debt Crisis and Development: A Comparison of Major Competing Theories," *Geoforum*, vol. 18 (1988), pp. 3–28.

# 9 Gearóid Ó Tuathail, 'The Effacement of Place? US Foreign Policy and the Spatiality of the Gulf Crisis'

Reprinted in full from: *Antipode* 25(1), 4–31 (1993)

On January 15, 1991 a world-wide audience watched a geopolitical crisis become a war live on their television screens. The Iraqi invasion of Kuwait on August 2, 1990 was the ostensible cause of the crisis. With the approval of the United Nations Security Council, Iraq was subject to a comprehensive economic blockade in the weeks that followed, while the United States coordinated an international military coalition of thirty-six nations against

Iraq. The US state itself moved half a million soldiers into the region and coordinated 250,000 soldiers from other nations together with millions of tons of food, fuel and equipment to support this offensive force. On January 15, 1991, after the failure of diplomatic initiatives, the US-led coalition launched an intense aerial bombardment of Iraq and Kuwait which lasted six weeks. After a short 100 hour ground war to recapture Kuwait, a cease-fire was declared by President Bush at midnight on the 27th of February 1991. An estimated 100,000 to 150,000 combat and civilian Iraqis and Kuwaitis were killed in the war. United States combat losses were remarkably low at 148 dead (38 from so-called "friendly fire"), making the Iraq-to-United States loss ratio roughly a thousand to one (Draper, 1992b). Iraq, Kuwait and the Persian Gulf all suffered acute environmental devastation. Thousands of Kurdish and Shi'ite Iraqis died after the cease-fire in shortlived uprisings. Hundreds of Iraqis, mostly children, are still dying as a consequence of the allied assault, continued UN sanctions, and the rule of Saddam Hussein.[1]

With such a grim balance sheet for a war that is still killing people (and may not yet be completely over), geographers have a moral and political obligation to confront and challenge the strategies by which the war was given to be seen by politicians, "experts" and the mass media within the "Western" world. As intellectuals sensitive to the social production of space and place, geographers have a special role in the documentation and deconstruction of the spatiality of the Gulf crisis, a spatiality which created the conditions of possibility for a techno-frenzied slaughter (which does not deserve the name "war") and the consummation of militarist fantasies that had long been harbored under the Reagan–Bush administrations. This paper is an attempt to open up the spatiality of what was projected as "the Gulf crisis" to empirical investigation and theoretical reflection. Its object of focus is the US role in the crisis and not the regional origins of the crisis, nor the motivations of the international coalition that participated in the war (see Brenner, 1991; Bullock and Morris, 1991; Cooper, *et al.* 1991; Draper, 1992a, 1992b; Halliday, 1991; Simpson, 1991 on these issues). Proceeding in an approximate chronological way from the origins of the "crisis" to the war itself, the paper addresses four general issues. First, it seeks to construct an (not the) explanation of why the Iraqi invasion of Kuwait came to be narrated as "the Gulf crisis," an event of supposed global significance and danger that necessitated a military response. Second, it seeks to document how the crisis was inscribed with meaning and a particular stage set of historical and geographical backgrounding by the Bush administration. The third section of the paper moves away from consideration of the official US inscription of the crisis to explore, using Paul Virilio's notion of "chronopolitics" as its point of departure, the related but less traditional problematic of speed and the management of time during the crisis and war. Finally, the fourth section reflects briefly on the nature and moral consequences of the electronic imaging of place by the military war machine and the mass media. Although there is sometimes an uneasy relationship between official, chronopolitical and electronic constructions of place, all these forms of inscription can be described as abstracting place and eviscerating its social dimensions and human character.

Deconstructing the discursive strategies by which the Gulf crisis was geographed necessarily requires one to develop a narrative of intelligibility about the spatiality of the Gulf crisis. Such a narrative should be premised on the impossibility of separating our inherited language and discourse from the "reality" of the Gulf slaughter; our very categories of reading and writing are already infested with those that made the Gulf war possible. This very fact should make us particularly self-conscious of the narratives of intelligibility we use to write the Gulf as critical geopoliticians (Dalby, 1991). The one that implicitly informs this paper (and other works of critical geopolitics) is the effacement of place, the erasure of the sociality of place by official geopolitical inscriptive strategies, speed and technology. The paper concludes by explicitly turning on this narrative and demonstrating how it itself has its limits. Narratives which summarize the spatiality of the Gulf crisis as characterized by an effacement of place should not reify place and treat it as a natural, innocent object outside of any relationship to signification and inscription.

## US foreign policy, the end of the Cold War, and the origins of the Gulf crisis

While the Iraqi invasion of Kuwait in August 1990 had complex regional origins, the subsequent narration of this event as the globally significant and dramatically threatening "Gulf crisis" must be traced back to the dilemmas created for the United States and Atlanticist security structure by the end of the Cold War. As a geopolitical system, the Cold War gave a distinct organizational structure to global politics and helped bring into existence not only two distinct geoeconomic and geopolitical orders but two permanent military-industrial complexes in the West and East. The reality of US–Soviet antagonism helped legitimate US military hegemony amongst the advanced capitalist states while the ideological use of Cold War discourses of danger shaped the very nature of political life within these states. Articulations of Cold War geopolitics helped constitute and secure a field of geographical identities ("the West," "the Soviet Union," the "United States") while serving to discipline domestic social and cultural difference within these spaces (Campbell, 1992a; Kaldor, 1990; Ross, 1989).

The Reagan rerun of classic Cold War geopolitics in the early eighties helped secure a conservative domestic order within the United States but the contradictions of Reaganomics left the US state seriously crippled with huge trade and budget deficits (Corbridge and Agnew, 1991). The relative economic standing of the United States vis-à-vis Japan and Germany declined and became an issue of national concern. Gorbachev's "new thinking" in Soviet foreign policy explicitly sought to deconstruct the Cold War image of the Soviet Union in Western society. In the late 1980s arms agreements, revolutions in Eastern Europe, the collapse of the Berlin Wall, and the manifest economic failures of communism rendered the US national security community's standard script of world politics obsolete and anachronistic (Ó Tuathail, 1992b). The collapse of the Cold War (due to the economic failures of Soviet communism) provided the new Bush administration, an administration elected because of its successful manipulation of Cold War signifiers, with the challenging task of moving beyond Cold War reflexes, attitudes and institutions. The possibility of this being realized was not good considering the fact that the overwhelmingly white, male conservatives that made up the Bush administration owed their careers, political position and very subjectivities to the Cold War. Many were veterans of the Nixon and Ford administrations and followed Kissinger in viewing post-Cold War "euphoria" as dangerous.[2] The discursive functioning of "euphoria" (which had a narcotic sub-text) revealed the persistence of a gendered Cold War subjectivity amongst this community, particularly those in key foreign policy positions (especially National Security Advisor Brent Scowcroft and Secretary of Defense Richard Cheney). "Euphoria" was the position of the idealist, the softheaded, loose, feminine reader of international politics whereas "prudence and realism" was the position of the seasoned Cold Warrior, the hard-headed, masculine analyst of international affairs (who had seen it all before).

From the perspective of the Bush administration, the expectations released by the end of the Cold War, therefore, were dangerous aspirations in need of containment. Domestically, there was a clamor for a "peace dividend" and renewed sentiment for a more "isolationist" role for the United States in world affairs. The possibility of an American *glasnost* after the Cold War also threatened to loosen the cultural hegemony of Cold War understandings of "America" and "the West" (already under contestation in debates over the canon of Western civilization) not to mention the lock of the Republican Party on the White House. Internationally, the end of the Cold War also raised the awkward issue of whether the subordination of Western Europe (principally Germany) and Japan to US military authority would continue. The traditional identity, authority and role of the "United States of America" was becoming less distinct and more uncertain. In Deleusian and Guattarian terms, the implosion of Cold War discourse "de-territorialized" the "United States" as an identity (Doel, 1993; Bogue, 1989; the globalization of the 1980s had already de-territorialized its economic identity).

"Re-territorializing" the role of the "United States" and the "Western" security system was an implicit, reflex priority for the Bush administration in 1990. Traditional anti-communist threat narratives had to be rewritten as general threat narratives in order to re-secure and re-discipline a domestic as well as an international order. The United States still lived in a dangerous world and instability itself, as Bush noted in early 1990, was now the enemy. Deeply imbedded discourses of danger concerning Third World otherness (Islam, despotism, anarchy, narco-terrorism) were re-asserted independent from Cold War narratives and activated to justify the US invasion of Panama in December 1989. This Theodore Roosevelt-style hemispheric police action, however, was a regional affair and incapable of serving as the means of re-asserting the "United States" as a purposeful state in world affairs. Saddam Hussein's invasion of Kuwait less than a year later, however, offered an opportunity to do this in a very dramatic and forceful way. The invasion became a global platform for a comprehensive re-articulation of the US role in world politics. The "United States" became meaningful again and the role of the "Western" security system in a changing world was invested with a new purpose. This purpose was to safeguard the "Western" way of life against disruptive anarchies which threatened geopolitical order and the stability of the global economy.

The Iraqi invasion of Kuwait was a disruptive anarchy in the flows of the global economy. As a place "Kuwait" always had a somewhat problematic identity. Like all states, its existence was a product of global historical forces and dependent upon discourses of imagined community and sovereignty. Unlike the paradigmatic Western nation-states, however, the precariousness of these discourses in the case of Kuwait was transparent. A few days after the Iraqi invasion the *New York Times* (August 5, 1990) described Kuwait as "less a country than a family owned oil company with a flag and a seat at the United Nations." Ostensibly a sovereign nation, its population of over 2 million comprised 1.5 million expatriate workers and only 570,000 legal Kuwaiti citizens. Only 60,000 males from this group had a legal right to vote if allowed to by the Al-Sabah family who controlled the state's enormous wealth. Much of this wealth was non-territorial in form. Kuwait's overseas assets were estimated at $100 billion or more, and were invested largely in the United States (where it owned Santa Fe International), Britain (where it had a 9.8% stake in British Petroleum, a 10.5% stake in Midland Bank and complete ownership of St Martin's Property), Japan (10.3% stake in Arabian Oil), Spain (72% ownership of Torras Hostench) and Germany (25% ownership in Hoechst) (*Business Week*, 1990a). "Iraq" also was a problematic territorial entity and a lesser presence than Kuwait in the stream of international capital flows (Luke, 1991). It had an international debt of $80 billion, half of which was owed to Arab Gulf states, Saudi Arabia and Kuwait in particular. The rest was guaranteed trade debts to assorted Western countries (including France, Hungary, Yugoslavia, Poland and Ireland) and military debt mostly to the USSR (Karsh and Rautsi, 1991: 29). It had holdings in banks in London and the Netherlands among other places (Economist Intelligence Unit, 1990: 13).

In invading the territory of Kuwait the Iraqi army did not, however, gain access to Kuwait's non-territorial transnational cybernetic wealth (Luke, 1991). The ability of the Al-Sabah family to escape the invading army and relocate their corporate headquarters to the luxury of Sheraton Al-Hada Hotel in Tiaf, Saudi Arabia, allowed them to mount an informational war against Iraq across the international airwaves. Their counter-offensive began immediately with the founding and financing of "Citizens for a Free Kuwait" to lobby the US Congress for military intervention. The "citizens" dimension to the organization was purely fictional since funding came overwhelmingly from the Kuwaiti government in exile (MacArthur, 1992: 49). This group hired Hill and Knowlton, a Washington, D.C. public relations firm, to act as the coordinators of their informational air-war against Iraq. Deploying both the powerful discourse of sovereignty and equally powerful gendered discourses about territory and the body (the "rape" of Kuwait and the killing of babies; the "eye-witness" account of the removal of babies from incubators was subsequently shown to be unfounded; MacArthur, 1992), the lobbying battle achieved a series of strategic victories on Capitol Hill and on the screens of the mass media. The contested reality of "Kuwait" came to be articulated in terms of sovereignty not in terms of sociology, history or even justice. Discursive fictions of the sovereignty narrative, like "the Kuwaiti people," usually achieved aerial supremacy over competing political ("feudal monarchy"), sociological ("migrant labor" colony) and economic ("blood for oil") narrations of the territory (and crisis).

The effacement of the material realities of Kuwait in the public relations efforts of "Citizens for a Free Kuwait" was decisively aided by the inscription of Kuwait as a global symbol of a "victimized nation" by the Thatcher and Bush administrations. The decision to mobilize an international coalition against the invasion reversed the previous US policy tilt towards Iraq that began in 1979 after the Iranian revolution. The United States had removed Iraq from its list of countries sponsoring terrorism in 1982 and restored full diplomatic relations in 1984 (Gigot, 1990).[3] From 1985 to 1990 the United States authorized over $4 billion in guaranteed agricultural exports to Iraq (Ridgeway, 1991: 13). During the Iran–Iraq war the United States shared intelligence information with Iraq on the movements of the Iranian army and intervened to re-flag Kuwaiti tankers carrying Iraqi oil, thus enabling Iraq to continue fighting. In the fall of 1988 when evidence of Iraqi use of chemical weapons against the Kurds came to light, the US Senate passed tough unilateral sanctions against Iraq by a vote of 87–0. Through persistent lobbying by the Reagan administration and US business interests the sanctions never became law (Gigot, 1990: 6). In the geopolitical calculus of the Reagan and Bush administrations a "strong Iraq" provided a buffer against a "fundamentalist Iran."[4]

On August 2, 1991 George Bush and his advisors had a two-hour meeting with Margaret Thatcher in Aspen, Colorado.[5] Thatcher, evoking Churchill and World War II, urged Bush to make a stand against Saddam Hussein.[6] Revealing both her own exaggerated masculinized subjectivity and gendered understanding of diplomacy she reportedly stated: "Remember, George, this is no time to go wobbly" (J. E. Smith, 1992: 63). At a National Security Council meeting the next day Bush evidently steeled hiself to embody the recommended stiffness. CIA Director William Webster, interpreting the texts generated by spy satellites, suggested that Iraq was about to invade (penetrate) Saudi Arabia. Though certain military and political advisors were unconvinced, Bush (with the strong support of Brent Scowcroft) concluded that this was the case and that the United States needed to send forces to Saudi Arabia as a (steel) shield against potential "naked aggression." Through telephone diplomacy and the dispatch of a high-level US delegation (led by Secretary of Defense Cheney) to meet with King Fahd, the United States secured permission from the Saudi monarchy to send troops there under the contingency plan 1002-90 (Woodward, 1991). This recently updated contingency plan was the result of a war game simulation with Iraq as enemy that had been designed and run by General Schwartzkopf in the summer of 1990. On August 8, 1991 125,000 United States troops from the 82nd Airborne Division and units of the US Air Force began arriving in Saudi Arabia.

**Inscribing the Gulf crisis: the practical geopolitical reasoning of the Bush administration**

The critical study of geopolitical ideology can be usefully divided into the study of formal and practical geopolitical reasoning (Ó Tuathail and Agnew, 1992). The former involves the study of the intellectuals and texts of the geopolitical tradition. The latter involves the study of the pragmatic practice of statecraft by national security elites and is concerned with the strategies of inscription and re-inscription of identity and difference upon international politics. This takes the form of the relentless construction of imaginary geographical boundaries between the self and the other, the domain of freedom and the domain of danger, the inside realm of community and the outside realm of anarchy with the former always privileged over the latter. In times of crisis this process of geopolitical scripting rigidly designates (in a Kripkean sense) the map of international politics. Places become rigidly inscribed with sets of identities, descriptions, histories and intentions. The boundary between community and anarchy becomes a Manichean divide. In the process the complex and ambiguous human geography of places and peoples becomes eviscerated. States lose their quality as socially constructed geographical places (locations for the sustainment of life), and become abstractions in a geopolitical power game (Dalby, 1990; Ó Tuathail and Agnew, 1992).

The speeches and policy statements of the Bush administration on the Gulf crisis are performances of practical geopolitical reasoning. Using the public record of speeches and policy statements by top officials in the Bush administration, as chronicled in the US State Department's weekly publication *Dispatch*, one can identify certain recurring inscription strategies by which the crisis was rendered meaningful to the United States and world public. Only a few strategies can be examined here, so I have concentrated on the scripting strategies used in explanations of the ostensible reasons for US force deployments (oil and the "new world order"), and explications of the historical and geographical meaning of the crisis (World War II and Vietnam as systems of signification). Other differentiation strategies involving technology (Oriental primitivism versus Western high-tech "smart" weapons), gender (the Iraqi "rape" of Kuwait versus the protective and socially "liberated" armies of the West), and religious morality (Islamic barbarism versus Western "just war" morality) are also important but are not considered here.

*Oil: geopolitical hyperrealism*

Unlike many Americans, President Bush, as a former oil executive, knew where Kuwait was located on the world map. Oil gave the Iraqi invasion of Kuwait a general geopolitical significance but the assumption that the US state rationally intervened in the region to safeguard the West's supply of oil is highly contentious. From a realist geopolitical perspective, one which purports to read potential threats in the context of empirical strategic realities,

US military intervention was not particularly necessary. The defense of Kuwait was not in the vital interests of the United States and administration policy before the invasion reflected this fact (see the Glaspie-Hussein transcript in Sifry and Cerf, 1991, pp. 122–133; Assistant Secretary of State for Near Eastern and South Asian Affairs John H. Kelly's exchange with Representative Lee Hamilton on July 31, 1990 in Draper, 1992a, pp. 52–53). Even after US forces were committed, resort to the use of force was not necessary and many leading geopoliticians and other intellectuals within the United States argued against going to war (see the opinions of Admiral Crowe, Arthur Schlesinger, Zbigniew Brzezinski and others collected in Ridgeway, 1991; and Sifry and Cerf, 1991). In practice, the Bush administration's reasoning on oil was a form of geopolitical hyperrealism, a condition where geopolitical reasoning loses all but the most fantastic relationship to empirical strategic realities.[7] One can document this by considering three different dimensions to the US state's practical geopolitical reasoning concerning oil.

*The defense of Saudi Arabia*. Four days after the Iraqi invasion, Bush first stated that "this will not stand." Implicitly this suggested the rollback of the invasion. Inside accounts have suggested that from the beginning Bush's goal was to go beyond defending Saudi Arabia but this objective was initially shared neither with the public nor many officials outside his closest advisors (Woodward, 1991). Bush's first address to the nation on the invasion ("The Arabian Peninsula: US Principles," August 8, 1990) described the mission of the US troops dispatched to Saudi Arabia as "wholly defensive." "The sovereign independence of Saudi Arabia," Bush declared, "is of vital interest to the United States" (*Dispatch* September 3, 1990: 53). Bush's immediate goal was to protect the remaining elements of the pro-Western geopolitical order in the Middle East, the centerpiece of which was Saudi Arabia. This order comprised rich dictatorial regimes who ruled over sparse populations and vast oil reserves, states historically created by the West as client regimes (Brenner, 1991: 128). Iraq's military did appear to threaten this order yet the empirical evidence that they intended to invade Saudi Arabia was ambiguous at best (its deployment in Kuwait was entirely defensive; Bullock and Morris, 1991: 168–69). Furthermore, it was doubtful if Iraq had the capability to do so successfully since Saudi Arabia, the largest importer of arms in the world, was far from being defenceless. Finally, it was not obvious that the Iraqi invasion of Kuwait necessarily threatened "Western interests" in the region since Iraq had historically been understood as serving these interests rather well. It was Iraq, not Kuwait, that acted as the main check on the expansion of Iranian power in the region.

*Oil reserves and supplies*. Once the Bush administration decided on the necessity to deploy US forces to defend Saudi Arabia it also began to claim that Iraq posed a serious threat to the West's oil reserves and supplies. In his August 8, 1990 address Bush contrasted Iraq and the United States. Iraq is "a rich and powerful country that possesses the world's second largest reserves of oil and over a million men under arms. It's the fourth largest military in the world." The United States, by contrast, "now imports nearly half the oil it consumes and could face a major threat to its economic independence. Much of the world is even more dependent upon imported oil and is even more vulnerable to Iraqi threats" (Bush, *Dispatch* September 3, 1990: 52).

Bush's description of the Iraqi threat and the supposed vulnerability of the world to Iraqi control over oil bore little relation to strategic realities. First, Iraq's military force was largely composed of poorly trained conscripts adept at little beyond defensive strategy (Freedman and Karsh, 1991).[8] Second, stressing control over reserves of oil was misleading since figures on oil reserves vastly overstated the importance of Middle Eastern oil to the Western economy (Bandow, 1991). Third, even with control over Kuwaiti oil Iraq could not pose a serious threat to world oil supplies. Before the invasion Kuwait and Iraq together produced between four and five million barrels a day (b/d) for the world oil market. Their combined production was 4.7m b/d in July 1990, 20 percent of OPEC crude oil production that month and almost 9 percent of the global supply in the third quarter of the year (Shelley, 1992: 166). This oil was stopped from reaching the world market by a UN embargo after the invasion, not by Iraqi action. The price of oil on the international spot market did soar and fall in response to the vicissitudes of the crisis. In supply terms, however, by the end of September 1990 the 11 other OPEC members had boosted their joint production by 3.5m b/d. A brief shortfall in global supply was soon overcome. There was to be no oil shock for the structures of the global oil market were quite different from those of the 1970s (Johnson, 1989; Shelley, 1992).

*The chokehold image*. The reasoning of the Bush administration on oil was ultimately based on deep-rooted strategic fantasies (Ó Tuathail, 1992a) rather than on existent strategic realities. Before the Gulf crisis Robert Johnson (1989), a former member of the National Security Council staff, argued that US strategy in the Persian Gulf was based on the erroneous assumption (rooted in a "misunderstanding of events in the seventies") that the greatest threat was one of a severe oil supply crisis. Despite the outdated and mistaken nature of this assumption Bush and other officials made frequent recourse to images of the choking hands of Saddam Hussein on the lifeline of the West (again corporeal and geopolitical territoriality are made mutually indistinguishable). In announcing US troop increases on November 8, 1991 Bush declared that the world community must prevent Saddam Hussein from "establishing a chokehold on the world's economic lifeline" (*Dispatch* November 12, 1990: 258). Geopolitical reasoning blended with demonology as Saddam Hussein was represented as a sadistic madman who threatened the strangulation of the vulnerable West. The Bush administration could thus represent itself in this unconscious fantasy drama as a guardian of law and order, a doctor/police power with responsibilities for the health and welfare of the "Western way of life." Speaking at the Pentagon on 15 August, 1990 Bush described how the US role was about:

> maintaining access to energy resources that are key – not just to the functioning of this country but to the entire world. Our jobs, our way of life, our own freedom, and the freedom of friendly countries around the world would all suffer if control of the world's great oil reserves fell into the hands of Saddam Hussein (Bush, "Against Aggression in the Persian Gulf" *Dispatch* September 3, 1990: 54).

The irony of this hyperbolic geopolitical reasoning is that it was *not* shared by countries who had more reason than the United States to be concerned about oil supplies. At the time of the invasion Japan imported 99% of its oil and received two-thirds of that from the Middle East. It was the largest customer for Kuwaiti and Iraqi petroleum (*Business Week*, 1990b). However, Japan was reluctant to back the use of military force in the Gulf. In a heated parliamentary debate over the invasion most leaders did not show notable concern about a potential oil crisis (Tamamoto, 1991). The Japanese consul general in New York was reported as stating in January 1991 that "[e]xperience tells us that whoever controls oil will be prepared to sell it. We are prepared to pay" (*Business Week*, 1991). Van Wolferen (1991: 28) claims this geo-economic perspective was quietly encouraged by the Japanese bureaucracy. Yet by the war's end the Japanese state had pledged nearly $13 billion towards the costs of the war while Germany, another reluctant warrior, promised $10.7 billion to the US-led coalition (Hamilton and Clad, 1991).

### *The Bush administration's "new world order"*

The second, and more significant, motivation for US interventionism in the Gulf was what President Bush described as the "new world order." The phrase itself had twentieth century associations with Woodrow Wilson and the Nazis among others. In recent times, however, it was Mikhail Gorbachev who revived the ideal as an expression of his vision of a post-Cold War world (Der Derian, 1992: 162–63). Within the Bush administration the concept gradually gathered favor as it became obvious that the Cold War world order had indeed ended. Both Bush and Robert Gates used the phrase in speeches in April and May of 1991. However, it was in an address before a joint session of Congress on September 11, 1990 that Bush first began to actively promote the concept as the new "big idea" in US foreign policy thinking. From the evidence of this speech the "new world order" was that which could be achieved once Iraq's threat to the peace was resolved: "The crisis in the Persian Gulf . . . offers a rare opportunity to move towards a historic period of cooperation. Out of the troubled times . . . a new world order can emerge" (Bush, "Towards a New World Order," *Dispatch* September 17, 1990: 91). Implicitly Bush's speech suggested it would have five elements: American global leadership, East–West co-operation against threats to the peace, proper functioning of the United Nations, allied support for US leadership and economic renewal within the United States itself. Three points are worth noting about the Bush administration's articulation of its vision of a "new world order."

First, the "new world order" was envisaged as a *post*-Cold War world order more than a fundamentally *new* world order. Bush's first means of defining it was to point to the US–Soviet joint statement at Helsinki a few days earlier condemning Iraqi aggression. This is significant. Though the Cold War as a geopolitical system had ended, the superpowers were assumed to still hold central authority in world politics. The Bush administration's "new world order" did not countenance a delegation of authority to geo-economic powers like Japan and Germany. Both states were to remain "prisoners" of geopolitical alliances established during the Cold War which subordinated them militarily and diplomatically to the United States (they appeared willing "prisoners" in the short-term). World order was to be established with the Soviets and the other permanent members of the UN Security Council, a group that did not include Japan and Germany, in a way originally envisaged by Franklin D. Roosevelt and modified by Churchill and Stalin. Co-operation between the former geopolitical rivals offered the new possibility of converting the US into a global instrument for the promotion of order as the permanent members of the Security Council saw it. The Bush administration's vision of the "new world order" was thus a traditional backward-looking geopolitical rather than a new futuristic geo-economic one. Threats were defined, first, in terms of the sovereignty of existing states (invasions, terrorism, ethnic violence), second, in terms of geopolitical order (nuclear proliferation, territorial settlements, terrorism) and only third in economic and environmental terms (global debt, environmental degradation, economic decline).

Second, the traditional geopolitical cast given to the "new world order" was reinforced by the claim that its realization was threatened by disruptive regional conflicts in the Third World. Though the "new world order" vision claimed to address the global geopolitical system its articulation did not, in practice, transcend the symbolic case of the Iraqi invasion of Kuwait. This event enabled the Bush administration, firstly, to define the global system as a system in transition (without new rules) and, secondly, to define this transition as dangerous. US Secretary of State James Baker stated before the House Foreign Affairs Committee (September 4, 1990):

> Iraq's unprovoked aggression is a test of how the post-Cold War world will work. Amidst the revolutions sweeping the globe and the transformation of East–West relations, we stand at a critical juncture in history. The Iraqi invasion of Kuwait is one of the defining moments of a new era – an era full of promise but also one replete with new challenges. . . . we must respond to the defining moments of this new era, recognizing the emergent dangers lurking before us. We are entering an era in which ethnic and sectarian identities could easily breed new violence and conflict. It is an era in which hostilities and threats could erupt as misguided leaders are tempted to assert regional dominance before the ground rules of a new order can be accepted (Baker, "America's Stake in the Persian Gulf," *Dispatch* September 10, 1990: 69).

Such a discourse of danger echoed colonialist discourses from the past. "Ethnic and sectarian identities" breeding new violence and conflict was code for non-Western and non-rational spaces where dictators ruled and anarchy lurked. Throughout the US justifications for interventionism in the Gulf the issue was inscribed as a conflict between "the rule of law" and "the law of the jungle" (e.g. Bush, *Dispatch* September 3, 1990: 54; October 22, 1990: 205; January 21, 1991: 38). The latter expression recalls colonial images of wild, untamed spaces where "Western" traditions of law and order were obscured or no longer applied. Iraq's invasion was "a ruthless assault on the very essence of international order and civilized ideals" (Bush, *Dispatch* September 3, 1990: 55), a "relentless assault on the values of civilization" (Baker, *Dispatch* February 11, 1991: 82). The crisis was an opportunity to reinforce "standards of civilized behavior" (Baker, *Dispatch* September 10, 1990: 69). Bush's frequently expressed admiration for Theodore Roosevelt is worth remembering for Roosevelt's Corollary of 1904 ("Civilized society . . . impotence . . . international police power") and whole subjectivity (the restless male hero who carries a big stick!) was reenacted by Bush during the Gulf crisis. Implicated in such drawing of international political space are long-standing racist and orientalist dispositions that can be found in US foreign policy, dispositions brought into sharp relief by the emergence of Iraq's nuclear program as a further reason for US intervention and the necessity of war. Unlike the United States and other responsible "Western" nations (such as Israel), the Iraqis could not be trusted with nuclear, chemical or biological weapons. Iraq's potential military potency, therefore, needed to be emasculated (see Quayle, "The Gulf: In Defense of Moral Principle," *Dispatch* December 24, 1990: 350).

Third, the Bush administration's vision of a "new world order" rested on the traditional myths of American exceptionalism and Manifest Destiny. In his State of the Union address on January 29, 1991 Bush declared that what was at stake was a "big idea: a new world order" in which "America" has a unique role and responsibility:

> For two centuries, America has served the world as an inspiring example of freedom and democracy. For generations, America has led the struggle to preserve and extend the blessings of liberty. And today, in a rapidly changing world, American leadership is indispensable. Americans know that leadership brings burdens, and requires sacrifice. But we also know why the hopes of humanity turn to us. We are Americans; we have a unique responsibility to do the hard work of freedom (Bush, "State of the Union," *Dispatch* February 4, 1991: 65).

This rhetoric was traditional American universalism and the fact that it privileged idealism over strategic realities, and did not distinguish a hierarchy of vital places, brought criticism from generally sympathetic geopoliticians like Kissinger (1991). America's triumph in the Cold War, he noted, required adjustment of traditional concepts. But the vision of the Bush administration looked towards the past rather than towards the future. Bush's call to Americans "to prepare for the next American century" in the same speech recalled Henry Luce's similar appeal in 1941. The discursive horizon for Bush's proclamations was World War II.

### *World War II and Vietnam as inscription strategies*

The mythic narratives of World War II are a deeply entrenched Anglo-American vision of a "just war" (Luke 1989, 1991). Socialization, the inscriptive power of the mass media (particularly cinema and television, where history is black and white footage), made these mythic scripts available to all. Bush, himself a World War II veteran, was steeped in these. He frequently cited Winston Churchill and his view that World War II need not have been fought if Hitler had been thwarted in his 1936 push into the Rhineland (e.g. *Dispatch* December 10, 1990: 312). This "historical lesson" was considered a timeless verity of statecraft. Ruthless dictators should be confronted and stopped. Appeasement eventually leads to war. "Naked aggression" must not be allowed to succeed. Bush's first address had set the tone: "if history teaches us anything, it is that we must resist aggression or it will destroy our freedoms. Appeasement does not work. As was the case in the 1930s, we see in Saddam Hussein an aggressive dictator threatening his neighbors" (*Dispatch* September 3, 1990: 53).

In Bush's nostalgic script World War II was a feel-good war, one where a dictator was stopped, nations fought together to resist aggression, American soldiers and citizens enjoyed their finest hour and the promise of a new world order was secured. Complex histories and the realities of place were overwhelmed and effaced by a script which lifted the conflict from its actual geographical and historical co-ordinates. In the Bush administration's re-make of World War II, Iraq was cast as fascist Germany, Kuwait as the victimized small nation, the exiled Kuwaitis as the free French, Saddam Hussein as Adolf Hitler, and the allies as their true heroic selves (Luke, 1991). Iraqi atrocities were like those of the Nazis in Poland. Addressing the troops in the Saudi desert at Thanksgiving reminded him of November 23, 1944 when he and "another group of Americans far from home" also fought for freedom. "Once again, Americans have stepped forward to share a tearful goodbye with their families before leaving for a strange and distant shore" (*Dispatch* September 17, 1990: 94). *Life* magazine helped codify and circulate this nostalgic structure of feeling by producing a special weekly edition under the headline "In Time of War" where new and old photographic images of war mixed in mutual legitimation (see, for example, the March 11, 1991 issue "Heroes All" and the March 18, 1991 issue "Coming Home"). Lost in the reasoning of the Bush administration and its circulation in civil society was the real World War II, a war Fussell (1989: 132) writes was "indescribably cruel and insane," a "savage, insensate affair, barely conceivable to the refined imagination and hardly comprehensible without some theory of mass insanity and inbuilt, inherited corruption."

The most threatening system of signification for war in US public life was Vietnam. The moral and political ambiguities of that war did not sit well with the Manichean moral universe created around the Gulf crisis. The Vietnam war became represented as a "syndrome," an impotence in the face of war, that had to be overcome by a demonstration of resolve and power (Bush stated: "Saddam is going to get his ass kicked"). Whereas the Arabian desert became simulated as scenes from a World War II movie in official reasoning, the same landscape had to be exorcised of flashback scenes from Vietnam:

> ... this will not be another Vietnam. This will not be a protracted, drawn-out war. The forces arrayed are different. The resupply of Saddam's military would be very different. The countries united against him in the United Nations are different. The topography of Kuwait is different. And the motivation of our all-volunteer force is superb (Bush "The Gulf: A World United Against Aggression," *Dispatch* December 3, 1990: 296).

Amazingly, considering the enormous free range pulverisation and ecological sabotage of that country, Vietnam was cast as a slow war, one where America's troops fought "with one hand tied behind their backs" (Bush, *Dispatch* January 21, 1991: 38). Yet it was the mythic narratives of World War II, as worked by Eisenhower, Kennedy and Johnson, which led the United States directly into Vietnam (Luke, 1989: 168–69). Rewrites of the same narratives and representation of caution in the face of war as a medical condition eventually secured the Bush administration the war in the Gulf it ultimately appeared to want.

**Speed and space: chronopolitics and the Gulf crisis**

The evisceration of the real (economic, historical, geographical, topographical, etc.) by the Bush administration's practical geopolitical reasoning was amplified by the role of technologies of time–space compression and destruction during the Gulf crisis. The case for the effective disappearance of space given the speed of military technologies is provocatively overstated by the French critical strategist Paul Virilio (Virilio and Lotringer, 1983; Virilio, 1986; 1989). Space, for Virilio, is no longer in geography but in electronics. "Unity is in the terminals. It's in the instantaneous time of command posts, multi-national headquarters, control towers, etc. Politics is less in physical space than in the time systems administered by various technologies . . . the distribution of territory becomes the distribution of time" (Virilio and Lotringer, 1983: 115). Speed is taken to be the essence of war. Whereas strategy in the past was dominated by geopolitics (defined as the control of territory), today it is dominated by chronopolitics, the politics of time and acceleration (Der Derian, 1990; 1992). "Territory," Virilio asserts, "has lost its significance in favor of the projectile. In fact the strategic value of the non-place of speed has definitely supplanted that of place . . ." (1986: 133).

Virilio's style is a mixture of hyperbole and richly suggestive insight (his arguments recall the themes of Lewis Mumford, and E. P. Thompson on "exterminism"). Like Guy Debord (1983) and David Harvey (1989) he seeks to politicize space, time and speed, but whereas the latter ground their arguments in the turn-over time of production, space, time and speed for Virilio are ultimately military phenomena which are grounded in the means of destruction, in the technological weapons ("vehicles" and "projectiles") of the military. In a statement typical of his style he asserts: "Stasis is death, the general law of the world . . . history progresses at the speed of its weapons systems" (Virilio 1986: 68). Extending this general law into an analysis of history, Virilio provides a suggestive re-reading of conventional historical processes in terms of the speed of movement – the dromos – they allowed. For

Virilio "there was no 'industrial revolution' but only a 'dromocratic revolution,' there was no democracy only dromocracy; there is no strategy, only dromology" (1986: 46). In his scheme it is the pace of production, the mobility of a population and the management of speed that are crucial.

The problematic Virilio engages – the relationship between weapon systems, speed and space in militarized states – is one that deserves careful consideration by geographers. The course of the Gulf crisis provides provocative evidence for the eclipse of place by pace. First, it is worth recalling that George Bush's original reason for going to Aspen, Colorado on August 2, 1990 was to deliver a major statement on the reshaping of US power projection capabilities after the Cold War. In a homage to military technology, Bush pointed to how "the United States has always relied upon its technological edge to offset the need to match potential adversaries' strength in numbers." Speed and flexibility are explicitly recognized by Bush as crucial:

> we must focus on rapid response. As we saw in Panama, the US may be called on to respond to a variety of challenges from various points on the compass. In an era when threats may emerge with little or no warning, our ability to defend our interests will depend on our speed and our agility. And we will need forces that give us global reach. No amount of political change will alter the geographic fact that we are separated from many of our most important allies and interests by thousands of miles of water . . . A new emphasis on flexibility and versatility must guide our efforts (Bush, 1990: 678).

Holding a military lead in the logistics of acceleration and flexibility was clearly recognized in US strategy as vital to power projection. Under the Carter administration the United States had established a Rapid Deployment Force to enforce the Carter Doctrine in the Middle East. Bush's proposals amounted to an extension of the principles of rapid deployment into a new policy of "flexible containment" (Luke, 1991). Rather than the static postures of traditional Cold War containment, military force would be ready for use against shifting and variable enemies in a variety of locations (NATO had recently re-structured its forces in this way). In December 1989 the location was Panama and in August 1990 it was the Arabian desert. The domination of pace not place concerned the military. The air and sea-lifting capacity of the US military soon negated the significance of Saudi Arabia's distance from the United States. Geography was overcome by logistical power.

Second, subsequent debate within the Bush administration about Gulf strategy became a debate about pace and time, not distance and geography. For those who advocated the "containment" and "strangulation" of Iraq the key issue was one of letting sanctions have the time to work. The United States needed to be a patient power according to this line of reasoning. For Bush and Scowcroft, time was on the side of the Iraqi army for the longer it remained in Kuwait the more difficult it would be to dislodge it. Woodward (1991: 42) provides an account of the crucial October 23, 1990 meeting when the containment option was rejected by the Bush administration. In this account the president rejected sanctions by stating "I don't think there's time politically for that option." Elections were approaching in November 1990 (the offensive deployment announcement was delayed until after these) and public opinion was considered unlikely to tolerate a long drawn-out war. The result was the establishment of a firm deadline for Iraq to disengage (measured according to Eastern Standard Time not local time in Kuwait) and a series of diplomatic battles over the timing of meetings (Baker-Aziz) and withdrawals (Bush's 24 hour deadline of February 22). The control of time during the crisis was all important.

Third, the US foreign policy debate beyond the administration also divided on questions of time rather than questions of whether or not the United States should be involved in a region so distant (see the opinions in Ridgeway, 1991 and Sifry and Cerf, 1991). Sam Nunn, the leading Congressional expert on defense, opposed the Bush administration's policy of no rotation for the troops and its pursuit of the offensive military option. In a *Washington Post* article Nunn (1991) described his opposition in the following manner:

> What guarantees do we have that a war will be brief and that American casualties will be light? . . . Our policy cannot be based on an expectation that the war will be over quickly and easily . . . I am afraid too many recall our most recent conflicts in bumper-sticker terms:
> – "Vietnam: long, drawn out – bad"
> – "Grenada-Panama: quick, decisive – good."
> . . . We in America like instant results. We want fast food and fast military victories.

For the administration a fast military victory was vital to sustain its public relations image of the war. In chronopolitical terms Vietnam was a failure because it was not over fast enough. As Virilio would have put it, the Bush administration and the US military were dromomaniacs.[9]

Finally, the course of the actual military conflict provides further evidence for the significance of chronopolitics. The strategy of the Iraqi army was essentially one of static defense in the hope they could draw allied forces into killing zones and raise casualties to unacceptable levels (Freedman and Karsh, 1991). Schwartzkopf's strategy was one which placed its emphasis on the technological strengths of the allied forces, particularly the allied monopoly of the means of military acceleration. Because SCUDS were relatively slow missiles most were rendered ineffective by the faster Patriot missiles. Because allied air power was not only more numerous but faster than Iraq's, the allies soon gained air supremacy and a monopoly of sight over the battlefield. Schwartzkopf's "Hail Mary" play was a rapid strike across the desert deep into Iraq. The military goal was not to hold and secure territory but to swiftly engage the Republican guard and quickly cut off the escape route for the Iraqi army retreating from Kuwait. Lastly, once the war was called to a halt, the overriding concern of US policy was to get the troops home as quickly as possible. Significantly, the ground war was measured in terms of time (100 hours) not territory.

The question remains whether these and other examples are evidence of the domination of chronopolitics over geopolitics as Virilio suggests. The opposition between the two seems misplaced, a consequence of Virilio's (1986) totalizing style and refusal to play by the traditional representational rules of strategic writing. (Lotringer describes Virilio's writing as telescopic, writing in a state of emergency; Virilio and Lotringer, 1983: 38–42.) This is both the weakness and strength of Virilio's project for, like Baudrillard, the style of exposition is part of the argument. That argument challenges the rules of representation (including the temporal ones) that are assumed in geopolitical discourse. In exploding our received understanding of geopolitics, it opens up a general field of geo-politics (the politics of geographical inscription) and problematizes a spatiality implicated with but also beyond the language games of formal and practicing intellectuals of statecraft. A fragment of this beyond is the seductive spatiality produced by electronic screens of power (Der Derian, 1992; Luke, 1989).

## The electronic spatiality of the Gulf war

One can argue that on January 15, 1991 the textual spatiality of the geopolitical crisis, the spatial understanding of the crisis produced by the discursive strategies of the Bush administration, gave way to the televisual spatiality generated by coverage of the actual war itself (Der Derian, 1992: 175). Verbal scripts were superseded by visual images and active debate by passive watching. The mass media coverage of the war was duplicitous, both a reportage yet also a covering of the war's actuality, both a transmission of its immediacy yet also a screening of its murderous results. The instantaneous coverage of the crisis and war by transnational television networks such as CNN (whose slogan – "Anytime. All the Time" – is pure chronopolitics) reinforced the disappearance of space and distance. World leaders in Baghdad, Moscow, London and Washington all received CNN and could watch each other talk and events unfold as the CNN's panoptic danced around different locations on the globe. The war began in prime-time (7:00 EST) and was brought to the world by US reporters (CNN supplies its news feed to 103 countries) who represented the US-led attack by drawing upon exclusively American experiences ("sky lit up like the Fourth of July").

The spatiality generated by how the war looked on television had paradoxical features. Television presented the war as live yet distant, as instantaneous yet remote, as dramatically real yet comfortably televisual. A real war easily became a television mini-series war, its "live" drama edited into video tape highlights every night for the entertainment of the watching audience. The mass media itself demonstrated a certain awareness of the war's televisual spatiality and began to describe it as a "video game" or a "Nintendo war." The war seemed to unfold in the unreal dematerialized space and time of the video game, a discrete, self-contained, electronically generated space of colorful graphics, reflex action shots and an absorbing aesthetics of destruction (Der Derian, 1992; Levidow and Robins, 1991a).

Confining analysis of how the war came to be seen to ruminations of its video-game nature, however, is hardly adequate. Not everyone experienced the war through the reception of televisual images. Nor should one assume these images were received passively and produced the same insidious effects throughout the world (imagine how CNN was received in Jordan). Tracing the origins of the electronic spatiality of the war allows one to move below the realm of media appearances into the institutional and material structures which produce particular ways of seeing places in wars. Inevitably this takes one into the overwhelmingly male microworld of the militarized state, its gaze, cybernetic technologies and techno-strategic manner of reasoning (Levidow and Robins, 1991b). Physically such microworlds are far removed from the social world, in deep underground bunkers, in mountain cores, ocean waves, or high in the atmosphere. How the dematerialization of place is produced by these features of the military microworld is worth investigation.

First, the importance of seeing potential places of war is vital to the militarized state. Beside the "war machine" that has developed in certain states in the twentieth century there has always existed an ocular "watching machine" to provide military commanders with a visual perspective on potential and actual fields of battle (Virilio, 1989: 3). The modern form of this geopolitical panopticon (key-hole spy satellites, AWACS, photo reconnaissance [PHOTINT], listening posts [COMINT], radar sweeps [SIGINT]: this panopticon extends well beyond the visible spectrum; De Landa, 1991) played its part in the origins of the Gulf crisis (the interpretative analytics of the National Security Agency and Webster's CIA). Its logic, both historically and now, has been to produce representations of places (maps, photographs, videos, digitized images, radar traces, GIS packages; Schulman, 1991; N. Smith, 1992) which have come to efface the materiality of the places themselves. The photographic images, GIS contours, and cyberspace grids appear thoroughly objective but they produce what Virilio (1989: 3) terms an "obliviousness to the element of interpretative subjectivity that is always in play in the act of looking." In the cybernetic optics of this watching machine, places become dematerialized into electronic traces and blips on a military screen (Der Derian, 1990). This representational dematerialization often portends their actual dematerialization as images become targets and whole countries "target-rich environments."

Second, the dematerialization of place is also produced by the mediating and distancing effects of modern technologies of war. For the Western military, at least, the Gulf war was the most comprehensively mediated war this century, the culmination of a process that reached a critical historical stage during World War I with the first sustained use of long-range artillery. This allowed the sustained separation of soldiers from the carnage their weapons inflicted on an unseen and therefore abstract enemy (Bauman, 1989; 1990). Since then long-range technological warfare has become the norm for most militarized states. Sherry's (1987) history of the US Air Force documents how the compartmentalization of destruction into routine technical tasks, and the remoteness of Air Force personnel from what their bombs actually did on the ground, sustained a moral detachment from the destructiveness of war. Modern fire-and-forget and over-the-horizon cruise missiles used by the United States in the Gulf war took this detachment even further. War was reduced to the remote controlled destruction of places whose only existence to military personnel was as electronic target coordinates on a screen. For those flying over the "KTO" (the Kuwaiti Theater of Operations; place as a stage for performing war) the quiddity of the place was derealized by the pace of their vehicles and the electronic eyes they relied upon to see. Even "frontline" (just where this is located in modern cybernetic warfare is uncertain) special forces and infantry soldiers did not always have unmediated vision for they often fought in the dark with night-vision goggles.

Third, the dominance of techno-strategic reasoning within the microworld of militarized states further embellishes the dematerialization of place. Techno-strategy has its origins in the nuclear deterrence models of the RAND corporation in the 1960s when young specialists, trained in game theory, systems analysis and rational choice modelling, displaced older classical forms of strategic thinking (Klein, 1988). The institutionalized dominance of techno-strategy (which reduced politics to technical questions) has since led to the pervasive use of simulations within the US military microworld. Simulations construct an internal environment of meaning that is insulated from actual historical events, actors and places yet, as Der Derian (1990: 301) suggests, "they have demonstrated the power to displace the 'reality' of international relations they purport to represent." Frequent representations of the Gulf war as a "game" (Schwartzkopf's strategy was based on a war game) by both military and public made it easy to view places within that "game" as obstacle course objects for "attriting" within the field of play.

The ability of the television media to challenge the US military's structured way of seeing the war was circumscribed by military censorship and self-censorship. There was a complicity of technologies between both for the eye of the military's watching machine and the eye of the television camera both converted material places into intangible electronic images. Yet the free movement of the media's camera could not be countenanced lest it puncture the official vision of the war. The video feeds of the major networks had to be cleared by the military, reporters were policed, live shots were carefully disciplined and only selected aesthetically engaging video kills were deemed fit for public release. But despite repeated attempts to destroy Iraqi television (and tag its vision as "censored") the harrowing carnage at the Al-Amiriya air-raid shelter on February 13, 1991 was recorded. For once the US military was faced with the real human consequences of its "operational success" and shown to be trapped in its own techno-strategic language (Aksoy and Robins, 1991). In most cases, however, the real carnage of the war was edited out or framed within a general heroic narrative such as those now evident in the photography books and videos instantly produced to commodify the war and sustain its after-image.

Coverage of the Gulf war by the mass media as a whole positioned citizens as spectators at an event over which they had no control. The mass media became a site for a politics of psychological participation. Patricia Mann (1990: 182) has argued:

> As media viewers, we are politically passive, but we are often at the same time constituted as emotionally active observers. Our judgements of these political events and processes are entangled with our reception of them as spectacles. Our

> emotions are engaged by spectacles, we clap or boo in response to the behavior of our football team, or our Congress, or our military forces in Saudi Arabia. We do not expect to influence the actions or the behavior of any of these entities.

Michael Mann (1988: 183–187) terms this form of passive participation yet active emotional engagement by a citizenry in war spectacles "spectator-sport militarism." Geopolitics, normally the preserve of the few, became a public sport during the Gulf war with colorful maps, studio terrain models, computer graphic simulations, and the array of defense "experts" all props in the televised game (Der Derian, 1992; MacArthur, 1992).[10] In the phantasmagoria of the war, escaping the geopolitical framing of place (the conditions of possibility for the deadly pulverization of Iraqi and Kuwaiti locations) was only possible in moments of disorientated shock, brought on by the rare screening of the charred bodies left in the wake of "smart" weapons.

## Conclusion: the effacement of place?

Implicated in the very construction of this essay is a theoretical problem of how one describes the spatiality of the Gulf crisis. Although the official geopolitical spatialization of the crisis and its electronic spatiality are different, both can be described as effacing the "reality" of the places involved in the Gulf crisis and war. The Bush administration's foreign policy texts can be described, in terms inspired by Edward Said, as reducing the complexity of place to abstracted, Westernized, Manichean images. The spatiality of the Gulf crisis is thus read as a distorted Western writing of the reality of place and politics in the "Middle East" (de-pluralized as a region by the assumption that the West was the center of the globe). A nuanced and colored local identity was effaced and erased by a crude monochromatic Western work-up of the region.

Similarly, the chronopolitical and electronic refraction of place can be described as erasing an ordinary presence of identity. Speed blurs the sociality of place while the electronic screens of the military war machine and the mass media dematerialize its everyday life presence. Bauman's (1989, 1990) departures from the moral philosophy of Emmanuel Levinas to analyze the Holocaust and modernity's "effacement of the face" provide a certain intellectual grounding for such arguments. Read spatially, Bauman's reading of Levinas is a chronicle of the effacement of place as much as it is the effacement of the face. Levinas describes humans as having an innate morality which is prior to ontology and subjectivity (Levinas, 1985). The "I" is always for the "Other." Responsibility for the Other is the essential defining characteristic of subjectivity. "Ethics does not follow subjectivity; it is subjectivity that is ethical" (Bauman, 1990: 18). From this perspective, the key question raised by the Holocaust (or the Gulf war) is not the social production of a murderous practice but the social suppression of moral responsibility. The "accomplishment of the Nazi regime consisted first and foremost in neutralizing the moral impact of the specifically human existential code" (Bauman, 1989: 185).

Responsibility arises out of the proximity of the Other as a face. "Proximity means responsibility, and responsibility is proximity" (Bauman, 1989: 184). Although described as both physical and mental, Bauman has a strong understanding of proximity as a geographically defined concept. Our inborn moral impulses can be blocked when we can no longer physically *see* and meet the face of the Other. Distance de-ethicalizes our responsibility for the Other: "moral inhibitions do not act at a distance. They are inextricably tied down to human proximity" (Bauman, 1989: 192). We can add, as Bauman tentatively does, that "place" or "neighborhood" is a condition of possibility for moral practice (1990: 23–30). Explaining such events as the Gulf war, therefore, is a matter of explaining the social production of distance which inhibits moral responsibility and pure seeing.

The concept of morality outlined by Levinas philosophically and translated into sociological terms by Bauman offers an attractive narrative of intelligibility which geographers could well use to explain events such as the Gulf war. It is flawed, however, in a number of ways. Very briefly, its notion of morality relies on an essentialist view of human nature. Social proximity is no guarantee of moral behavior (as the fighting in Somalia, Armenia, and the former Yugoslavia is graphically demonstrating). It also assumes the possibility of a pure, unmediated seeing of the face (Derrida, 1978: 99). In its geographical form, it holds out the danger of an innocenting of neighborhood or place. Place, like the face, is pure presence (Derrida, 1978: 100–109). Yet to hold to the view that place is an innocent, corrupted by the diabolical swirl of images in a postmodern world, is to refuse to acknowledge how place is always a contested and historically inscribed entity (a concept with its own historical geography). This fact is more apparent in "Iraq" and "Baghdad," with its didactic and totalitarian urban monuments, than in most places (Al-Khalil, 1991). Seeing the face or the sociality of place never reaches the possibility of transparency for seeing is already infested with codes of recognition. The effacement of place narrative, therefore, must always be followed with a question mark for "place" must not rest unexamined, an entity signified in and of itself, a concept independent of any relationship to language, inscription and signification. Critical geopolitics can only be constructed on "hollowed ground" (Doel, 1993).

## Notes

1 The Pentagon's three-volume study of the Gulf war omits all references to the Iraqi dead. In March 1992 a US Census Bureau report on the post-war demographics of Iraq estimated 158,000 "excess" deaths in 1991. Military fatalities were estimated at approximately 40,000, civilian deaths during the war at 13,000. The violence surrounding the uprisings accounted for approximately 5,000 military and 30,000 civilian deaths. Deaths caused indirectly by the war were estimated at 70,000. The Census Bureau analyst who produced this report was initially fired from her position because of it but subsequently reinstated after threatened legal action (Ridgeway, 1992).

2 The danger of "euphoria" after the end of the Cold War was first influentially stated by Henry Kissinger in his attack on the Intermediate Nuclear Forces treaty

signed by Reagan and Gorbachev in 1987 (Blumenthal, 1990). Bush's National Security Advisor (Brent Scowcroft) and Deputy Secretary of State (Lawrence Eagleburger) were former members of Kissinger Associates, Henry Kissinger's high-profile lobbying firm.

3 Once the Gulf crisis was proclaimed Iraq was reinstated as a state which sponsors terrorism. See "Fact Sheet: Iraq's Support for Terrorists" (US Department of State's weekly *Dispatch*, November 5, 1990: 241–243).

4 The details of the US tilt towards Iraq are only now becoming part of the public record, thanks to the persistent efforts of Henry Gonzalez in the US House of Representatives. For an excellent exploration of the details of the illegal loans, credits and aid provided by the Bush administration to Saddam Hussein see Campbell (1992b).

5 An account of this meeting and other key moments during the Gulf crisis is provided by participants in an instant-history documentary called *The Washington Version*, a three-hour series sponsored and distributed by the right of center American Enterprise Institute. The documentary takes the form of heroic Western men telling how they made the tough decisions during the crisis and war. It was shown on British television in January 1992 to commemorate the first anniversary of the war.

6 It is worth recalling that Thatcher was under pressure at the time for her anti-European vision of "Britain" and "Britishness." She was likely to benefit domestically from any re-invigoration of the "special relationship" with the United States, a relationship which defined "Britishness" in Atlanticist not European terms. As it happened she was forced to resign soon afterwards because of her anti-European views.

7 The notion of a hyperreal is derived from the provocative and loose writings of Jean Baudrillard (1983). It is particularly useful for understanding contemporary geopolitical reasoning (see Der Derian, 1990, 1992; Luke, 1991; Ó Tuathail, 1992a). Baudrillard's two published pieces on the Gulf war (1991a, 1991b) have generated extended polemic (Norris, 1992) and outraged dismissal (N. Smith, 1992). While there is much (perhaps too much) to fault with Baudrillard's flippant totalizing and bleak hyperbole, it does serve the purpose of forcing one to confront the limits and foundations (however hollow) by which one is enabled to write. Norris's (1992) readable polemic against Baudrillard has the merit of exploring these issues. He can, however be accused of mischaracterizing Baudrillard's project (see Gane, 1991). His diagnosis of postmodernism as a "widespread cultural malaise" reveals an interesting disciplining commitment on his part. "Ultra-relativist" thought is a disease. Treating Baudrillard, neopragmatism and Lyotard as similar ignores crucial differences between their traditions. Most significantly, Norris's reading of the "truth" of the Gulf crisis and war is simply not adequate to explain why the crisis and war took the form they did.

8 On the sustained overestimation and exaggeration of Iraqi forces see Simpson (1991: 332–333) and Draper (1992b: 43).

9 According to Virilio (1986: 153) dromomaniacs is a name in psychiatry given to compulsive walkers. The compulsive energetic nature of Bush as a person has been noted by many. After the war the President was diagnosed as suffering from a hyperthyroid problem, the symptoms of which include impatience and restlessness.

10 The sports/war intertext (Shapiro, 1989) was most graphically illustrated during the 1991 Gulf War Super Bowl when television coverage kept switching from the war on the sports field to the field of war in the Middle East. The ceremonies saw fans waving small flags in the stands, "we're #1" gestures from players in the same

style as those of bomber pilots, a child singing "You are my hero . . . ," and Third Reich style flag marching. Bush called the Gulf War his Super Bowl (Larsen, 1991).

## References

Aksoy, A. and Robins, K. 1991: Exterminating angels: Morality, violence and technology in the Gulf War. *Science as Culture* 2: 322–336.
Al-Khalil, S. 1991: *The Monument: Art, Vulgarity and Responsibility in Iraq*. Berkeley: University of California Press.
Bandow, D. 1991: The myth of Iraq's oil stranglehold. In M. Sifry and C. Cert (eds) *The Gulf War Reader*. New York: Times Books, pp. 219–220.
Baudrillard, J. 1983: *Simulations*. New York: Semiotext(e).
Baudrillard, J. 1991a: The reality gulf. *The Guardian*. March 11.
Baudrillard, J. 199lb: La guerre du Golfe n'a pas Europe lieu. *Libération*, March 29.
Bauman, Z. 1989: *Modernity and the Holocaust*. Oxford: Polity Press.
Bauman, Z. 1990: Effacing the face: On the social management of moral proximity. *Theory, Culture and Society* 7: 5–38.
Blumenthal, S. 1990: *Pledging Allegiance: The Last Campaign of the Cold War*. New York: Harper Collins.
Bogue, R. 1989: *Deleuze and Guattari*. London: Routledge.
Brenner, R. 1991: Why is the United States at war with Iraq? *New Left Review* 185: 122–37.
Bullock, J. and Morris, H. 1991: *Saddam's War*. London: Faber and Faber.
Bush, G. 1990: United States defenses: Reshaping our forces. *Vital Speeches* 56 (22): 677–678.
*Business Week* 1990a: Kuwait's billions may really be in the deep freezer. August 20, p. 35.
*Business Week* 1990b: US–Japanese relations may slip on Mideast oil. August 27, p. 27.
*Business Week* 1991: Why shirking the burden isn't in Japan's best interests. January 28, p. 34.
Campbell, D. 1992a: *Writing Security*. Minneapolis: University of Minnesota Press.
Campbell, D. 1992b: Politics without principle: Narratives of the Persian Gulf War and the ethicality of Operation Desert Storm. Paper presented at the International Studies Association convention, Atlanta, April 1–4.
Cooper, A. F., Higgott, R. and Nossal, K. R. 1991: Bound to follow? Leadership and followership in the Gulf conflict. *Political Science Quarterly* 106: 391–410.
Corbridge, S. and Agnew, J. 1991: The US trade and budget deficits in global perspective: an essay in geopolitical-economy. *Environment and Planning D: Society and Space* 9: 71–90.
Dalby, S. 1990: *Creating the Second Cold War*. New York: Guilford.
Dalby, S. 1991: Critical geopolitics: discourse, difference and dissent. *Environment and Planning D: Society and Space* 9: 261–283.
Debord, G. 1983: *Society of the Spectacle*. Detroit: Black and Red.
De Landa, M. 1991: *War in the Age of Intelligent Machines*. New York: Zone Books.
Der Derian, J. 1990: The (s)pace of international relations: simulation, surveillance and speed. *International Studies Quarterly* 34: 295–310.
Der Derian, J. 1992: *Antidiplomacy: Spies, Terror, Speed, and War*. Oxford: Blackwell.
Derrida, J. 1978: Violence and metaphysics: An essay on the thought of Emmanuel Levinas. In *Writing and Difference*. Chicago: University of Chicago Press.
Doel, M. 1993: Proverbs for paranoids: writing geography on hollowed ground. *Transactions, Institute of British Geographers* 18, 3: 377–394.

Draper, T. 1992a: The Gulf War reconsidered. *The New York Review of Books*, January 16: 46–53.

Draper, T. 1992b: The true history of the Gulf War. *The New York Review of Books*, January 30: 38–45.

Economist Intelligence Unit 1990: *Iraq*. Number 3, 1990.

Freedman, L. and Karsh, E. 1991: How Kuwait was won: Strategy in the Gulf war. *International Security* 16, 2: 5–41.

Fussell, P. 1989: *Wartime: Understanding and Behavior in the Second World War*. New York: Oxford University Press.

Gane, M. 1991: *Baudrillard: Critical and Fatal Theory*. London: Routledge.

Gigot, P. 1990: A great American screw-up: The US and Iraq, 1980–1990. *National Interest* 22: 3–10.

Halliday, F. 1991: The Gulf War and its aftermath: first reflections. *International Affairs*, 67: 223–234.

Hamilton, D. and Clad, J. 1991: Germany, Japan, and the false glare of war. *Washington Quarterly* Autumn: 39–49.

Harvey, D. 1989: *The Condition of Postmodernity*. New York: Basil Blackwell.

Johnson, R. 1989: The Persian Gulf in US strategy. *International Security* 14: 122–160.

Kaldor, M. 1990: *The Imaginary War: Understanding the East–West Conflict*. Clifford: Blackwell.

Karsh, E. and Rautsi, I. 1991: Why Saddam Hussein invaded Kuwait. *Survival* 33: 18–30.

Kissinger, H. 1991: A false dream. In M. Sifry and C. Cerf (eds) *The Gulf War Reader*. New York: Times Books, pp. 461–465.

Klein, B. 1988: After strategy: The search for a post-modern politics of peace. *Alternatives* 13: 293–312.

Larsen, E. 1991: Gulf War TV. *Jump Cut* 36: 3–10.

Levidow, L. and Robins, K. (eds) 1991a: *Cyborg Worlds: The Military Information Society*. London: Free Association Books.

Levidow, L. and Robins, K. 1991b: Vision wars. *Race and Class*, 32, 4: 88–92.

Levinas, E. 1985: *Ethics and Infinity: Conversations with Philippe Nemo*. Pittsburgh: Duquesne University Press.

Luke, T. 1989: *Screens of Power*. Urbana: University of Illinois Press.

Luke, T. 1991: The disciplines of security studies and the codes of containment: learning from Kuwait. *Alternatives* 16: 315–344.

MacArthur, J. 1992: *Second Front: Censorship and Propaganda in the Gulf War*. New York: Hill and Wang.

Mann, M. 1988: *States, War and Capitalism*. Oxford: Basil Blackwell.

Mann, P. 1990: Representing the viewer. *Social Text* 27: 177–184.

Norris, C. 1992: *Uncritical Theory: Postmodernism, Intellectuals and the Gulf War*. London: Lawrence and Wishart.

Nunn, S. 1991: War should be very last resort. *Manchester Guardian Weekly*. January 20, p. 19.

Ó Tuathail, G. 1992a: Foreign policy and the hyperreal: The Reagan administration and the scripting of "South Africa." In J. Duncan and T. Barnes (eds) *Writing Worlds: Discourse, Text and Metaphors in the Representation of Landscape*. London: Routledge, pp. 155–175.

Ó Tuathail, G. 1992b: The Bush administration and the "end" of the Cold War: A critical geopolitics of US foreign policy in 1989. *Geoforum* 23.

Ó Tuathail, G. and Agnew, J. 1992: Geopolitics and discourse: Practical geopolitical reasoning in United States foreign policy. *Political Geography* 11: 190–204.

Ridgeway, J. (ed.) 1991: *The March to War*. New York: Four Walls Eight Windows.

Ridgeway, J. 1992: The cover-up of Desert Storm's civilian deaths and environmental damage continues. *Village Voice*. April 21, 19.

Ross, A. 1989: Containing culture in the Cold War. In *No Respect: Intellectuals and Popular Culture*, pp. 42–64. New York: Routledge.

Schulman, R. 1991: Portable GIS: From the sands of Desert Storm to the forests of California. *Geo Info Systems* 1: 24–33.

Shapiro, M. 1989: Representing world politics: The sports/war intertext. In J. Der Derian and M. Shapiro (eds) *International/lntertextual Relations*. Toronto: Lexington Books, pp. 69–96.

Shelley, T. 1992: Burying the oil demon. In H. Bresheeth and N. Yuval-Davis (eds) *The Gulf War and the New World Order*. London: Zed, pp. 166–180.

Sherry, M. 1987: *The Rise of American Airpower*. New Haven: Yale University Press.

Sifry, M. and Cerf, C. (eds) 1991: *The Gulf War Reader: History, Documents, Opinions*. New York: Times Books.

Simpson, J. 1991: *From the House of War*. London: Arrow.

Smith, J. E. 1992: *George Bush's War*. New York: Henry Holt.

Smith, N. 1992: History and philosophy of geography: real wars, theory wars. *Progress in Human Geography* 16: 257–71.

Tamamoto, M. 1991: Trial of an ideal: Japan's debate over the Gulf Crisis. *World Policy Journal* 8: 89–106.

United States Department of State, Dispatch 1 and 2, various numbers, Washington, D.C.: Government Printing Office.

Van Wolferen, K. 1991: No brakes, no compass. *The National Interest* Fall: 26–35.

Virilio, P. and Lotringer, S. 1983: *Pure War*. New York: Semiotext(e).

Virilio, P. 1986: *Speed and Politics*. New York: Semiotext(e).

Virilio, P. 1989: *War and Cinema: The Logistics of Perception*. London: Verso.

Woodward, B. 1991: *The Commanders*. New York: Simon and Schuster.

# SECTION FOUR

# *GEOGRAPHIES OF POLITICAL AND SOCIAL MOVEMENTS*

## Editor's introduction

The politics of collective action is largely about mobilizing groups to obtain either collective goods or redress of grievances from political and economic organizations, of which the most important is the state. Collective goods could be policies or the provision of regulations and resources to specified groups and places. Social movements arise spontaneously around particular issues at specific locations. Sometimes they expand to encompass like-minded and similarly organized people elsewhere. Often they either die out once a cycle of activity is past, become formal interest groups and political parties or are incorporated into more formal organizations such as existing political parties (Tarrow 1994). Over time and in the process of state-building the 'repertoires' of collective action used by movements have shifted from the localized and sporadic to the national and specific; from burning the hay ricks of landlords to mass demonstrations in capital cities, strikes and boycotts (Tilly 1986). But movements must still have roots somewhere even when the issues they promote and the strategies they use are non-local. They cannot successfully mobilize if they are entirely 'top-down' in organization. Of course, this creates problems when local memberships opt for different strategies or choose to pursue different goals. This is more likely to happen, however, when the state is less centralized and local autonomy provides alternative outlets to a focus on the centre (Tarrow 1994, 62). Then, of course, as Tocqueville pointed out in his classic studies *Democracy in America* and *The Old Regime and the French Revolution*, people may well have both less to complain about and more opportunity to do so without needing to form movements.

The period since the 1960s, particularly during the years 1965–84, has been one of intense social movement activity around the world (Scott 1990; Eyerman and Jamison 1991). Given that the material incentives for individuals to join movements are usually insufficient to prompt active participation, this often elicits surprise in some quarters. A popular argument is that one can always wait for others to take on the task in the hope that you will benefit without much exertion. This 'free rider problem' should mean minimum participation in large groups that offer little prospect of immediate personal rewards (Olson 1965). But people affiliate with movements for a large number of reasons other than personal material advantage and they use external resources such as institutional opportunities and social networks to sustain themselves when resources cannot be provided from within the movement (Tarrow 1994). The shared experiences and social interaction of living together in places also undermines the 'isolated' actor premise upon which the free rider problem rests (Barnes and Sheppard 1992). Consequently, the flowering of movements is not totally irrational if nevertheless beyond the intellectual reach of Olson's theory of 'rational choice'. The recent past, like other periods such as the turn of the twentieth century, has obviously been a period in which many people have chosen to act with others to further goals that might in a different historical context have appeared unrealizable. One result of this has been an explosion in research on social movements, both present-day ones and ones from the past (see, e.g., Jenkins and Klandermans 1995).

Although this period of intense social movement activity has coincided with the processes of globalization, as yet states provide the main 'opportunity structures' within which and against which many social movements organize. For example, within Europe social movements remain largely state-orientated even with the growth of European-wide institutions associated with the European Community (Tarrow 1995). Just as the territorial state arose to manage issues whose spatial scope extended beyond the range of existing political institutions, however, so collective action cannot be expected to remain embedded at the state level when so many issues (from environmental problems to human rights questions) are international or global in nature. There are changing 'political economies of scale', to use Cerny's (1995, 602) apt phrase. As **Ó Tuathail** (**Reading 9**) points out, in an age of global communications there is also a sense in which 'moral proximity' is no longer localized in face-to-face interactions. 'Caring' is not a distance-decay function in the way it once was when one had less information about the needs of distant strangers. The dilemma for movements is that even with the expansion of electronic communication (e-mail and the World Wide Web) they are still drawn into the politics of particular states because of relative ease of access, the continuing ability of states to coopt as well as coerce their populations and the difficulties of organizing transnationally. There are truly transnational movements (the environmental movement Greenpeace is a good example), but they are relatively few in number. So far, the environmentalist slogan 'think globally, act locally' could appropriately be supplemented with 'but still organize nationally'.

Many political parties, particularly progressive ones and ones on the extreme right, begin life as social movements, so there is a less clear-cut distinction between social and political movements than appears at first sight (see, *inter alia*, Alapuro 1976; Agnew 1987; Weitz 1992).

The 'new' social movements of the recent past (such as environmental and feminist ones), however, have often been dismissive of established parties because of the fear of cooptation into 'politics as usual' and because of the statist-orientation of political parties. But this was and is not always the case. Parties can be powerful if dangerous allies for social movements. Greater opportunities can also open up for movements that turn themselves into parties (for example, the German Greens). Political parties have been attractive to students of collective action because they often participate in elections so the results can be used to make inferences about the nature of support (which social groups support which parties) and the incidence of political ideologies among populations. Electoral geography is largely devoted to this pursuit (and to 'improving' the conduct of elections through better districting, etc.) There is now a large body of research detailing the co-variation between political parties, political ideologies, social groups and specific places (see Agnew 1987; Johnston *et al.* 1990). An important part of this research has concerned the ways in which places (or contexts) mediate the impact of membership in social groups through the specifics of everyday life (e.g. Eagles 1995; O'Loughlin *et al.* 1995; Agnew 1996). Distinctive geographies of political parties, therefore, are the result not simply of a coincidence between where certain social groups reside and votes for different parties but of the way places structure political ideologies and affiliations.

This kind of insight about the structuring role of place in politics is the inspiration for the article by **Sari Bennett** and **Carville Earle** (**Reading 10**), which offers a geographical interpretation of an important issue in American political history: the failure of a socialist party to take root at the turn of the twentieth century, the last great period of social movement activity in the United States. Previous interpretations have isolated the role of ethnic divisions among American workers, the general prosperity of American workers or the faulty political tactics of socialist leaders. Bennett and Earle prefer an approach that focuses on the places where the Socialist Party was relatively successful in US presidential elections between 1892 and 1920, tracing success to the sedimentation of trade union or 'labour power' in these places in the years after the Civil War. In a statistical analysis amply illustrated by maps, they show how two factors operating differentially over space undermined the prospects for the Socialist Party in national elections: the increasing gap between skilled and unskilled wages in large cities which divided the working class and the industrial diversity of large cities which reduced the relative numbers of unskilled workers in the heavy industries that stimulated militancy. The Socialist Party was successful only in smaller cities. A base in the larger cities eluded it and undermined its long-term prospects.

This spatial-analytic perspective is fairly typical of the reasoning used in much electoral geography. The major criticism of it is that it is reductive, searching for potential causal variables that operate differentially across space. In contrast, political-economic approaches frame spatial variation in political party or social movement activity in terms of an overarching theory of political economy. Party success or failure is interpreted as reflecting the cycles of the economy and the balance between social forces at any point in time. One good example of this type of structural reasoning is provided in an article by **Peter Osei-Kwame** and **Peter J. Taylor** (**Reading 11**). They use a world-systems framework to argue that political parties in Ghana in the period 1954–79 were constrained by the country's location within the global division of labour to compete over which economic strategy best served the country (loosely, import substitution versus basic commodity-export orientation) while appealing to different ethnic clienteles for electoral support. They use a quantitative analysis of election results to identify a number of different sub-periods in which different parties prevailed largely through isolating in opposition the party most representative of the Ashanti ethnic group, closely identified with cocoa growing and, hence, most opposed to the 'semi-periphery' economic strategy (import substitution) of the governing parties. But the 'successful' parties have no core areas of support (there is no 'normal vote' indicating local stability in voting throughout much of the country over time) indicating that they must 'shop around' the ethnic groups to mobilize support. Osei-Kwame and Taylor interpret this as a pattern likely to be found in other 'peripheral states' in which parties must work with a limited range of economic policy alternatives in the presence of major ethnic divisions.

The possibility of spontaneous agency in political mobilization is excluded from a structural account such as this (Johnston 1991; Katznelson 1992). Irrespective of claims to the contrary within the article, the structural position of Ghana within the world economy and its significant ethnic divisions are identified as the only determinants of political activity. This would not sit well with either those who identify the local intersection of social and environmental causes for the growth of movements (e.g. Howard and Homer-Dixon 1995) or those students of 'new' social movements who see multiple opportunities for political action for even the most disadvantaged of groups wherever they are within the world economy. A 'resistance' politics is possible even when resources and opportunity structures are not richly present (Pred 1990; Staeheli 1994). A postcolonial strain of argument has grown up in political geography concerning the oppositional movements that have appeared in many countries in recent years drawing largely from Castells (1983), Scott (1976, 1985) and Guha (1989). After a swift but useful review of dominant resource-mobilization and identity-orientated theories of new social movements which contain little or no explicit geographical content, **Paul Routledge** (**Reading 12**) gives a geographical account of one contemporary social movement in India that sees it as emanating from a 'terrain of resistance' (a whole set of local conditions) against the coercive, cooptive and seductive powers of the state (also see Routledge 1993).

The movement in question has been devoted to opposing the Indian government's attempts at establishing a number of military facilities (particularly the National Testing Range for missiles) in Orissa, north-east India. Routledge identifies a series of local and locational conditions that have contributed to the mobilization of people against state plans. Key among these is a local 'sense of place' that cannot be reduced to either a set of local material interests or local 'externality fields'. The idiom of the movement is strongly connected to this feeling for or identity with place through the songs and dramas used in political activities. The case strongly suggests that collective action (at least in India) involves cultural codes that are strongly place specific. Other movements in India are given a similar interpretation, linking their efflorescence to an incipient democratization of Indian society against the depredations of an overbearing state.

Tocqueville's claim that there are greater opportunities for democracy when society is stronger than the state and when there is considerable 'local patriotism' seems strongly supported by this example from the old continent. Perhaps it is the absence of local cultural codes and the reduction of politics to the pursuit of personal interests that accounts for the decline of civic involvement in the United States. Or perhaps it is only television that is responsible (Putnam 1995). Whatever the precise causes, there is a burgeoning literature in political theory which suggests the importance of high levels of social communication and shared 'cultural codes' for successful democratic practice (see, e.g., Walzer 1983; Kymlicka 1991; Bauman 1995; Sandel 1996). Whether these can be achieved in cyberspace is still an open question (see, e.g., Turkle 1996). But the contemporary impoverishment of public life in liberal democracies such as the United States is leading more scholars to ask if the geography of political organization (the ways in which structures of governance, the political parties and social movements are organized over space) is not out of balance with the cultural complexity and shifting economic geography of the times.

## References

Agnew, J. A. 1987: *Place and politics: the geographical mediation of state and society*. London: Allen and Unwin.

Agnew, J. A. 1996: Mapping politics: how context counts in electoral geography. *Political Geography*, 15: 129–46.

Alapuro, R. 1976: Regional variations in political mobilization: on the incorporation of the agrarian population into the state of Finland, 1907–1932. *Scandinavian Journal of History*, 1: 215–42.

Barnes, T. J. and Sheppard, E. 1992: Is there a place for the rational actor? a geographical critique of the rational actor paradigm. *Economic Geography*, 68: 1–21.

Bauman, Z. 1995: *Life in fragments*. Oxford: Blackwell.
Castells, M. 1983: *The city and the grassroots: a cross-cultural theory of urban social movements*. Berkeley CA: University of California Press.
Cerny, P. 1995: Globalization and the changing logic of collective action. *International Organization*, 49: 595–625.
Eagles, M. (ed.) 1995: *Spatial and contextual models in political research*. London: Taylor and Francis.
Eyerman, R. and Jamison, A. 1991: *Social movements: a cognitive approach*. University Park PA: Penn State Press.
Guha, R. 1989: *The unquiet woods*. New Delhi: Oxford University Press.
Howard, P. and Homer-Dixon, T. 1995: *Environmental scarcity and violent conflict: the case of Chiapas, Mexico*. Washington DC: American Association for the Advancement of Science.
Jenkins, J. C. and Klandermans, B. (eds) 1995: *The politics of social protest: comparative perspectives on states and social movements*. Minneapolis MN: University of Minnesota Press.
Johnston, R. J. 1991: *A question of place: exploring the practice of human geography*. Oxford: Blackwell.
Johnston, R. J., Shelley, F. M. and Taylor, P. J. (eds) 1990: *Developments in electoral geography*. London: Routledge.
Katznelson, I. 1992: *Marxism and the city*. Oxford: Clarendon Press.
Kymlicka, W. 1991: *Liberalism, community and culture*. Oxford: Clarendon Press.
O'Loughlin, J., Flint, C. and Shin, M. 1995: Regions and milieux in Weimar Germany: the Nazi Party vote in geographic perspective. *Erdkunde*, 49: 305–14.
Olson, M. 1965: *The logic of collective action*. Cambridge MA: Harvard University Press.
Pred, A. 1990: *Making histories and transforming human geographies: the local transformation of practice, power relations and consciousness*. Boulder CO: Westview Press.
Putnam, R. D. 1995: Tuning in, tuning out: the strange disappearance of social capital in America. *PS: Political Science and Politics*, 28: 664–83.
Routledge, P. 1993: *Terrains of resistance: nonviolent social movements and the contestation of place in India*. Westport CT: Praeger.
Sandel, M. J. 1996: *Democracy's discontent: America in search of a public philosophy*. Cambridge MA: Belknap Press of Harvard University Press.
Scott, A. 1990: *Ideology and the new social movements*. London: Unwin Hyman.
Scott, J. C. 1976: *The moral economy of the peasant: rebellion and subsistence in Southeast Asia*. New Haven CT: Yale University Press.
Scott, J. C. 1985: *Weapons of the weak: everyday forms of peasant resistance*. New Haven CT: Yale University Press.
Staeheli, L. 1994: Empowering political struggle: spaces and scales of resistance. *Political Geography*, 13: 387–91.
Tarrow, S. 1994: *Power in movement: social movements, collective action and politics*. Cambridge: Cambridge University Press.
Tarrow, S. 1995: The Europeanisation of conflict: reflections from a social movement perspective. *West European Politics*, 18: 223–51.
Tilly, C. 1986: *The contentious French*. Cambridge MA: Belknap Press of the Harvard University Press.

Turkle, S. 1996: Virtuality and its discontents: searching for community in cyberspace. *The American Prospect*, 24: 50–7.

Walzer, M. 1983: *Spheres of justice: a defense of pluralism and equality*. New York: Basic Books.

Weitz, E. 1992: *Popular communism: political strategies and social histories in the formation of the German, French, and Italian Communist Parties, 1919–1948*. Ithaca NY: Western Societies Program, Occasional Paper 31.

# 10 Sari Bennett and Carville Earle, 'Socialism in America: a Geographical Interpretation of its Failure'

Reprinted in full from: *Political Geography Quarterly* 2(1), 31–55 (1983)

This paper offers a geographical interpretation of a central issue in American politics: the failure of socialism in the United States. Between the Civil War and the end of World War I, communities throughout the United States embraced socialist ideas and the Socialist Party bid at becoming an integral part of the American political tradition. But by 1920, the Party was in shambles; socialism had failed; and the United States stood alone as the only industrial capitalist nation lacking an effective labor party.

Why did socialism fail here? Why did the American political tradition eschew class politics in favor of a multi-class, two-party system? These questions have baffled four generations of scholars, and we would be presumptuous in claiming that geography promises the definitive interpretation. Yet geography offers an angle of vision, a spatial perspective, usually lacking in previous interpretations. Heretofore, the failure of American socialism has been studied at the most general, aggregate level. Failure has been attributed either to the prosperity of the American worker or to tactical strategies of socialists, trade unionists, and government. Rarely do these interpretations descend to the level of the locality and the enormous spatial variations among them. We attempt to do just that and, consequently, our analysis highlights two critical factors previously neglected by students of American socialism.

Socialism's failure, in our view, is best explained by focusing on its modest local successes. The latter occurred in communities having two geographical attributes: a converging ratio in the wages of skilled and unskilled workers and a condition of transition from a small-scale pre-industrial economy to an industrial one. Although socialism appealed *where* the material interest of labor came together and *where* industrial transition provoked a reformist response, the fact remains that neither process was sufficiently ubiquitous to ensure a broad geographical base for socialist politics. In the industrializing United States, the failure of socialism can be viewed as a product of the spatial selectivity of wage convergence and industrial transition. Consequently, the Socialist Party, though achieving some stunning local victories, was incapable of unifying the ranks of labor over an extensive geographical territory. Its successes were always limited, and therein lay the principal cause of failure.

## Prevailing interpretations and the geographical perspective

Previous interpretations of the failure of American socialism divide into two schools: one stresses structure, the other tactics. The structural school regards

American society as inhospitable for socialism. The tactical school, on the other hand, emphasizes the poor quality of socialist leadership and strategy (Laslett and Lipset, 1974).

One of the earliest proponents of the structural school was the German scholar Werner Sombart. Sombart (1906) attempted to document the prosperity of the American worker at the beginning of the twentieth century, and he inferred that their economic well-being led them to an affirmation rather than a critique of industrial capitalism. His 'roast beef and apple pie' interpretation of socialism's failure has since been elaborated by others, notably Seymour Martin Lipset. 'In both Canada and the United States,' writes Lipset, 'relatively egalitarian status structures, achievement oriented value systems, affluence, the absence of a European aristocratic or feudal past, and a history of political democracy prior to industrialization have all contributed to produce cohesive systems that remain unreceptive to proposals for major structural change' (1978: 90–104). Socialist reform, according to the structuralists, contradicted the interests of prosperous, socially mobile American workers.

The tactical school, on the other hand, stresses the leadership and strategy of the socialists and their adversaries – the trade unions and the government. The socialists have received especially harsh criticism. Daniel Bell (1967) maintains that the movement suffered because it was more religious than political. Socialists had a certain millennial quality that made it difficult for them to accept the kinds of political compromise essential for a broad electoral appeal. Socialism, in the view of Kolko (1970), was further troubled by sectarianism: self-flagellation and doctrinal division sapped strength from the movement and were principal causes of failure.

While some scholars have stressed socialism's internal difficulties, others have focused on the tactical wisdom of its conservative adversary, the trade unions (Ulman, 1955; Taft, 1957). These unions, especially the American Federation of Labor, have been regarded as practical and pragmatic. They organized a limited group – the skilled trades – and sought limited objectives, notably improvements in wages and working conditions. They eschewed party politics, accepted the reality of industrial capitalism, and used its organizational forms (business unionism) on behalf of skilled workers.

The wisdom of the trade unions and the incompetence of the socialists certainly possesses a kernel of truth. but some members of the tactical school have sensed a tautological problem, i.e. that winners are wise, losers inept. These sceptics suggest two further considerations for explaining socialist failure – intolerance and repression. American attitudes, says Kenneth McNaught (1966), invariably stiffen against ideas that are exotic, bizarre, or heterodox – all of which apply to socialism in the early twentieth century – and this intolerance paved the way for the enthusiastic repression of socialism during World War I. The vehicle of repression, as documented by James Weinstein (1967, 1968), was the corporate state. Following a series of local socialist victories between 1910 and 1916, the government working through the courts and federal investigation savaged the ranks of the Socialist Party. The Party, although advocating the democratic reform of capitalism, was branded as alien, radical and revolutionary.

Our summary of the structural and tactical interpretations, though desperately compressed, provides a point of take-off for a geographical interpretation of socialism's failure. Such an interpretation, in our judgment, must address the alleged prosperity of American workers and the division between reform socialists and conservative trade unions. The remainder of our essay provides in three parts a reformulation of the structuralist and tactical interpretations. First, we offer a critique of the alleged well-being of nineteenth century American workers. Theoretical and empirical evidence indicates, contrary to the structuralist view, that unskilled American workers were not well paid (Earle and Hoffman, 1980). Second, we argue that the failure of socialism was premised on yet another structural feature – the remarkably wide wage differential among skilled and unskilled workers. This wage differential divided the working class and contributed to the political cleavage between socialists and trade unions. We maintain that this division in the ranks of American labor had deep roots in the American economic experience; that it rested on the contrary material and ideological positions of the skilled and unskilled; and that these positions stemmed from the peculiar geography of the American labor market and its manner of determining wages.

The third part of this paper examines the geography of the American labor movement during the Gilded Age and the Progressive Era. Spatial analysis identities the wage ratio and city size as two critical factors in the failure of socialism. During these periods, the socialists and the trade unions established distinctive geographical loci. The trade unions gathered strength in the large industrial cities where, after 1890, wage ratios diverged sharply. This process amounted to a reversal of earlier trends of the 1870s and 1880s when city wage ratios converged, labor presented a more united front, and strike actions multiplied. Meanwhile, the socialists gained support principally in the smaller cities of industrializing America. These communities, experiencing the transition from pre-industrial to industrial society, had been virtually powerless during the strike-torn 1880s. After 1890, socialism promised a program, if not a panacea, for the strains caused by modernization. In large measure, socialism failed because of its spatially concentrated appeal. Radical politics proved incapable of penetrating the large industrial cities where diverging wage ratios divided working-class interests and a diversified economy diluted the socialist constituency.

A geographical interpretation suggests that the failure of socialism lodged in the large city. There, wage divergence and trade union growth progressed simultaneously between 1890 and 1903. But that is not all of the story since large city wage ratios converged steadily between 1907 and 1920. Our thesis would predict a convergence of class interests and improvements in socialist performance. That this did not happen, we believe, can be explained only by a combination of factors, notably: the solidification of the trade unions by 1903; their alliance with the Wilson administration; and the wartime repression aimed at the heterodox political left. Although structural conditions after 1907 favored class convergence, the trade unions were already well established. After 1907, they struggled successfully in maintaining their privileged position. The persistence of the skill differential in the United States stands as testimony to their success.

## The labor market in industrializing America

### *The unskilled workers*

A geographic perspective on American labor economics does not concur with the rosy estimates of worker welfare made by Sombart, Lipset, and other structuralists. We maintain that industrialization was less beneficial for the American worker than the 'roast beef and apple pie' interpretation would have it. Contrary to this version of American history – a version that is reinforced by the economist's doctrine of labor scarcity – American industrialization before and after the Civil War was based on the use of cheap and unskilled labor. The wages of the unskilled in the North were low by comparison with those of the antebellum South or England. American wages could be judged as an improvement only by comparison with the exceedingly low earnings that workers made in seasonal rural employment. Yet industrialization was not immizerizing. As economic growth and development took place, primarily as a result of entrepreneurial investments in cheap labor, the wages of the unskilled edged upwards. They rose owing to a combination of factors that took effect in the second half of the nineteenth century: the emergence of competitive urban labor markets; the development of more intensive agriculture which helped raise the bidding level for urban workers; and labor's aggressive use of strikes and other tactics that aimed at increasing wages. But before detailing those mechanisms that modestly improved the wages of unskilled labor, we should briefly examine why so many previous interpreters have been misled about the alleged prosperity of the unskilled American worker.

The benevolence of the American economy toward the worker is a long-established view dating back at least until the second half of the nineteenth century. According to this view, the economy's capacity for equitably rewarding the worker rested on the most important of material conditions – a favorable land–labor ratio. Since land was abundant relative to labor, all forms of labor were scarce and, hence, all workers earned high wages. Scholars trotted out a good deal of evidence in favor of the benevolence of the capitalist economy. Farmers lamented the high cost of wages paid out to farm laborers; industrialists complained of the rarity, expense, and extreme independence of skilled workers throughout the nineteenth century. Since literary evidence of this sort comported so nicely with abstract theory, it was only natural that scholars have regarded all American labor as scarce and dear. Consequently, it was alleged that, as the American economy grew, labor reaped its fair share in the form of high wages.

This sanguine view of American labor underlies the principal historical interpretations of American economy and society. Frederick Jackson Turner's frontier thesis (1920), perhaps the most influential interpretation of this nation, maintained that free land on the frontier served as a 'safety-valve' for city dwellers and immigrants. The rural frontier alleviated class pressures in the East by drawing down the supply of urban labor and, thus, increasing urban wages. Although historians have challenged Turner's interpretation, they have less successfully disputed his sanguine view of the urban wage-earner. Similarly, in economic history the favorable land–labor ratio underpins the principal interpretation of American industrialization. H. J. Habakkuk (1962) contends that American machine technology was a direct response to labor scarcity. Machines economized on dear labor and increased labor productivity. The worker's economic position was enviable. Technology raised marginal productivity to the already high level of wages; America got industrialized and the worker retained his high wages.

The doctrine of labor scarcity presents an optimistic view of the economic status of the American worker; yet, certain evidence contradicts these views. Numerous studies of American income and wealth indicate a highly inegalitarian society, particularly in the cities. Equally at odds with labor scarcity doctrine is the trend in nineteenth century factor shares. Edward Budd (1960) has shown that labor's share of output was modest in mid-century. Thereafter, labor's rising share corresponded with the growth of labor organization. Additional evidence on labor's share of value added in manufacturing confirms the trend identified by Budd. This evidence, fragmentary though it is, suggests that antebellum wages may not have been so high as is normally believed. Furthermore, the high wages implied by the doctrine of labor scarcity have been challenged by Peter Temin (1966, 1971). His theoretical analysis arrived at the paradoxical conclusion that American wages were probably cheaper than British wages during the antebellum period.

What we require then, is a theory that reconciles this contradictory evidence and, at the same time, explains the absence of class politics in the industrializing United States. We submit that labor scarcity theories are inadequate because they misunderstand the nature of the urban economy and its labor market during the process of economic development. More precisely, the most misunderstood market is the one that prevailed for unskilled urban labor. On that score, nineteenth century observers were more acute than later scholars. James Montgomery, an authority on the antebellum cotton industry, observed that American capital was motivated by the 'great object' which meant paying 'their help just such wages as will be sufficient inducement for them to remain at the work. Hence the greater the quantity of work produced, the higher the profits, because paid at a lower rate of wages' (1840: 97–98). But what was the source of this cheap labor to which Montgomery alluded? The source was no mystery as it had been clearly identified in the 1790s by none other than Alexander Hamilton in his classic *Report on Manufactures* (McKee, 1934: 206–208).

It is a matter of small irony that historians of economic thought have ignored Hamilton's central concern with labor economics. Hamilton knew that any persuasive argument on behalf of industrialization had to meet the theoretical critique of political economists. Development, in their judgment, had as an indispensable condition the existence of cheap labor and the wage fund, and since the United States suffered from labor scarcity and high wages, industrialization was practically impossible. Hamilton took pains to counter the myth of American labor scarcity. He modestly suggested that unskilled American labor was not so dear as was normally supposed, and he proceeded to identify four sources of cheap labor that nullified the critique of political economists and which opened up industrial possibilities for the United States. American manufacturers, argued Hamilton, could tap the labor market for women, children, immigrants, and farmers and farm workers idled by the seasonality of agriculture. Indeed, American urban entrepreneurs were to do precisely that. After the War of 1812, cotton textile firms installed the most advanced machinery such as the power loom and the throstle frame, thereby enabling the employment of unskilled operatives – women and children, and, in the Middle States, men (Earle and Hoffman, 1980).

Textiles, then, offered a model of new economic possibilities. The industry concurrently wed the machine with cheap, unskilled labor, and it introduced a new locational strategy. The new strategy involved the location of new factories and their unskilled workforce on the outskirts of large cities, in the first industrial 'suburbs' or in new autonomous towns and villages proximate to a rural labor supply. Such was the locational pattern for Waltham and Lowell, Massachusetts as well as Manayunk outside Philadelphia, and Hampden north of Baltimore. These unskilled workers were geographically isolated from the urban artisans who might have provided articulate leadership and political 'savvy'. In consequence, the workingmen's political parties and union organizations of the 1820s and 1830s flourished among the skilled trades in the cities but had modest effect on the lives of the unskilled workers in the industrial satellites (Laurie, 1971).

But geographical location merely reinforced other divisions of a deeper nature. Unskilled workers had little power. Wages were low and they were easily replaced by other unskilled workers. Given their vulnerability, unskilled workers, generally speaking, were docile, politically apathetic, and a malleable labor force in the hands of industrialists. To make matters worse, the factory operatives rarely gained the support of their class allies, the skilled workers, because the two groups stood in direct competition. In one industry after another, cheap labor and the machine threatened the displacement of skilled workers. The interests of labor remained deeply divided so long as there existed *a wide spread between the wages of the skilled and the unskilled* – a spread that encouraged the substitution of the new technology based on machines and of the unskilled for the old technology of skill. So long as these geographical and economic divisions persisted, the potential for a united, radical working class was nil (Ober, 1948; Lydall, 1968; Phelps Brown and Browne, 1968; Lindert, 1978; Earle and Hoffman, 1980).

### *Wage differentials*

The economic geography of nineteenth century labor must assume considerable importance in any explanation of the failure of class politics. Cheap unskilled labor was not only an inducement to industrialization, it was also a wedge driven into the ranks of labor splitting the highly paid skilled workers from the unskilled. Since the operation of the antebellum labor market has been examined elsewhere, we present here a spare outline of its functioning as a basis for understanding the political division in the labor movement after the Civil War (Earle and Hoffman, 1980).

In a developing economy such as the antebellum United States, wage determination results from the interaction of two sectors, the rural and the urban. The wages of *unskilled* workers are determined principally by the earnings workers could make in alternative employment. In a developing economy, the principal alternative is the rural, agricultural sector. The level of urban wages, or the transfer wage, is determined by assuring that net urban earnings equal or exceed net rural earnings. Or more precisely, where $(W_u – C_u)=(W_r – C_r)$, where $W_u$ is the transfer wage, $W_r$ is rural earnings, and $C_u$ and $C_r$ are the costs of living in city and country, respectively. The urban entrepreneur must then pay an unskilled wage ($W_u$) that is determined principally by the nature of rural production and employment opportunities.

Cheap, unskilled labor prevailed in one important antebellum region, the grain belt of the northern United States. There, a low transfer wage reflected the acute seasonality of wheat and corn production and their constricted employment opportunities. Wheat farmers usually hired labor for 10 days to two weeks at harvest, while corn farmers hired labor for planting and tillage between April and early July – a period that also encompassed the wheat harvest and some haying activity. This seasonal regime required a unique rural labor force responsive to the demands of the 'crop season'. These rural laborers, often comprising 20–30 per cent of the adult male labor force, worked on farms for about a third of the year. During the rest of the year, they were unemployed or underemployed. Their principal rural earnings, thus, were circumscribed by the seasonality of grain culture.

The plight of the rural laborer was a bonanza for industrial and urban entrepreneurs. Because of the hybrid rural labor market – a perfectly competitive market in the spring and early summer and imperfectly competitive in the remainder of the year – the urban investor in the grain belt faced a low transfer wage. Cheap, urban labor stretched from Pennsylvania to Illinois. For example, during the 1820s, Philadelphians hired unskilled urban labor for about $110 to $150 per year, while English firms in the Midlands and Lancashire paid about $225. Similarly, during the 1850s, entrepreneurs in Chicago paid workers about $190 per year while their counterparts in southern cities paid $450 or thereabouts. In each case, the contrast in wages was rooted in divergent agricultural systems, their seasonality, and the employment opportunities provided by staples.

Unskilled American labor was cheap and readily available from the farm or foreign immigration, but the same was not true of skilled labor. Artisanal skills were scarce and dear – a fact noted in Eli Whitney's claim that interchangeable parts would 'substitute correct and effective operations of machinery for that skill of the artist which is acquired only by long practice and experience; a species of skill which is not possessed in this country to any considerable extent' (Blake, 1894: 122). In consequence, the introduction of machines and factory production was encouraged as much by the low wages of unskilled workers. By combining unskilled labor with machines, grain belt economies manufactured sophisticated products at lower costs than the artisans.

The rapid acceptance of the power loom offers a case in point. Although the productivity of the new technology (the power loom) was about the same as the old artisanal technology (the hand loom), the combined cost of the new machine and the unskilled operative fell substantially below the wages of the skilled weaver. Precisely the opposite occurred in England. The higher costs of the unskilled (a reflection of the intensity of English agriculture, its longer season, and the higher annual earnings of English rural laborers) retarded the diffusion of the power loom.

The nature of the American labor market did more than encourage industrialization through the introduction of machines and factories; equally important, the market drove a wedge in the growing ranks of the wage-earning classes. As industrialization progressed, the labor market divided the political interest of the skilled and the unskilled because of the wide disparity in their wages and economic interests. In England, conversely, a much narrower wage ratio facilitated the convergence of class interest, first by reducing the competitive threat of unskilled workers displacing the skilled and second by structuring a convergence rather than a divergence of class interests.

In bold outline, the antebellum labor market consisted of cheap unskilled labor, expensive skilled labor, and a large wage differential which separated these two components of the working class. This American wage structure provided the foundation for postbellum developments in the economy and, more particularly, in the American labor movement. During the period of rapid industrialization after the Civil War, subtle geographical shifts in the wage structure appeared. In 1865, many portions of the northeastern states reported large ratios in the wages of skilled (machinists) and unskilled (common laborers) workers. These areas predictably occupied zones within the American grain belt such as central Pennsylvania, Illinois, western Indiana, southern Michigan, and Wisconsin. On the other hand, the narrowest wage ratios in 1865 occurred among the Atlantic seaboard cities and in the upper Midwest. During the next 15 years, the depressed years of the 1870s resulted in a general broadening of the wage ratio, as is evident in the map of 1880 (Figure 10.1). Yet, two important areas bucked this trend. Narrow or converging wage ratios characterized the seaboard cities from Philadelphia to Boston, much of Illinois, and southern Ohio and Indiana. Among the seaboard cities, the antebellum wage structure seems to have remained intact after the Civil War; while in the Midwest, sharply converging wage ratios reflected a steady rise in unskilled wages (United States Congress, 1886).

While we do not as yet know the precise causes of these changes in wage structure, we can appreciate the remarkable spatial variation. Moreover, these data permit a test of their political implications for the labor movement. As attention shifts to the political geography of labor, we highlight the decisive role played by the geography of labor economics. For an instant in the 1880s

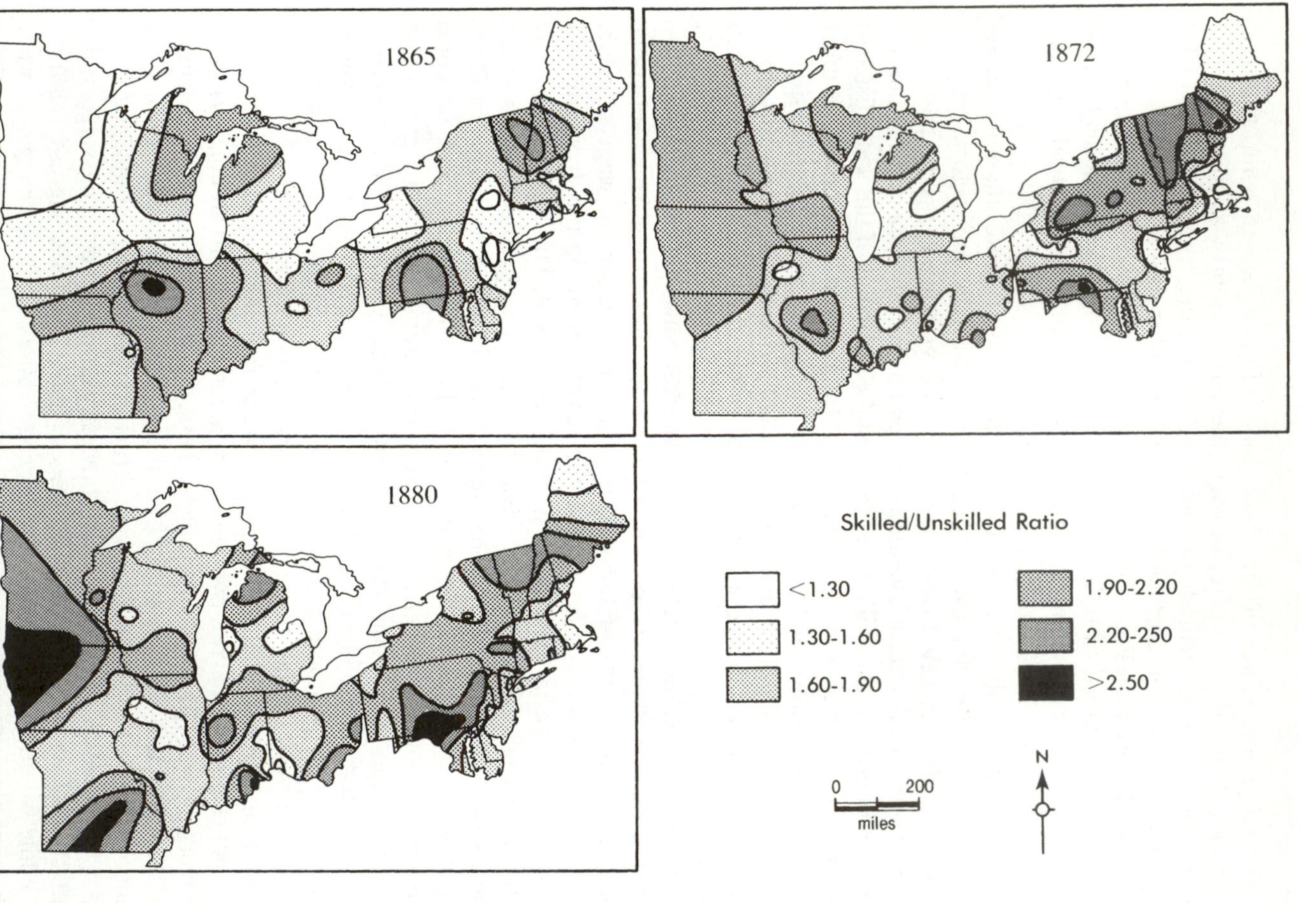


**Fig. 10.1** Skilled/unskilled wage ratios, 1865, 1872 and 1880

and 1890s, a narrowing wage ratio in certain key areas portended the socialist moment; yet this wage convergence was transitory. As wage ratios and the material interests of workers diverged toward the end of the nineteenth century, the promise of class politics gave way to division in the working class.

## The labor movement and the failure of socialism, 1865–1920

The progress of socialism was, of course, tied to a much broader labor movement. Within that movement socialists worked for a unified working class that was organized in itself and for itself (that is a class unified by its class consciousness and its opposition to capitalism). Although the socialist moment never arrived in the United States, this philosophy made considerable inroads into the labor movement during the half century after the Civil War.

The first surge of socialism having any mass appeal came with the Knights of Labor during the turbulent 1880s. Founded in 1869, this semi-secret order, aiming at the unification of all wage-earners in opposition to capitalism, achieved the pinnacle of its influence in 1886. One year later, Frederick Engels (1887, preface) envisioned the deliverance of socialism during the Gilded Age:

> The Knights of Labor are the first national organization created by the American working class as a whole; whatever be their origin and history, whatever their platform and their constitution, here they are, the work of practically the whole class of American wage-earners, the only national bond that holds them together, that makes their strength felt to themselves no less than to their enemies, and that fills them with the proud hope of future victories . . . to an outsider it appears evident that here is the raw material out of which the future of the American working-class movement, and along with it, the future of American society at large, has to be shaped.

But Engels' optimism was unjustified. The decline of the Knights was as meteoric as its rise.

The second surge of socialism came at the turn of the century and took the form of the Socialist Party of America. Unlike the Knights, who eschewed political parties in favor of educational campaigns, the Socialist Party put its faith in political organization and the electorate. At its peak, just before World War I, the Party counted over 100 000 members, elected over 1000 public officials, and garnered nearly 900 000 votes in the presidential election of 1912. During the next eight years, Party support ebbed away to the point that the feeble campaign of 1920 was its last of national significance.

These socialist attempts at remaking America in the Gilded Age and the Progressive Era must be accounted failures. They misled those like Engels who dreamt of a socialist millennium. Yet it is important to remember that, in the context of failure, socialists achieved success in varied local communities throughout the nation. By examining these geographical variations we may more fully understand the environments that favored the growth of socialism and, as importantly, those that did not.

We begin our inquiry by focusing on the outburst of labor unrest that served as the nineteenth century context for the Knights of Labor and later the Socialist Party. During the 1880s, strikes and lockouts rose from a handful a year to nearly a thousand; the conflict between labor and capital, once confined to older Atlantic port cities, spread throughout the northeastern quarter of the nation. The 1880s marked a conjuncture of industrial conflict, socialist programs, and growth in the Knights of Labor. The background for this conjuncture requires a brief review of the earlier historical geography of the American labor movement.

*Strikes: the temporal pattern, 1790–1880*

The nineteenth century was truly the century of the strike. Although workers in western Europe and, to a lesser extent, the United States, had employed the strike before 1800, this strategy became more widespread and more effective after that date. In the United States, the number of strikes steadily accelerated up until the 1880s when strike actions achieved unprecedented levels. Contemporaries and historians usually dated the upsurge in labor unrest at 1877, the year the extensive railroad strikes broke out all over the northeastern United States. Yet a careful examination of the strike curve (Figure 10.2) suggests a turning-point between 1865 and 1880 (compiled from Third Annual Report 1888: 1029–1108; and Tenth Annual Report 1896: 34–1373).

The Civil War marks a great divide in the American labor movement. Before the Civil War, strikes were sporadic and the curve hovered below 15 strikes per year. The pace accelerated to about 20 per year during the War, and then, between 1867 and 1877, the curve rose from 20 to over 50 per year. Although the railroad strikes of 1877 may have galvanized the American labor movement, that year does not stand out as a sharp break or discontinuity with preceding strike activity. The inflection occurred in 1880 when over 700 strikes were reported, and again in 1886 when strikes topped the 1000 mark.

The groundwork for the turbulent 80s seems to have been laid in the years between the Civil War and 1875 – a decade characterized by agitation for the eight-hour day, by the severe depression of the early 1870s, and by efforts at forming labor organizations. Indeed this crucial decade marked the organization of 26 new national unions to go along with the six already in existence. Most of these organizations consisted of trade unions, but at least one attempt was made at forming a broader organization of workers. The National Labor Union, aiming its efforts at the organization of all workers, was unfortunately short-lived and it had expired for all practical purposes by 1870. During the next decade, the Knights of Labor, founded in 1869, sustained the thrust of organizing the entire class of wage-earners; yet the Knights made few inroads at first (Commons *et al.*, 1918: 42–48; Montgomery, 1967).

A curious thing happened after this crucial decade, 1865–1875. Labor organizations that had grown rapidly to over 30 unions and perhaps 300 000 members rapidly lost ground. By 1878, just eight or nine national unions

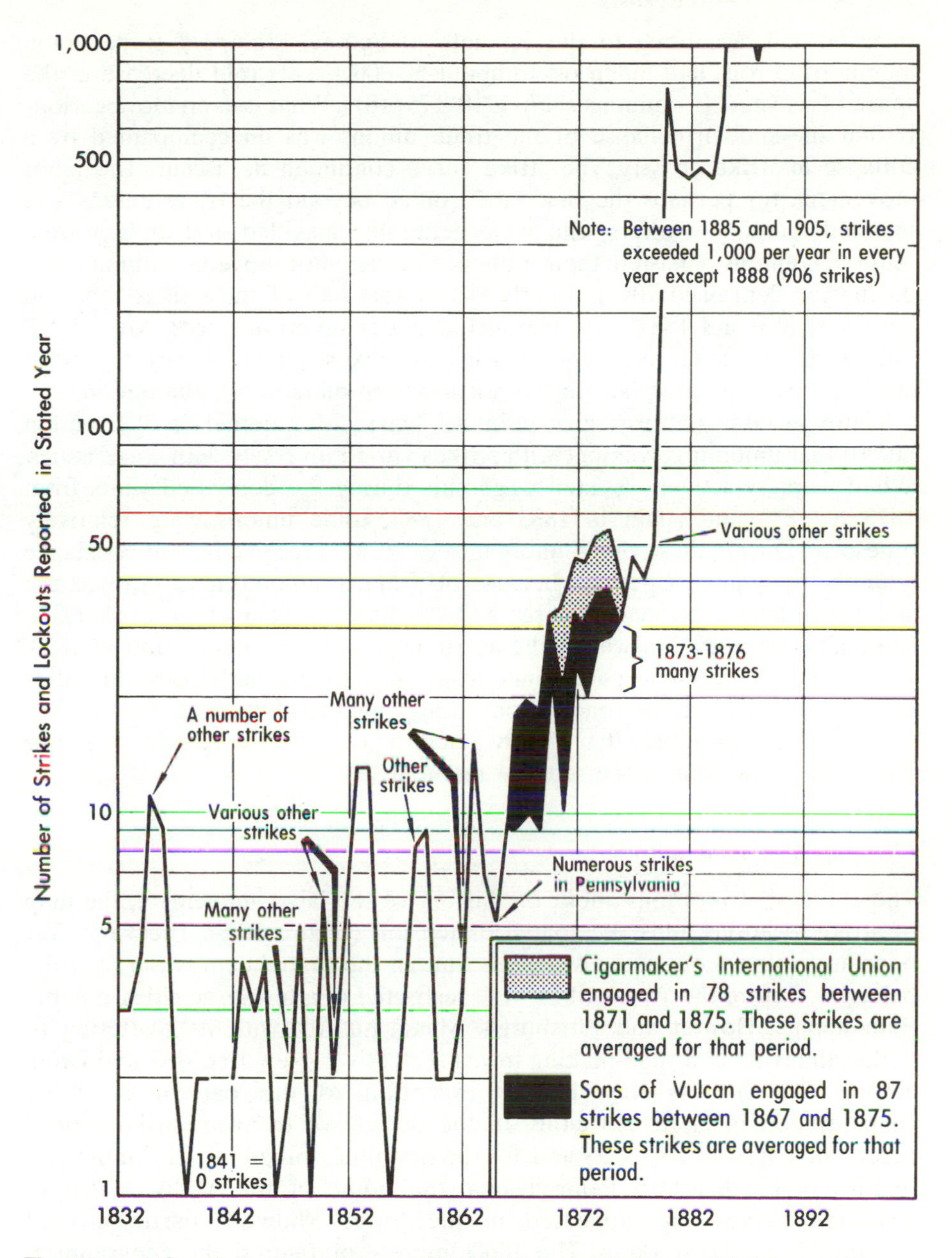


**Fig. 10.2** US strikes and lockouts, 1832–1885 (compiled from Third Annual Report 1888: 1029–1108; and Tenth Annual Report 1896: 34–1373)

survived and, according to the estimate of Samuel Gompers, trade union membership rolls had fallen off to about 50 000 – a six-fold decrease in the space of six years (Commons *et al.*, 1918: 175–181). What is even more curious is that the sudden collapse of the trade unions was unaccompanied by a collapse in strike activity. The strike curve continued its ascent. The labor movement, for perhaps the first time, spilled beyond the trade unions and infected ordinary workers who were generally unskilled and unorganized. Although unions persisted, their influence in the labor movement diminished during the depressed 70s and early 80s. Barely half of the strikes reported between 1881 and 1887 were conducted under union auspices. After 1887, unions experienced a resurgence. During the next seven years they organized 60–70 per cent of all strikes in the northeastern quadrant of the nation.

These variable patterns give point to David Montgomery's observation (1980) that union involvement with strikes varied inversely with wage issues. When workers actively resisted wage cuts during the depressed years from 1873 to 1879 and again in 1883 and 1884, trade unions were relatively quiescent. During these times, unorganized workers received slight assistance from the trade unions, perhaps because of strapped union finances or because skilled workers experienced lesser wage reductions than other workers. In sum, labor organization was in the ascendancy until the depression of 1873. As wage issues became paramount from 1873 to the mid-1880s, the labor movement staked out a broader front. It appealed to all workers, particularly the unskilled – an appeal that greatly aided the cause of the Knights of Labor – and gave a portent of the socialist moment.

### *Geography of labor unrest*

The curve of strikes tells about the history of the labor movement; the map of strike locations adds the spatial dimension (Figure 10.3). The Civil War once again stands out as a divide, this time in the spatial expression of strike behavior. Before the War, strikes were restricted to a few large cities, notably Philadelphia, Boston and Pittsburgh. Massachusetts, with its scattering of strikes in textile and shoe-making towns (Fall River, New Bedford, and Lynn, among others), was the singular exception to the pattern of strike concentration in large, old cities. In the decade of the war, strike actions accelerated in New York City and, for the first time, spread into the anthracite regions of northeastern Pennsylvania, the home of the Molly Maguires. Scattered strikes also appeared in the lower Midwest, particularly at Cincinnati and at St Louis. The 1860s merely prefigured the geographical strike explosion of the 1870s when strikes spread into the bituminous coal-mining and iron centers of western Pennsylvania and eastern Ohio, and intensified in Cincinnati and St Louis. These two decades irrevocably shattered the pre-industrial pattern of localized strikes in the seaboard cities of the Northeast. Strikes were not only more numerous, they were becoming a familiar part of the American social landscape from large cities to mining camps to midwestern towns.

The 1880s perpetuated the geographical extension of labor unrest, as strike

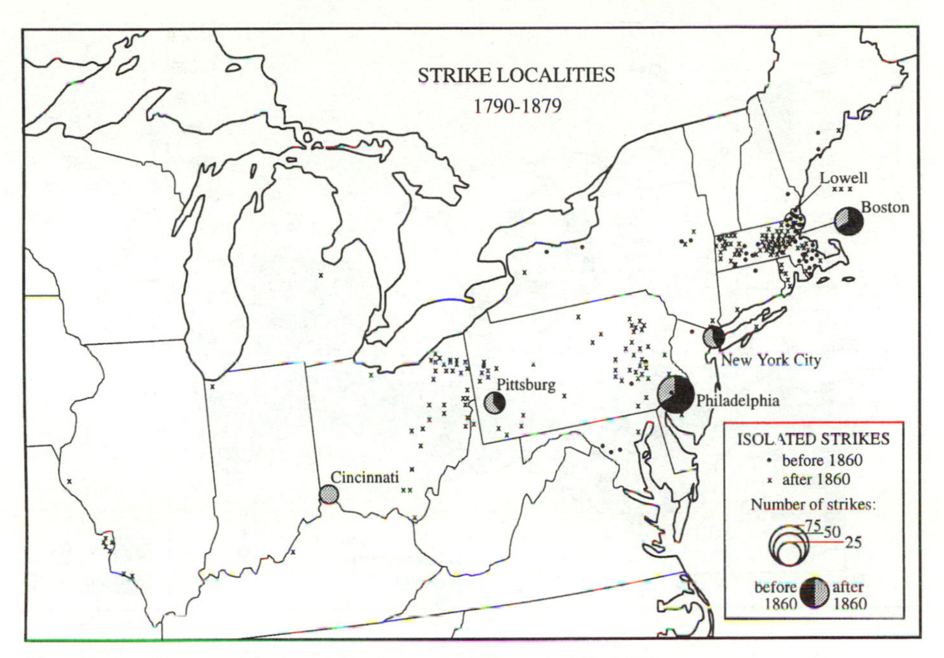


**Fig. 10.3** Geographical distribution of strikes, 1790–1879

activity marched westward to Indiana, Illinois and Iowa, and southward to West Virginia and Kentucky. The strike, though not yet a national phenomenon, was firmly established in the industrializing northeastern quadrant of the nation. Between 1881 and 1894, the quadrant reported strikes in 99 per cent of all cities with 20 000 or more people, in 58 per cent of all towns and cities above 2500 persons, and in over 50 per cent of all 973 counties within the region. The older areas along the seaboard were an important region of labor conflict, but the geographic center gradually shifted toward the Midwest (Bennett and Earle, 1982a).

The strike had become labor's chief weapon for exerting power during the 1880s and 1890s. It was a metric of labor's influence within the workplace and the community. Examination of the geography of strikes for the period 1881–1894 allows us to bring together certain strands of our interpretation of the labor movement. First, the labor unrest of this period was marked by the rapid growth of the Knights of Labor (Figure 10.4), the first occasion in which a socialist philosophy enjoyed a mass appeal (Knights of Labor, 1883; Garlock, 1974). Second, the strike record permits analysis of the interplay between strikes, the Knights of Labor, union activity, and the material interests of labor.

The geography of labor unrest during the Gilded Age reveals several patterns (Bennett and Earle, 1982a and 1982b):

1. In the northeastern states, labor unrest was concentrated in just 112 'strike-prone' counties where 85 per cent of the strikes occurred.
2. Within these strike-prone counties, labor power (as measured by the

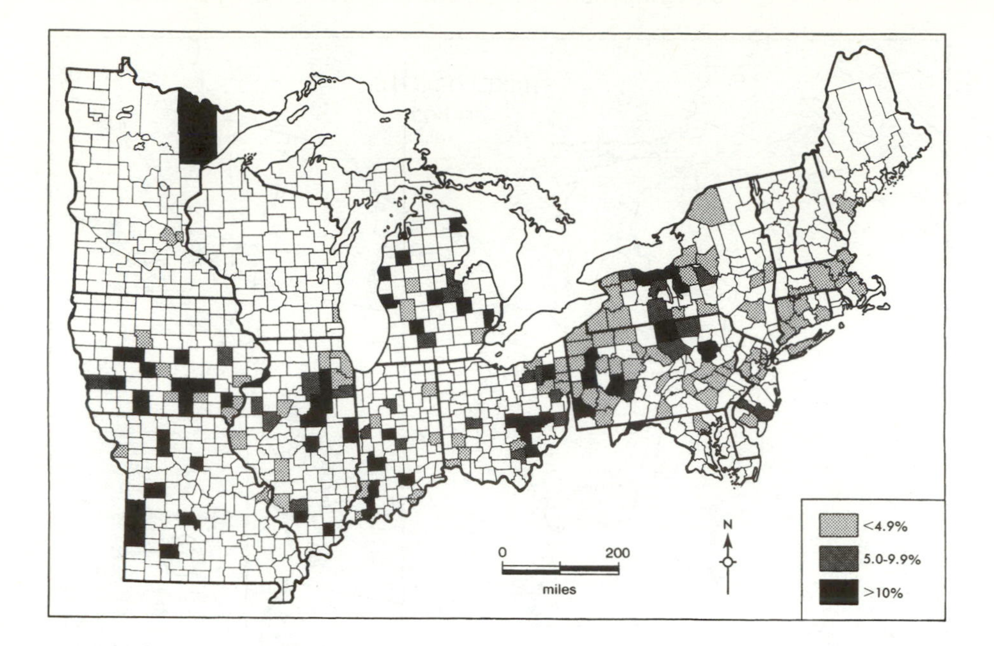


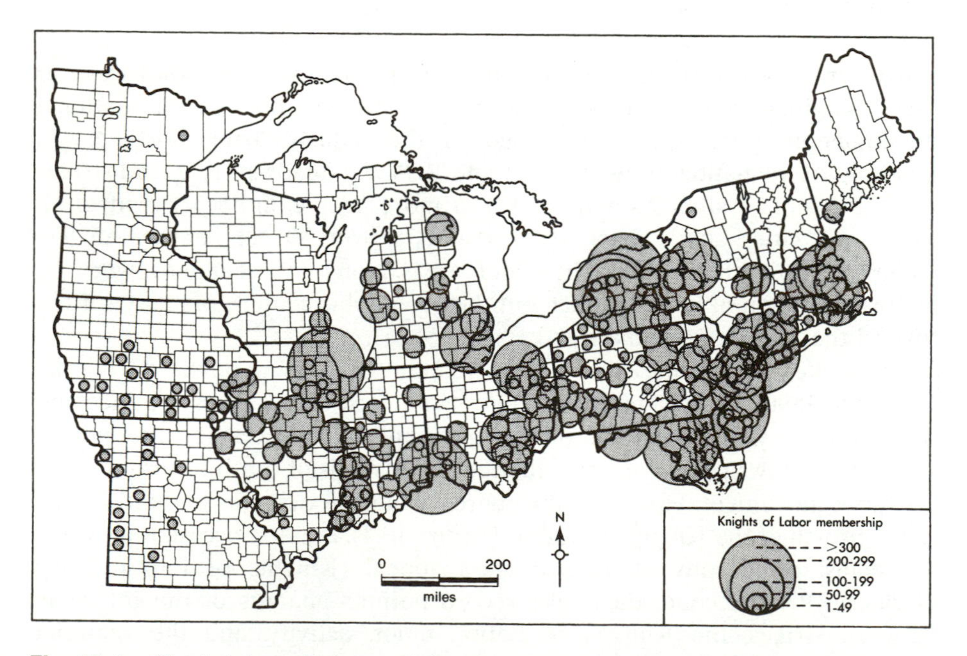


**Fig. 10.4** The knights of labor membership by county, 1883 (top); as a percentage of manufacturing labor force, 1880 (bottom)

strike rate per 1000 population) varied in a curvilinear fashion, declining in smaller communities (85 000 or less), and then inflecting and rising sharply in the larger industrial cities. This scale reversal of labor power departs from the prevailing historiography which argues that labor power declined as industrialization occurred (Gutman, 1970).

3. Formal labor organizations (the Knights of Labor and the American Federation of Labor) had limited effect on labor power in the large industrial cities. Far more decisive was the wage ratio of skilled and unskilled labor. During the 1870s, this ratio narrowed in the Midwest, particularly in Illinois, and was attended by a burst of strike activity that previously had been confined to the East. As wages in the large midwestern cities narrowed, labor power increased – a process that suggests an alignment of the material interests of skilled and unskilled workers, and thus a broadening base for labor unrest.
4. In addition to the positive impact of community scale and narrowing wage ratio, labor's power among large communities of the northeastern states was further augmented by high wages of manufacturing workers and a high degree of industrial concentration. The favorable effect of high wages on labor power suggests that better-paid workers were more likely to strike because of their greater ability to absorb income lost as a result of a strike. A high degree of industrial concentration perhaps implies that Gilded Age workers were more willing to strike if there was an opportunity to find a job with another firm in the same industry if they were fired for strike activity (Kerr and Siegel, 1954). Economic security and a convergence of material interests thus seem the decisive factors in the exercise of labor power in large industrial cities.
5. Although labor's power waned as communities grew into cities, labor mustered some strength where wages were good, where the Knights of Labor were active, and where large firms paid labor a small share of income. The grievance of the small wage share, perhaps more easily perceived in the small town than in the industrial city, was one of the principal laments of the Knights of Labor, who seem to have capitalized on this issue in pre-industrial communities. No issue rallied more support to the socialist perspective than the issue of the inequity with which capitalism apportioned the fruits of labor.

Labor power in the Gilded Age reflected, in sum, the transitional nature of the late nineteenth century – an age caught between pre-industrial and industrial worlds. Labor power declined in the smaller communities though descent was mitigated somewhat by the grievance of capitalist inequity and the expression of that grievance through the Knights of Labor. Among the industrial cities, labor power was reconstituted not by formal labor organizations, as might be expected on the basis of Gilded Age historiography, but by large populations and economic factors including high wages, industrial concentration and, most importantly, a convergent wage ratio (these findings are consistent with the aggregate analyses of Edwards, 1981: 12–133).

*Socialism's Gilded Age antecedents*

The labor turbulence of the 1880s coincided with the spectacular rise of the Knights of Labor – the first socialist program enjoying mass appeal. Yet the modest successes at the Knights were localized and transitory. They displayed surprising strength in unexpected places, notably the small towns and cities of the Middle West, and their enlistment of over 700 000 members in 1886 buoyed up the socialist movement. Yet the Knights lost control of the labor movement soon after 1886, and by the early 1890s the organization moved swiftly toward obscurity.

Yet the Knights bequeathed a socialist legacy to the Progressive Era – a legacy assumed by the Socialist Party of America. Unlike the Knights, who disdained politics in favor of inculcating class consciousness, the socialist politicians aimed at undoing American capitalism through the ballot box. In 1892, the Socialist Labor Party put forth its first national ticket, and in 1896, under Daniel DeLeon, the Party received three-tenths of a per cent of the presidential vote – its best ever. More successful was the Socialist Party, founded in 1901. Running Eugene Debs for President, the Party gathered six per cent of the popular vote in 1912 and at the local level elected over 1200 candidates, including 79 mayoral candidates. Although Debs's showing was modest, he in fact did quite well in various parts of the nation. In the northeastern states, Debs scored above the regional mean vote in western Pennsylvania, eastern and southwestern Ohio, southwestern Indiana and Illinois, and the northern portions of Michigan, Wisconsin, and Minnesota. Similarly, the Socialists, including the Public Ownership Party in Minnesota and the Social Democrats in Wisconsin, polled surprisingly well in the congressional campaign of 1912. The Socialists won over 15 per cent of the vote in Schenectady and parts of New York City, and to the west in places like Pittsburgh, Chicago, Terre Haute (Debs's home town), Dayton and Columbus. In Milwaukee, Socialist candidates polled over 30 per cent while the Public Ownership Party received about 17 per cent of the Minneapolis vote and about one-third of the vote in the bonanza wheat country of the Red River Valley (Figure 10.5 and Figure 10.6) (Robinson, 1970; Congressional Quarterly's Guide, 1975; Dubin, 1980).

Contemporaries and historians of the Socialist Party in the Progressive Era have rightly detected continuities with the labor movement of the Gilded Age. Engels quite clearly envisaged the Knights of Labor as a forerunner of socialism if not socialism itself. And then, too, was it coincidence that the Knights of Labor drew support from the small communities of the Midwest – precisely the kind of place where Eugene Debs was raised? Although the Knights and the Socialist Party differed in their strategic assault on capitalism, both shared the commitment to its gradual destruction through 'civilized means'. In this concluding section, we examine the Gilded Age antecedents of the Socialist Party. We do this through a multiple regression model that attempts to explain the variation in Socialist Party voting in 1912 (presidential) using 19 independent variables previously used in our analysis of Gilded Age labor power. We conclude with some thoughts on the failure of socialism as a consequence of industrialization, diverging wage ratios, and the alliance of trade unions with the Democratic Party.

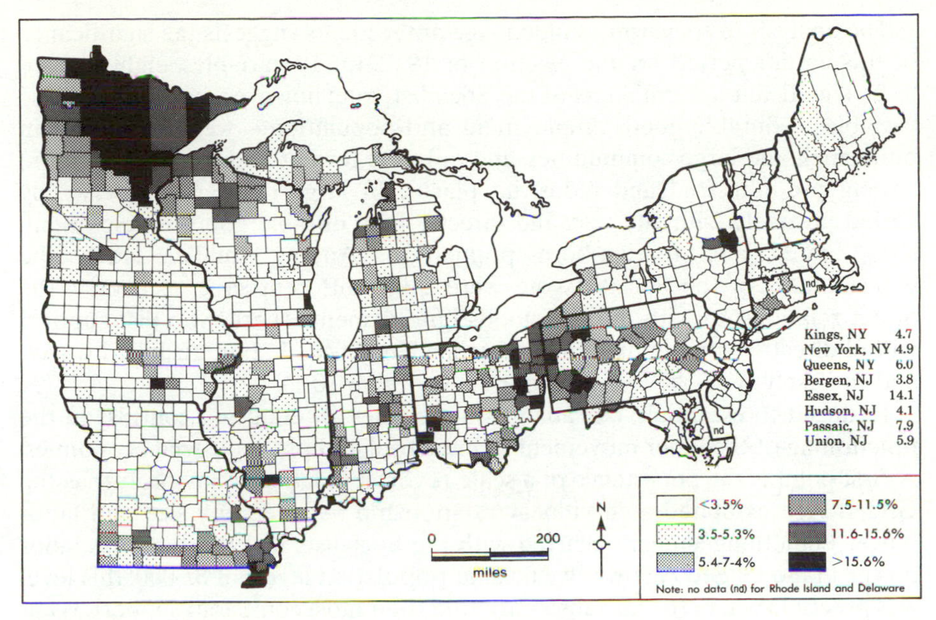


**Fig. 10.5** Socialist Party vote in the presidential election of 1912

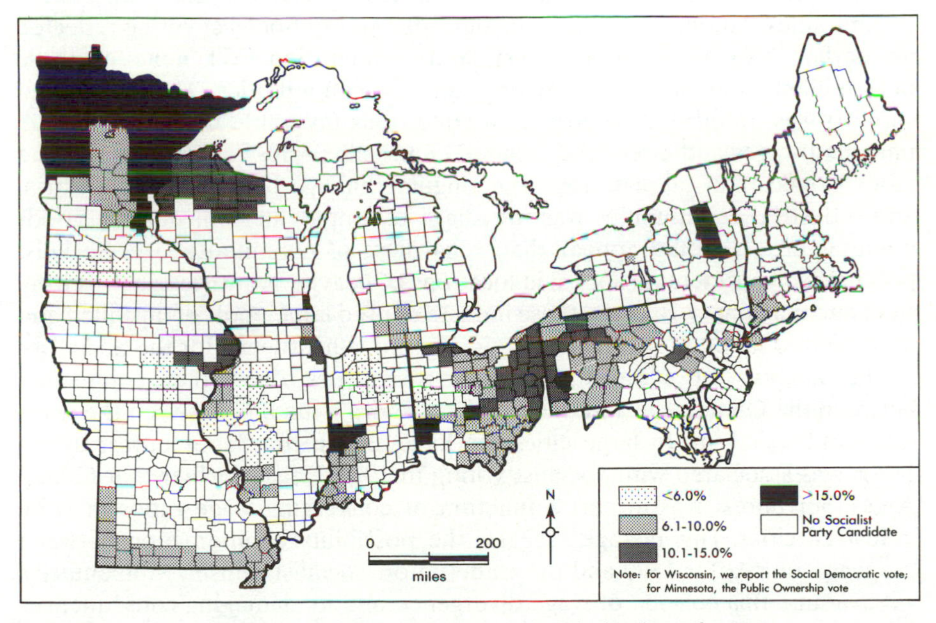


**Fig. 10.6** Socialist Party vote in the congressional election of 1912

The analysis of socialism's Gilded Age antecedents suggests the significance of this earlier period for the election of 1912. Of 19 variables, eight proved to be significant determinants of the Socialist vote; however, just two of these variables – durable goods employment and population – were significant in both small and large communities, and only the former had the same sign (+) among pre-industrial and industrial places. In large cities, Socialist voting varied inversely with city size, and directly with durable goods employment, strike rates, and foreign-born population. Among smaller cities, the determinants of the Socialist vote were quite different save, of course, the direct relationship with durable goods employment. There, Socialist support varied directly with population, the earlier Knights of Labor, unskilled wages, and inversely with the capital–labor ratio (Table 10.1).

Reduced to essentials, our analysis underlines an enduring contrast in the functioning of the labor movement in industrial and pre-industrial economies. A first point is the persistence of a scale reversal in labor behavior. Just as the Gilded Age exhibited a curvilinear relationship between city size and labor power, something similar occurred with the Socialists in 1912. Whereas labor power in the Gilded Age was weakest at population levels of 85 000, this level was precisely where the Socialists garnered their most consistent support. What appears to have happened is that the Socialists filled a power vacuum amongst those local economies undergoing industrial transition. A second point is the remarkable contrast in Socialist determinants between small and large communities. Among large cities, the determinants of Socialist voting reflected an earlier history of labor unrest, and immigration. Yet none of these determinants was significant among smaller communities. There, Socialist support was predicated on structural conditions favorable to labor, e.g. high unskilled wages and a tendency toward economic parity between capital and labor (low capital–labor ratios). The singular linkage between pre-industrial and industrial communities was the share of employment in durable goods manufacture. It would appear that, regardless of community size, Socialist politics had considerable appeal in locations of heavier industries, such as iron, steel and machinery. Because these industries used large pools of unskilled and unorganized labor, they provided an important constituency for the Socialists.

The analysis of socialist antecedents indicates one other connection between the Gilded Age and the Progressive Era – the link between the wage ratio and socialism. In large cities, we have shown that labor unrest in the 1880s was associated with Socialist voting in 1912. Recall further that Gilded Age labor unrest was in part a function of converging wage ratios in large industrial cities. This linkage suggests the possibility of interaction between the wage ratio after 1890 and the trade union–socialist schism. Momentarily, we examine this episode of wage divergence and its damaging consequences for large city socialism.

### *Urban wage structure and the failure of socialism 1890–1920*

It would be presumptuous to press these conclusions too far since our analysis of socialism's antecedents is intended as suggestive rather than definitive.

**Table 10.1** Multiple regressions, strike-prone counties, 1881–1894 (dependent variable=socialist vote for President, 1912)

| *Counties with <85 000 population (n = 58)* | | *Counties with >85 000 population (n = 51)*[a] | |
|---|---|---|---|
| *Variables*[b] | *Std reg. coefficient* | *Variables* | *Std reg. coefficient* |
| DG* | 0.444 | DG* | 0.537 |
| UNW* | 0.268 | PP** | –0.357 |
| CLR** | –0.212 | SR** | 0.306 |
| PP*** | 0.180 | FB** | 0.283 |
| KL*** | 0.170 | | |
| Multiple $R$ = 0.60 | | Multiple $R$ = 0.52 | |
| Multiple $R^2$ = 0.36 | | Multiple $R^2$ = 0.27 | |

* Coefficients significant at 5 per cent; ** coefficients significant at 10 per cent; *** coefficients significant at 15 per cent.

[a] Due to incomplete data, Providence, Rhode Island; New Castle, Delaware; and Washington, DC were omitted from the analysis.

[b] To avoid problems of autocorrelation we conducted a step-wise regression with 19 independent variables for small and large communities. That analysis indicated six independent determinants which were entered in the multiple regression equations here. The 19 independent variables are listed below:

| | |
|---|---|
| (PP) | County population, 1890 |
| (SR) | Strike rate/1000 population, 1890 |
| | Strike success rate/1000 population, 1890 |
| (FB) | Percentage of county population foreign-born, 1890 |
| | Wages as a percentage of value added by manufacturing, 1890 |
| | Skilled wages, 1880 |
| (UNW) | Unskilled wages, 1880 |
| | Ratio of skilled to unskilled wages, 1880 |
| (KL) | Knights of Labor membership (1883) as a percentage of 1880 manufacturing labor force |
| | Value added per manufacturing employee, 1890 |
| | Wages per manufacturing employee, 1890 |
| | Capital investment per manufacturing employee, 1890 |
| (CLR) | Capital/labor ratio, 1890 |
| | Average number of employees per manufacturing establishment, 1880 |
| | Average number of employees per manufacturing establishment, 1890 |
| | Police expenditure per capita in principal city of the county, 1890 |
| (DG) | Percentage of manufacturing employment in durable industries, 1880 |
| | Percentage of manufacturing in largest industry, 1880 |
| | Percentage of total strikes union ordered, 1881–1894 |

What does seem clear is that American radicalism, whether in the 1880s or in 1912, enjoyed its greatest successes in the small and intermediate cities of pre-industrial America. Yet even in pre-industrial America there were changes in the locus of labor radicalism as it shifted in three decades from the small-town affiliation of the Knights to the intermediate cities-in-crisis affiliation of the Socialist Party. Although more detailed analyses using structural data from 1900 and 1912 will be required before we adequately comprehend the geographical bases of the Socialist Party, our preliminary work leaves little doubt that Herbert Gutman and Robert Wiebe were right in drawing a sharp contrast in culture, society and economy in the pre-industrial and the industrial worlds (Wiebe, 1967; Stave, 1975; Gutman, 1976). Radicalism in our analysis appears as a response to painful structural change, to transition from one way of life to another, and its locus rested in the middling city. Consequently, given the meteoric growth of American cities in this class, radical philosophies such as socialism were destined for a brief but useful life.

If socialism flourished in pre-industrial America, its failures lodged in the large cities of the northeastern United States. In these cities, the Socialists were pitted against the trade unions, whose organizing tactics blunted Party efforts at unifying all workers. The singular socialist success among cities larger than 100 000 occurred in Milwaukee where the Party received 27 per cent of the presidential vote in 1912. Conversely, skilled trade unions fared well in these cities. In the period 1880–1886, when the Knights of Labor organized local assemblies of skilled tradesmen, the trade assemblies in large cities (greater than 100 000) accounted for 26.5 per cent of all assemblies in the Knights. By 1900, the American Federation of Labor reported that 32.5 per cent of all local labor unions were situated in large cities (American Federation of Labor, 1900: 9–32; Garlock, 1974). Moreover, trade union membership data for Illinois in 1901 indicated the congeniality of the large city. Chicago, though reporting just 24 of the state's 169 AFL locals, contained nearly 116 000 trade unionists, or 82 per cent of the state total (Illinois, 1904: 312–316). Although much additional work on trade union locales and their membership must be completed before we can make definitive statements on the connection between large cities and the trade union, some basis for this connection can be gleaned from aggregate time series data (Table 10.2).

We earlier showed the association between Gilded Age labor power and the wage ratio of skilled and unskilled workers. The argument, to recapitulate, held that a convergence of wages implied a consciousness of class unity, thus contributing to intensified conflicts with capital. If we pursue this line of thought, certain linkages between the wage ratio and the labor movement can be detected for the period 1865–1920. The wage ratio of skilled labor (carpenters) and unskilled labor (common laborers) follows two distinct cycles. The ratio widens to 1870, narrows into the 1880s, widens to a maximum around 1903, and then converges rather sharply until 1918–1920. These movements of the wage ratio can be linked with changes in the labor movement. As the ratio widened after the Civil War, trade unions grew vigorously in number and membership. But after 1870, as the ratio converged and depression hung over the nation, trade unionists suffered substantial losses. Meanwhile, as worker interest converged, more radical organizations such as the Knights of Labor gained influence. But in the 1890s, the wage ratio rose and the interests of trade unionists diverged sharply and irrevocably from other workers. As the ratio widened toward its maximum in 1903–1907, the trade unions reported their most vigorous growth. In the 1890s, meanwhile, the Knights collapsed and their advocacy of socialism shifted gradually to DeLeon's Socialist Labor Party and Debs's Socialist Party.

Yet the most fascinating aspect of Table 10.2 has to do with the period 1907–1910, a period characterized by a narrowing ratio of wages in large cities, the concomitant growth of Socialist Party support to 1912, and its decline thereafter, despite ongoing wage convergence. By 1912, the socialist vote had

**Table 10.2** Wage ratios and labor, 1865–1920

| Year | Wage ratio | Knights of Labor membership | Union membership | Socialist Party vote | Socialist Party vote as a percentage of trade union membership | Number of AFL national unions |
|---|---|---|---|---|---|---|
| 1865 | 152 | | | | | |
| 1866 | 171 | | | | | |
| 1867 | 169 | | | | | |
| 1868 | 167 | | | | | |
| 1869 | 166 | | | | | |
| 1870 | 171 | | 300 000 | | | |
| 1871 | 167 | | | | | |
| 1872 | 164 | | | | | |
| 1873 | 168 | | | | | |
| 1874 | 167 | | | | | |
| 1875 | 165 | | | | | |
| 1876 | 167 | | | | | |
| 1877 | 163 | | | | | |
| 1878 | 161 | | | | | |
| 1879 | 161 | 13 400 | | | | |
| 1880 | 161 | 40 900 | 200 000 | | | |
| 1881 | | 31 800 | | | | |
| 1882 | | 52 800 | | | | |
| 1883 | | 85 800 | | | | |
| 1884 | | 104 200 | | | | |
| 1885 | | 158 600 | | | | |
| 1886 | | 638 400 | | | | |
| 1887 | | 1 708 000 | | | | 17 |
| 1888 | | 1 204 000 | | | | 7 |
| 1889 | | 616 000 | | | | 8 |
| 1890 | 176 | 288 400 | 372 000 | | | 8 |
| 1891 | 177 | 352 800 | | | | 6 |
| 1892 | 181 | 341 600 | | | | 4 |
| 1893 | 185 | 215 600 | | | | 7 |
| 1894 | 174 | 172 200 | | | | 6 |
| 1895 | 181 | 140 000 | | | | 7 |
| 1896 | 173 | 142 800 | | | | 8 |
| 1897 | 179 | | 440 000 | | | 5 |
| 1898 | 182 | | 467 000 | | | 9 |
| 1899 | 186 | | 550 000 | | | 8 |
| 1900 | 199 | | 791 000 | 94 768 | 12.0 | 11 |
| 1901 | 212 | | 1 058 000 | | | 8 |
| 1902 | 225 | | 1 335 000 | | | 13 |
| 1903 | 219 | | 1 824 000 | | | 16 |
| 1904 | | | 2 067 000 | 402 460 | 19.5 | 8 |
| 1905 | | | | | | 3 |
| 1906 | | | | | | 3 |
| 1907 | 226 | | 2 077 000 | | | 2 |
| 1908 | 204 | | 2 092 000 | 402 820 | 20.1 | 3 |
| 1909 | 202 | | 1 965 000 | | | 4 |
| 1910 | 188 | | 2 116 000 | | | 2 |
| 1911 | 192 | | 2 318 000 | | | 12 |
| 1912 | 194 | | 2 405 000 | 897 011 | 37.3 | 3 |

**Table 10.2** contd.

| *Year* | *Wage ratio* | *Knights of Labor membership* | *Union membership* | *Socialist Party vote* | *Socialist Party vote as a percentage of trade union membership* | *Number of AFL national unions* |
|---|---|---|---|---|---|---|
| 1913 | 226 | | 2 661 000 | | | 1 |
| 1914 | 187 | | 2 647 000 | | | 1 |
| 1915 | 195 | | 2 560 000 | | | 1 |
| 1916 | 189 | | 9 722 000 | 585 113 | 21.5 | 5 |
| 1917 | 172 | | 2 976 000 | | | 4 |
| 1918 | 140 | | 3 368 000 | | | 4 |
| 1919 | 149 | | 4 046 000 | | | 2 |
| 1920 | 146 | | 5 034 000 | 919 799 | 18.3 | 1 |

*Notes and sources*

Wage Ratio Series: This series contains the ratio of carpenters' wages (skilled labor) to the wages of common laborers (unskilled labor). Before 1881, the ratio is calculated on the basis of daily wages; after 1881, we use hourly wage rates. The pre-1881 wages ratio series is based on the Weeks Report. The post-1881 series is compiled from various publications: *Nineteenth Annual Report of the Commissioner of Labor 1904: Wages and the Hours of Labor* (Washington, DC, 1905: 444–471); US Department of Labor, *Union Scale of Wages and Hours of Labor*, Bulletins 131, 143, 194, 214, 245, 259, 274, 286 (Washington, DC, 1913–1921). The weakest link in the series covers the years 1907–1913 when our data-base is limited to four or five city wage observations. After 1913, the annual sample ranges from 12 to 26 observations. Between 1890 and 1903, the sample contains 19 urban observations. The strongest link in the series is the period before 1881 when sample size ranged from 22 in 1865, 41 in 1870, 57 in 1875, and 60 in 1880.

Knights of Labor Membership: Membership estimates are those made in Jonathan Garlock, "A structural analysis of the Knights of Labor: a prolegomenon to the history of the producing classes" (PhD dissertation, University of Rochester, 1974: 231). Garlock's projected membership offers an upper bound for the Knights. His procedure adds together the total reported membership for a particular year and the total of new members added during the succeeding year.

Union Membership: We use the estimates of the Bureau of Labor Statistics; for slightly higher estimates see those of the National Bureau of Economic Research. Both series may be compared in Albert Blum (1968), "Why unions grow", *Labor History* 9, 41–43.

Socialist Party Vote: Richard B. Morris, ed. (1970), *Encyclopedia of American History: Enlarged and Updated*, Harper and Row, New York.

AFL National Unions: This series lists the number of national unions that affiliated with the American Federation of Labor for the specified year. See, Gary M. Fink, ed. (1977), *Labor Unions*, pp. 423–447, Greenwood Press, Westport, CT.

risen to 37 per cent of the trade unionists in the country, but by 1916 it had fallen back to 21 per cent and then to 18 per cent in 1920. Meanwhile, trade union membership, which stagnated or declined between 1907 and 1914, grew rapidly between 1915 and 1920. Precisely when the socialists had the opportunity to broaden their audience, the trade unions were reinvigorated and made substantial gains.

Historians have offered several explanations for the socialist collapse. The factors usually cited include: the Wilson administration's friendship toward the trade unions; the passage of the Clayton Act in 1914; the establishment of the Department of Labor, all of which favored the trade unions; and the repression directed at labor radicals such as the Industrial Workers of the

World and the Socialist Party (Karson, 1958: 42–89; Adams, 1966: 168–175, 219–226; Weinstein, 1967; Montgomery, 1979).

These explanations of socialism's failure are, however, deficient with regard to spatial distribution. The critical failure of the Socialist Party, in our judgment, was its impotence in large industrial cities. Since Party strategy called for the reform of America through the 'civilized means' of democratic elections, it was imperative that the Party do well in the growing centers of population. By 1910, cities of over 100 000 contained over a fifth of the American population, and a decade later, over a quarter (Ward, 1971: 22). Although the Party won numerous local victories in 1910 and 1911, it fared poorly in all large cities except Milwaukee. Most of their triumphs came in 'small cities, villages and townships' (Hoxie, 1911: 613). Despite these local successes and Debs's improved showing in the presidential campaigns, a political base in large cities eluded the Socialists. The futility of their campaign became evident between 1911 and 1920. The Party, though electing 173 mayors and municipal officers across the nation, won just four of these elections – three in Milwaukee and one in Buffalo – in cities of over 100 000 population (Weinstein, 1967: 116–118).

To identify the geographical loci of socialist failure, of course, is not to explain the inhospitality of large American cities. Such an explanation will entail more research, yet our analyses suggest the importance of at least two structural conditions: the divergent wage ratio in large cities (1890–1903) which divided the political interests of the skilled and unskilled; and the increasingly diversified industrial base of large cities (a falling percentage of employment in durable goods manufacture). The former resulted in the siphoning off of the skilled workers from a working-class party; the latter, by lowering the proportion of heavy industry and unskilled workers, diluted the principal constituency of the Party. Each of these urban conditions weakened the Socialists and strengthened the trade unions. By 1910, trade unionism was well entrenched in the large cities, and not even wage convergence thereafter could dislodge it.

Subsequent events in large city labor markets seemed to justify the trade union triumph. During the prosperous 1920s, contrary to the predictions of economic theory, skilled wages rose disproportionately and wage ratios diverged. Not until the 1930s did the wage ratio converge as it had during World War I; and by the Great Depression, socialism as a force in American life was a moot issue (Ober, 1948; Lydall, 1968: 163–180, 182–185). The pure and simple trade unionism of the American Federation of Labor had prevailed.

## Conclusion

The failure of class politics and socialism has made the United States something of an exception among industrial–capitalist nations. The exceptionalism of the American labor movement, we tentatively suggest, owes a great deal to the nation's unique geographical structure. The rift between the skilled and unskilled, trade union and socialist, rests on the distinctive geography of American wage differentials which was rooted in the nineteenth century grain economy of the American manufacturing belt. In this region,

American wages historically maintained a wide skill differential which undermined class solidarity. By contrast, in Britain, skill differentials were considerably more narrow, thereby facilitating class consciousness and class politics (Phelps Brown and Browne, 1968: 47–48; Earle and Hoffman, 1980).

A second geographical explanation for socialist failure is the inhospitable environment of large industrial cities, particularly between 1890 and 1910. Sharply divergent wage ratios and an increasingly diversified industrial base worked to the advantage of the trade unions. The disaffection for socialism in large American cities proved irreversible despite later wage convergence and some stunning electoral victories in pre-industrial America.

By 1920, geographical considerations had played their part in the failure of socialism. Adverse structural conditions in America had largely negated class politics. Yet industrial managers in the 1920s still feared the resurgence of socialism, and they sought a mechanism to prevent it. Sam Lewisohn (1926: 199), Chairman of the Board of the American Management Association, grasped the solution: 'The problem of [income] distribution, which has so often been regarded as a drama, with labor and capital as the conflicting characters, turns out to be largely the prosaic task of using wage policies to increase national productivity.' And, we might add, to divide the ranks of labor in industrial America. Through artificial manipulation of the wage differential, industrial managers could perpetuate the historical rift in the labor movement which had, in the nineteenth century, arisen naturally from a grain economy.

## References

Adams, G., Jr 1966: *Age of Industrial Violence, 1910–1915: The Activities and Findings of the United States Commission on Industrial Relations*. New York: Columbia University Press.

American Federation of Labor 1900: *List of Organizations affiliated with the American Federation of Labor*, pp. 9–32. Washington, DC: American Federation of Labor.

Bell, D. 1967: *Marxian Socialism in the United States*. Princeton, NJ: Princeton University Press.

Bennett, S. and Earle, C. 1982a: The geography of strikes in the United States, 1881–1894. *Journal of Interdisciplinary History* 13, 63–84.

Bennett, S. and Earle, C. 1982b: Labor power and locality in the Gilded Age: the Northeastern United States, 1881–1894. *Histoire Sociale* 15, 383–405.

Blake, W. P. 1894: Sketch of the life of Eli Whitney, the inventor of the cotton gin. *Papers of the New Haven Historical Society* 5, 122.

Budd, E. C. 1960: Factor shares, 1850–1910. In *Trends in the American Economy in the Nineteenth Century*. National Bureau of Economic Research Studies in Income and Wealth, no. 24 (W. Parker, ed.) pp. 365–398. Princeton, NJ: Princeton University Press.

Commons, J. R., Saposs, D. J., Summer, H. L., Mittelman, E. B., Hoagland, H. E., Andrews, J. B. and Perlman, S. 1918: *History of Labour in the United States*, vol. 2. New York: Macmillan.

*Congressional Quarterly's Guide to U.S. Elections* 1975: Washington, DC: Congressional Quarterly.

Dubin, M. 1980: Congressional District Boundaries (unpublished).

Earle, C. and Hoffman, R. 1980: The foundation of the modern economy: agriculture and the costs of labor in the United States and England, 1800–1860. *American Historical Review* 85, 1055–1094.

Edwards, P. 1981: *Strikes in the United States, 1881–1974*. New York: St Martin's Press.

Engels, F. 1887: *Condition of the Working Class in England in 1844*. New York: J. W. Lovell.

Garlock, J. E. 1974: A structural analysis of the Knights of Labor: prolegomenon to the history of the producing classes. Unpublished PhD dissertation, University of Rochester.

Gutman, H. 1970: The workers' search for power: labor in the gilded age. In *The Gilded Age: A Reappraisal* (H. W. Morgan, ed.). Syracuse, NY: Syracuse University Press.

Gutman, H. 1976: *Work, Culture, and Society in Industrializing America*. New York: Vintage Books.

Habakkuk, H. J. 1962: *American and British Technology in the Nineteenth Century: The Search for Labour-Saving Inventions*. Cambridge: Cambridge University Press.

Hoxie, R. F. 1911: The rising tide of socialism: a study. *Journal of Political Economy* 19, 609–631.

Illinois State 1904: *Twelfth Biennial Report of the Bureau of Labor Statistics of the State of Illinois*, 1902. Springfield, IL: State of Illinois.

Karson, M. 1958: *American Labor Unions and Politics, 1900–1918*. Carbondale, IL: Southern Illinois University Press.

Kerr, C. and Siegel, A. 1954: The interindustry propensity to strike: interpretation. In *Industrial Conflict* (A. Kornhauser, R. Dubin and A. M. Ross, eds) pp. 189–212. New York: McGraw-Hill.

Knights of Labor 1883: *Proceedings of the General Assembly, 1883*.

Kolko, G. 1970: The decline of radicalism in the twentieth century. In *For a New America: Essays in History and Politics from Studies on the Left* (J. Weinstein and D. W. Eakins, eds). New York: Random House.

Laslett, J. H. M. and Lipset, S. M. 1974: *Failure of a Dream? Essays in the History of American Socialism*. Garden City, NY: Anchor Press/Doubleday.

Laurie, B. 1971: The working people of Philadelphia, 1827–1853. Unpublished PhD dissertation, University of Pittsburgh.

Lewisohn, S. A. 1926: *The New Leadership in Industry*. New York: E. P. Dutton.

Lindert, P. H. 1978: *Fertility and Scarcity in America*. Princeton, NJ: Princeton University Press.

Lipset, S. M. 1978: Marx, Engels, and America's political parties. *Wilson Quarterly*, Winter, 90–104.

Lydall, H. 1968: *The Structure of Earnings*. Oxford: Clarendon Press.

McKee, S., Jr 1934: *Alexander Hamilton's Papers on Public Credit, Commerce and Finance*, pp. 177–276. New York: Columbia University Press.

McNaught, K. 1966: American progressives and the great society. *Journal of American History* 53, 504–520.

Montgomery, D. 1967: *Beyond Equality: Labor and the Radical Republicans, 1862–1872*. New York: Knopf.

Montgomery, D. 1979: *Worker's Control in America: Studies in the History of Work, Technology, and Labour Struggles*. Cambridge: Cambridge University Press.

Montgomery, D. 1980: Strikes in nineteenth-century America. *Social Science History* 4, 81–104.

Montgomery, J. 1840: *Practical Detail of the Cotton Manufactures of the United States of America and the State of Cotton Manufactures of That Country with That of Great Britain*. Glasgow: J. Niven.

Ober, H. 1948: Occupational wage differentials, 1907–1947. *Monthly Labor Review*, 67, 127–134.

Phelps Brown, E. H. and Browne, M. H. 1968: *A Century of Pay: The Course of Pay and Production in France, Germany, Sweden, the United Kingdom and the United States of America, 1860–1960*. London: Macmillan.
Robinson, E. E. 1970: *The Presidential Vote, 1890–1932*. New York: Octagon Books.
Sombart, W. 1906: *Why is There No Socialism in the United States*? Tübingen, Germany: J. C. B. Mohr.
Stave, B. M. (ed.) 1975: *Socialism and the Cities*. Port Washington, NY: Kennikat Press.
Taft, P. 1957: *The A.F. of L. in the Time of Gompers*. New York: Harper and Row.
Temin, P. 1966: Labor scarcity and the problem of American industrial efficiency. *Journal of Economic History* 26, 277–298.
Temin, P. 1970–1971: Labor scarcity in America. *Journal of Interdisciplinary History* 1, 251–264.
*Tenth Annual Report of the Commissioner of Labor 1894.* 1896: Washington, DC: Government Printing.
*Third Annual Report of the Commissioner of Labor 1887.* 1888: Washington, DC: Government Printing.
Turner, F. J. 1920: *The Frontier in American History*. New York: H. Holt.
Ulman, L. 1955: *The Rise of the National Trade Union*. Cambridge, MA: Harvard University Press.
United States Congress 1886: House of Representatives, Misc. Doc. 42, vol. 13, Part 20, 47th Cong., 2nd Sess. *Report on the Statistics of Wages in Manufacturing Industries with Supplementary Reports* by Joseph D. Weeks. Washington, DC: Government Printing Office.
Ward, D. 1971: *Cities and Immigrants: A Geography of Change in Nineteenth Century America*. New York: Oxford University Press.
Weinstein, J. 1967: *The Decline of Socialism in America, 1912–1925*. New York: Monthly Review Press.
Weinstein, J. 1968: *The Corporate Ideal in the Liberal State*. Boston, MA: Beacon Press.
Wiebe, R. 1967: *The Search for Order*. New York: Hill and Wang.

# 11 Peter Osei-Kwame and Peter J. Taylor, 'A Politics of Failure: the Political Geography of Ghanaian Elections, 1954–1979'

Reprinted in full from: *Annals of the Association of American Geographers* 74(4), 574–89 (1984)

The recent interest in the geography of elections has found expression largely in studies of particular elections or series of elections employing cartographic or spatial analysis. The result has been the development of a strong empirical research tradition at the expense of any critical theoretical assessment. Elections are accepted as unproblematic, and the fact that most countries in the world do not hold regular elections has been of little or no concern to geographical researchers.

This paper has two purposes. First, we continue the empirical tradition of electoral geography by presenting a case study of elections in Ghana. Our case study is relatively unusual for electoral geography, however, because it deals with a country outside the First World. We capitalize on this feature for our second purpose: to provide a theoretical context within which to interpret such a case study. In Third World countries elections cannot be accepted as unproblematic; they can only be understood as one of several methods of government formation. Electoral geography must therefore fuse with a more general political geography set in a broad political economy framework.

The paper consists of four main sections. We begin by outlining some theoretical concepts that introduce a world-systems approach to studying elections. This is followed by a more specific treatment of the nature of the state in the periphery of the world-economy, identifying key paradoxes that afflict all such countries. We then use the case study of Ghanaian elections to illustrate these paradoxes through spatial analyses of voting returns from 1954 to 1979. Finally, we relate these findings to the world-economy framework.

## Theoretical preliminaries

The present research is part of a larger project aimed at placing electoral and political geography in a world-systems framework (Archer and Taylor 1981; Taylor 1981a, 1982a, 1982b, 1984a, 1984b, 1985). Here we shall discuss only the three fundamental issues that are directly relevant to our case study: the relation between economy and politics in the modern world-system, the relation between class conflict and ethnicity, and the relation between the ideological hue of a government and the location of the state in the world-economy.

### *World-economy and interstate system*

Wallerstein (1974, 1979) developed the world-systems approach to deal with the problem of large-scale change in societies. A change in the mode of production can introduce the kind of discontinuity that occurred in Europe after 1450 during the transformation from feudalism to capitalism. The capitalist mode of production took the form of what Wallerstein terms a "world-economy," initially based in Europe but gradually growing to incorporate the rest of the world by the twentieth century. The basic motor of contemporary change is, then, capital accumulation on a world scale. The main lesson to be drawn from the world-system approach is that it is futile to attempt to understand social change on a country-by-country basis; instead, each country must be analysed as part of a wider entity, the unfolding world-system that is the world-economy.

The capitalist world-economy incorporates two complementary structures. A world market for commodities exists to organize economic competition, and an interstate system exists to organize distortion of that competition. In Wallerstein's scheme both economic and political organization are necessary for the operation of the world-economy. In Chase-Dunn's (1982, 23) words:

> The system is not simply a free world market of competing producers. The interdependence of political-military power and competitive advantage in production for success in the capitalist world-system reveals that the logic of the accumulation process *includes* the logic of state building and geopolitics.

In short, Wallerstein offers a truly political-economic perspective.

*Government and structural location*

The spatial structure of the world-economy is expressed in two different ways. The economic structure consists of a dynamic pattern of core, semi-periphery, and periphery. This is complemented by a formal political compartmentalization of space into slightly more than 150 sovereign states. These two structures can be linked theoretically through a Marxist theory of the state.

At the most basic level, each state exists to relate its fragment of the world-economy to the larger system in such a way as to benefit those who dominate the government of the state, but the location of the state in the world-economy structure will constrain government strategy. This is clearly seen in tariff policy. Free trade arrangements enable the capitalist class in core countries, especially hegemonic core countries (Britain in the mid-nineteenth century, the USA in the mid-twentieth century) to exploit their relatively greater productivity on the world market. Semi-peripheral countries will want to protect their less efficient manufacturing industry from a free market; hence the protectionism of the USA and Germany in the previous century. Options for the peripheral countries are less clear; there are two basic positions: free trade policy, which may favor cash crops and raw material exports, or a more controlled trade policy, which would favor the development of industry. A free trade policy perpetuates dependence on the core; we shall call it the classic peripheral strategy. A controlled trade policy is then the semi-peripheral strategy in that it is directed explicitly at lessening core dominance of the local economy.

This simple materialist basis of government activity in the periphery is obscured by political and ideological competition from beyond the periphery. Young (1982), for instance, has investigated the economic growth of African states in terms of the self-proclaimed ideology of their governments. Hence the classic peripheral strategy is identified as "African capitalist," and the semi-peripheral strategy is a policy of "populist-socialist" governments. In world-systems terms both are capitalist; they merely represent different factions of capital controlling the state apparatus for different ends. Ideological labeling may be useful to local politicians and to those in the core interpreting the periphery in terms of "friendliness" toward the core, but to emphasize ideology over a country's location in the world-economy is to impress (inadvertently?) a cold-war mentality on political processes of the periphery.

*Politics of power and politics of support*

Our theory of the state is not only about relating the state to the world-economy. The state also contributes to the domination of one class over another, and, by allowing citizen participation in choice of government, elections operate to legitimate the processes of the state. Usually this involves competition between political parties, which mobilize the citizens as voters. This mobilization integrates the dominated class, sometimes only temporarily, into the national political system. In core countries parties have been particularly successful at political integration, producing states that we view as "stable democracies." In the context of Britain, Gamble (1974) has called this the politics of support. Most geographic studies of voting have been concerned with this kind of politics. There is, however, a second "politics" that Gamble describes; it corresponds to the second role we identify for elections. The politics of power is concerned with competition within the dominant class to use the state for particular interests. This is what Schattschneider (1960) means when he refers to Americans as "semi-sovereign people." Voters are "given" parties to vote for and have little direct say in the nature of the party system that circumscribes their electoral choice. In the politics of power, elections act as a means of conflict resolution between different sections of the dominant class. This politics has been neglected in electoral geography.

The main purpose of identifying these two "politics" is that the processes they describe need not be directly related. In the USA between the civil war and the New Deal, for instance, the political parties pursued distinctive economic policies: Republicans were the party of protection, Democrats the party of free trade. But this is only the politics of power because it concerns how the different sections of the dominant class organized policy when they obtained government office. In the country a politics of support operated in which these economic issues were not paramount. Members of the dominated class were mobilized behind these two parties on cultural, ethnic, racial, and religious lines (Taylor 1984a). Hence politics of power and politics of support were quite distinct in emphasizing different issues. In general we can note that it is advantageous for dominant classes to mobilize support along nonclass lines such as ethnicity because such a politics of support will conceal the class nature of the state. In Schattschneider's (1960, 71) terms potentially dangerous class issues are "organized out of politics." Hence the existence of a politics of support operating on nonclass lines in no way contradicts our class theory of the state. On the contrary it represents a highly successful, and hence very common, electoral arrangement for the dominant class.

These two processes, the politics of support and the politics of power, exist in all states throughout the world. However, the balance between the two and the degree to which elections are important mechanisms in these politics vary greatly between states. This variation is anything but random, and even a superficial knowledge of world political patterns indicates that the role of elections is related to material well-being. The liberal democracies are invariably rich countries, and liberal procedures such as electoral competition become increasingly the exception as we venture outside the "developed" world. Hence in order to understand fully the meaning of elections our studies must be conceptualized at the level of the world-economy. Core countries are relatively homogeneous in their adherence to liberal democracy, whereas in the periphery dominant classes have had to employ a wider range of political strategies with the result that the nature of the politics is much more heterogeneous. In these countries politics of support more often runs counter to the politics of power so that elections are relatively rarer. To understand this situation we need to consider the peripheral state in some detail.

## Paradoxes in the politics of peripheral states

Because the politics of peripheral states are heterogeneous, it is difficult to derive generalizations concerning the processes operating there. Whereas the politics of core states can be described by the simple phrase "liberal democracy," there is no equivalent, meaningful phrase to describe peripheral politics. Nevertheless, politics in poorer countries do operate in similar shared material circumstances and the politics of a state can be traced to the problems induced by these material circumstances. We can begin, therefore, to look for pattern in the heterogeneity by identifying the paradoxes inherent in the politics of peripheral states.

### *The paradox of the strong state*

Alavi (1979) identifies an "overdeveloped" political superstructure as a key characteristic of the peripheral state. Furthermore, he considers these strong states, represented by military-bureaucratic oligarchies, to be relatively autonomous. Alavi makes this claim on the basis of a class structure that includes not one but three "dominant" classes: an indigenous bourgeoisie, an external metropolitan bourgeoisie, and rural land owners. Although they will have different economic goals, all three classes share one overriding concern, the preservation of public order. Hence, according to Alavi, all three depend on military bureaucratic oligarchies that "assume . . . a relatively autonomous *economic* role which is not paralleled in the classical bourgeois state" (Alavi 1979, 42).

Alavi outlines a theory for the "strong" peripheral state but completely misses the paradox in the situation. The state may seem to be strong relative to its own society and economy, but it is notoriously weak in relation to the rest of the world-economy. It is not enough for Alavi to include an "external" dominant class in his analysis; he underestimates the most important feature of the peripheral state – the constraints imposed on it by the world-economy. As Hamilton (1982, 24) points out, state autonomy depends not only upon national class structure but also upon "the position of that social formation within the world capitalist system." The paradox of the strong state is that the "overdeveloped" superstructure does not indicate a meaningful autonomy, but rather reflects the state's inability to control its own destiny within the world-economy.

In the peripheral case, global weakness and the resulting poverty of the mass of the population means that state resources are concentrated upon maintaining order. Heeger (1974, 1) has succinctly summarized this position:

> Politics in underdeveloped societies has become preeminently a politics in search of order. Development, an intangible concept at best, has proved to be an elusive goal. Order, in contrast, is both more tangible, and, so it seems. more necessary.

O'Brien (1979) has shown how the "ideal" of political modernization has shifted away from representative democracy toward order and stability.

In our original terms, this shift represents the dominance of the politics of power over the politics of support. Shortly after independence, politics was based upon nationalist movements with development goals. The optimism of this politics of support has fallen victim to economic failure, leading to a more cynical politics of power centered on the three classes identified by Alavi. Peripheral states are therefore characterized by a particularly naked politics of power.

### *The myth of the nationalist mass party*

The politics of support, developed in the drive for independence, was based on mass political parties. The class alliance of the colonial state between metropolitan bourgeoisie and rural land owners was challenged by the indigenous bourgeoisie, who built mass parties to spearhead the nationalist movement. Despite the expectations of social scientists at this time (e.g., Wallerstein 1961) mass parties have not played a major role in the subsequent politics of peripheral states. Such parties have either been overturned by military coups or have succumbed to the politics of power and forsaken their mass base. This change could occur so easily because these political parties never had a national base. The myth of the national mass party has been exposed by Heeger's (1974) treatment of the role of parties in these new states.

The initial strategy of adopting a politics of support, although heartening for liberal-democratic theorists, was never as strong as it may have seemed:

> The nationalist movement . . . has been not so much a cohesive mass movement as a collection of movements in a society segmented by region, community, kinship and the pace of social change. In constructing a nationalist movement, nationalist elites have in a sense constructed a fiction, manipulating local protests and groups into concert with nationalist goals that are not necessarily widespread (Heeger 1974, 23).

Hence political parties in the periphery were never the same as the stable institutionalized parties found in core countries. The integrative nature of parties in core countries is replaced in peripheral states by a highly segmented party with segments linked together only by elites and by the party organization (Heeger 1974, 35). It was therefore fairly easy for this kind of organization to be converted from a vehicle of mobilization to a mechanism for allocating resources:

> Parties are seen as congeries of national, regional and local elites linked by a variety of personal relationships. Such relationships act as communication channels for the distribution of patronage and spoils (Heeger 1974, 21).

What emerges is a pattern of patronage, in some ways like old-style U.S. city machine politics. The emphasis on ethnicity in the politics of peripheral states does not reflect some return to a traditional society, but can be seen as an integral part of their modern politics of power. Early dreams of nation building have been destroyed and replaced by ethnic faction building in the competition for power. This is so important that even where ethnic cleavages do not exist, as in Kingston, Jamaica, political factions have had to be "artificially" created (Eyre 1984, 26).

If political parties have degenerated into mere tools for the politics of power, it follows that, contrary to liberal-democratic theory, the nature of peripheral states in which parties continue to flourish is not fundamentally different from peripheral states where military regimes have taken over:

> The almost classic distinction between those states with parties and those without, between those ruled by military or bureaucratic elites, may well be overstated. The various types of regime differ in terms of the institutional context in which elites organize the coalition network after they assume power (Heeger 1974, 31).

We can compare this distinction to that between plurality and proportional representation electoral systems in core states, an institutional difference that does not fundamentally change the nature of the state (Taylor 1984c).

### *The paradox of instability*

Hettne (1980, 173) has written that "Africa has since the mid-1960s been challenging the position of Latin America as the continent of military coups." We could add that Asia is also vying for this position. The paradox is that although social order is the prime concern of the peripheral state, governments have not been successful in achieving it. The politics are unstable as different elites fight for control of the state.

But why should the politics of power not run smoothly as it does in core countries? Heeger (1974, 79) is less helpful on this score; he concludes that "the political elites are themselves the major source of instability." This argument ignores the fact that the very frequency of the political upheavals undermines the legitimacy of the dominant classes. Although the political elites are, of course, the direct cause of the intermittent coups, we still need to understand why dominant classes act this way in peripheral states but not in core states. To answer this question we have to return to the scale of the world-economy.

The politics of all states revolves around the quantity of resources received from the world-economy for distribution within the state (Taylor 1982b, 1985). In core countries resources are abundant enough to allow a viable politics of support. In peripheral states there is uncertainty regarding how far resources will stretch even within the dominant classes. The quantity of resources depends on the state of the world-economy, and especially in a major recession, economic problems put the severest strains on every government, civilian or military. The resulting austerity makes all governments unpopular, but sufficient resources are never available to convert this opposition into a viable long-term politics of support. Hence, instead of heralding new periods of democracy, elections operate to replace one politics of power by another in much the same way as military coups: one set of elites is removed from power and a rival set is installed in government. Because the fundamental economic constraints on the country lie beyond its boundaries, changing the government does not lessen the crisis. Alternating civilian and military governments can occur, but the basic problems persist. We shall call this situation a politics of failure. Ghana has become a classic case of the politics of failure.

### The case of Ghana

Ghana epitomizes many aspects of the politics of peripheral states described above. The territory of the modern state of Ghana was put together by the British between 1844 and 1918. Initially the Gold Coast Colony was Britain's sole interest, but in the "scramble for Africa" Britain moved inland to defeat the Ashanti and take their territory in 1901; the Northern Territories were added in 1911 by annexation, and British Togoland was obtained from a defeated Germany in 1918. These colonial acquisitions produced a territory of many cultural groups, the four largest being the Akan including the traditionally dominant Ashanti, the Ewe from southern Togoland, the Ga and Adangbe from the central coastal area, and the Northern Group of largely Moslem ethnic groups (Osei-Kwame 1980, 84–85).[1] This cultural mixture has become the raw material for factional fighting in modern Ghana (Fig. 11.1).

The modern history of Ghana began with the return of Nkrumah in 1947 to become general secretary of the United Gold Coast Convention (UGCC). He soon broke with this elitist organization to form his mass nationalist party, the Convention People's Party (CPP). This initiated the basic cleavage in Ghanaian politics between Nkrumah on the one hand and his major rival, Busia of UGCC, on the other. The CPP was able to demonstrate its strength in a partial election in 1951 and in the first two full elections in 1954 and 1956 that preceded independence under Nkrumah's leadership in 1957. The 1960 elections ratified a new constitution and elected Nkrumah as president. In 1964 Ghana became a one-party state. From 1964 on there is a continual history of instability: Nkrumah was toppled by a coup in 1966; the military was replaced by a civilian government under Busia after his electoral victory in 1969; in 1972 the military overthrew Busia's government; in 1978 the military's attempt to produce a no-party ("union") government by referendum ultimately failed, and there was another military takeover in 1979; this led to new elections in 1979 producing a "pro-Nkrumah" government to be overthrown again in 1981 by the military.

The original cleavage between Nkrumah and Busia was, of course, more than a personal one. In 1957 Nkrumah represented the national centralizing forces within the country, and Busia came to head an opposition favoring

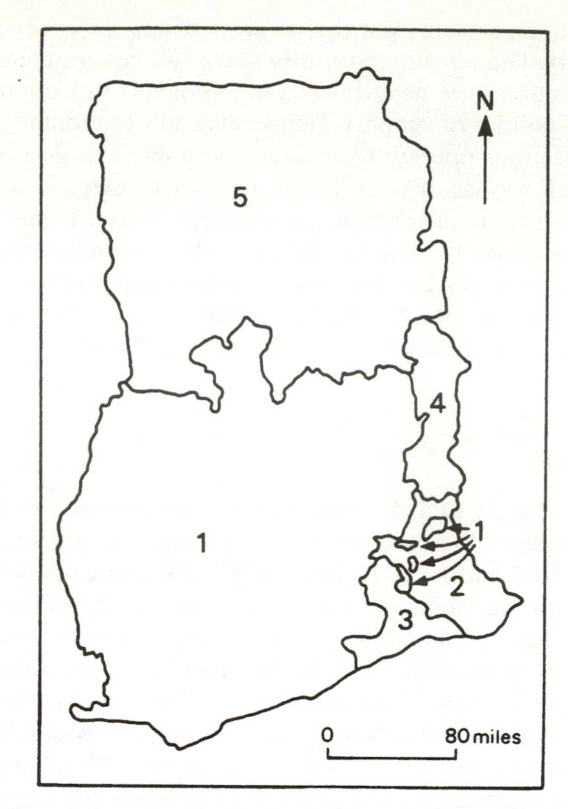


**Fig. 11.1** Major ethnic groups in Ghana: 1. Akan, 2. Ewe, 3. Ga-Adangbe, 4. Central Togo, 5. Northern (Mole-Dagbani)

decentralization in a federated state. These alternative political strategies were themselves based on rival economic interests. In the cocoa-growing areas (notably the Ashanti region) the landed interests supported a liberal trade policy aligned with federalism. They imagined that this would enable them to benefit fully from their cocoa-export earnings and allow them access to foreign industrial goods. In contrast, Nkrumah's government pursued a classic semi-peripheral approach. His policies promoted indigenous industry through protection and through the investment of cocoa-export earnings in the industrial-urban infrastructure.

The Nkrumah–Busia dispute can be directly related, therefore, to the fundamental issue of Ghana's role in the world-economy. In 1961 Nkrumah declared his government to be "socialist". His over-throw in 1966 thus became a defeat for socialism, with the military government returning the country to a market economy, a process continued by Busia after 1969. Busia's liberalism was in turn defeated in 1972. Hettne (1980, 177) suggests, however, that the differences between the two factions should not be overemphasized. The swings from "socialism" to "capitalism" are far less important to the country than was the world market price of cocoa. The brief periods when Ghanaian governments have been economically successful have coincided with good cocoa prices. Hence, according to Hettne (1980, 177) the demise of Nkrumah in 1966 and Busia in 1972 has more to do with the price of cocoa than with either one's political ideology.[2]

Even though both Nkrumah and Busia are now dead, faction building in Ghanaian politics, military and electoral, can be identified in terms of groupings that first emerged a generation ago around these two dominant figures. For want of a generally accepted terminology we shall refer to these two loose groupings as "centralists" and "liberals," respectively. These simple terms hide much complexity in both the politics of power and the politics of support; the rest of this paper consists of an empirical attempt to unravel centralist and liberal politics of support.

## The geography of a politics of failure

In the political geography of Ghana the paradoxes of the strong state and of instability are reflected in the succession of military coups and subsequent military governments. The myth of the mass party, on the other hand, is reflected in the rise and fall of political parties of both centralist and liberal tendencies. The geography of voting is a particularly sensitive index of this paradox. In this section we analyse maps of votes to illustrate the changing factionalism of Ghanaian politics.

### *Methodology*

We use the methodology previously described in Taylor (1981b) and extensively employed in Archer and Taylor (1981) to describe the "normal vote." This involves analyzing a time-series of votes for one party over a set of constant areal units in order to find continuing patterns in the votes; the continuing patterns are identified as normal votes. For instance the normal vote for the Democratic Party in the USA at the beginning of this century involved a three-region pattern of solid Democratic South, anti-Democratic North, and competitive West (Archer and Taylor 1981).

Common factor analysis is used to accomplish this task, with T-mode used to define the major patterns over time and S-mode used to delineate the overall regional structure (Archer and Taylor 1981, 216).[3] The approach requires a data matrix with elections along one axis and areal units along the other. The cells in the matrix give the percentage of votes for the party under analysis at particular elections in particular areas. There are two problems in producing this matrix for the Ghanaian case. First, there are no continuous parties from the first state-wide election in 1954 to the most recent in 1979. Second, the constituencies for elections to the legislature and for recording presidential and referendum results changed in 1969. We consider each problem in turn.

Although party labels have changed continually in Ghana the major contest has remained that between Nkrumah and his followers and Busia and his followers as we have seen. Therefore, despite changes in party names, the people supporting particular parties can be readily identified with either the centralist or liberal groups.[4] Similarly, for the referenda, the campaigners can be identified in the same way.[5] As the centralists of Nkrumah were the first to organize nationally, we analyse votes for this group in what follows.

The following votes are identified as the centralist grouping in Ghanaian elections:

(1) Convention People's Party vote in 1954 parliamentary election,
(2) Convention People's Party vote in 1956 parliamentary election,
(3) Yes vote in 1960 Republican constitutional referendum,
(4) Nkrumah vote in 1960 presidential election,
(5) National Alliance of Liberals vote in 1969 parliamentary election,
(6) Yes vote in 1978 Union Government referendum,
(7) People's National Party vote in 1979 parliamentary election, and
(8) Limann vote in 1979 presidential election.

These eight define the election axis of the data matrix.[6]

For the 1969 election the number of constituencies was increased from 104 to 140. This redistricting means that published election results cannot be compared directly across elections. We have constructed a set of 65 constant units by combining both sets of constituencies and maintaining consistent boundaries; in this way we have election results for the same units from 1954 through to 1979. Both sets of original political boundaries were drawn so that, as far as possible, ethnic groups were not mixed in the same constituency, and we have maintained this criterion in our new aggregations. The redrawn political boundaries are shown on Figure 11.2, and a summary of the old constituencies and the new constant units by regions is given in Table 11.1.[7] Figure 11.2 also incorporates a cartogram in which the 65 areal units are made equal in size. The advantage of using the cartogram for presenting the results

**Table 11.1** Summary of constituencies and constant units[a] for Ghanaian elections, 1954–1979

| *Regions* | *Number of constituencies in the 1954–1964 elections* | *Number of constituencies in the 1969–1979 elections* | *Number of constant units* | *Constant unit numbering as mapped in Figure 11.2* |
|---|---|---|---|---|
| 1 Ashanti | 14 | 22 | 7 | 1–7 |
| 2 Brong–Ahafo | 7 | 13 | 6 | 8–13 |
| 3 Central | 13 | 15 | 9 | 14–22 |
| 4 Eastern | 17 | 22 | 7 | 23–29 |
| 5 Greater Accra | 5 | 9 | 3 | 30–32 |
| 6 Northern | 10 | 14 | 5 | 33–37 |
| 7 Upper | 16 | 16 | 11 | 38–48 |
| 8 Volta | 13 | 16 | 10 | 49–58 |
| 9 Western | 9 | 13 | 7 | 59–65 |
| Total | 104 | 140 | 65 | |

[a] Constant units are combinations of constituencies and have the same boundaries for all elections from 1954 to 1979.
*Source:* Produced by the authors from unpublished material available in the Ministry of Local Government and the Office of the Electoral Commissioner, Accra.

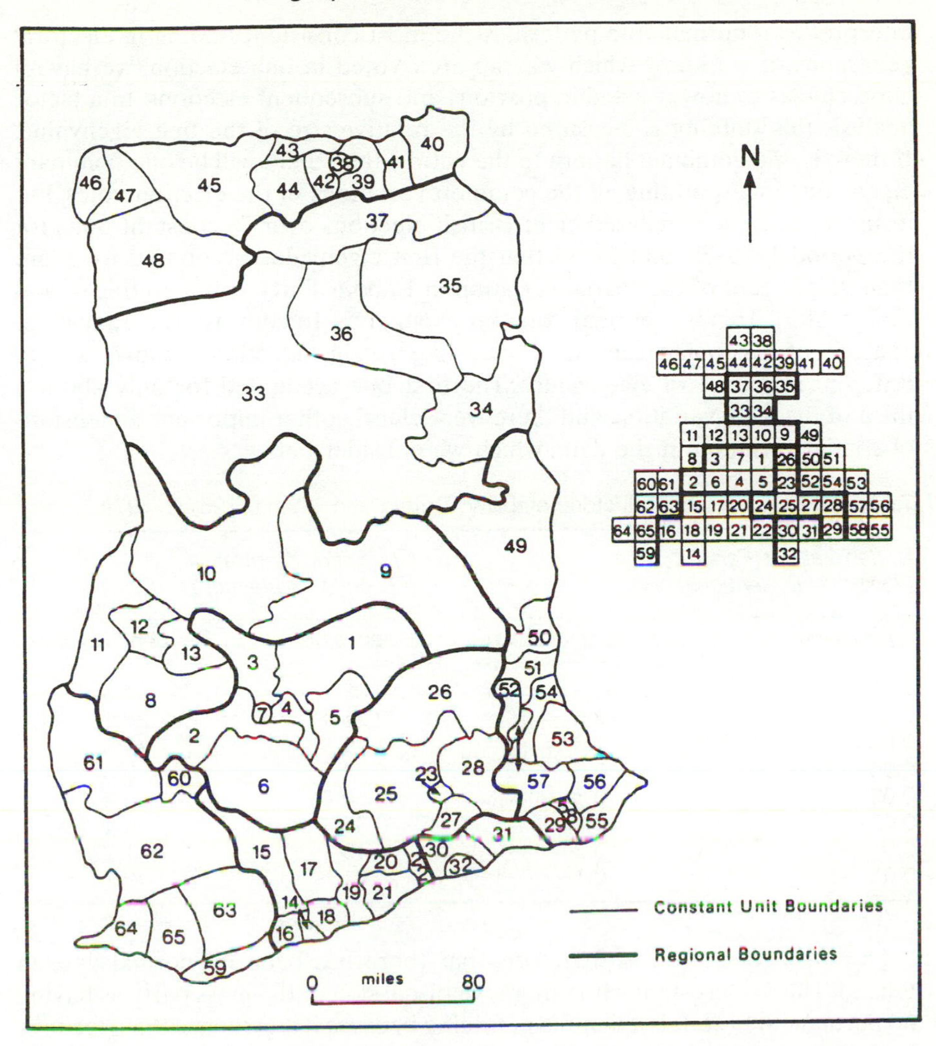


**Fig. 11.2.** Constant units of Ghanaian elections, 1954–1979. Constant units have been constructed by combining constituencies to produce a common area data base for 1954–79. The cartogram locates each constant unit as close as possible to its real geographical location. All constant units have the same areas on the cartogram, which is used to present the subsequent spatial analysis.

of the spatial analysis is that it portrays each areal unit as equal in the same way that the statistical techniques treat them as equal. The end result of this reorganization of the data is an 8 (elections) × 65 (constant areal units) data matrix, which constitutes the input to the analysis reported below.

*Measuring electoral stability*

In other studies of electoral geography, factor analysis has been used to measure the stability of voting patterns over time. This stability, which we interpret as a normal vote pattern, is the most consistent finding in electoral geography: if we know which way an area voted in one election, we have a good clue as to how it voted in previous and subsequent elections. In a factor analysis this stability is measured by the relative size of the first eigenvalue. If there is one dominant pattern to the voting, then there will be one dominant eigenvalue incorporating all the common variations in the election votes. For instance, we factor analyzed eight British elections over 78 constant units for the period 1955–79, and found that the first eigenvalue accounted for more than 95 per cent of the spatial variation in Labour Party voting in this period (Table 11.2). This is a normal vote par excellence. In contrast, the equivalent analysis of Ghanaian centralist votes over eight "elections" shows a very different sequence of eigenvalues. The first one accounted for only about a third of the total variation, and there were clearly other important dimensions of spatial variation in the data, which we consider below.

**Table 11.2** Comparing electoral stability: Britain and Ghana, 1954/5–1979

| *British Labour Party (78 units, 8 elections)* | | *Ghanaian "Centralist" (65 units, 8 elections)* | |
|---|---|---|---|
| *Eigenvalue* | *Per cent variance* | *Eigenvalue* | *Per cent variance* |
| 7.61 | 95.1 | 2.83 | 34.5 |
| 0.20 | 2.5 | 2.25 | 28.1 |
| 0.08 | 1.0 | 1.19 | 14.8 |
| 0.05 | 0.6 | 0.74 | 9.3 |
| 0.02 | 0.3 | 0.50 | 6.2 |
| 0.02 | 0.2 | 0.36 | 4.4 |
| 0.01 | 0.2 | 0.13 | 1.7 |
| 0.01 | 0.1 | 0.01 | 0.1 |

Our first conclusion is, therefore, that there has been no normal vote in Ghana. This is consistent with our above discussion of the mass party as having no permanent base. It is a significant finding because it is evidence that elections in peripheral states are different in kind from those we are familiar with in core states. There is, in fact, one other election analysis that supports our finding. In a study of Congress Party vote in six elections in the Punjab from 1955 to 1977 Dikshit and Sharma (1982) reported that the first two eigenvalues account for 35.2 per cent and 20.5 per cent of the spatial variation of the Congress vote. This confirms the lack of a Congress normal vote in Punjab and is very close to the first two eigenvalues for Ghana in Table 11.2.

We therefore suggest that the most consistent finding of electoral geography, the existence of the normal vote, reflects the ethnocentric bias of previous studies rather than any "general law." There are two types of electoral processes operating, and their spatial manifestations are stability and normal votes in core countries and instability with no normal votes in peripheral countries. Table 11.2 illustrates this contrast. But if there is no normal vote, what does the voting pattern of peripheral states consist of? To answer this we need to delve further into the results of our factor analysis.

*Consecutive mobilizations*

The eigenvectors in the Ghanaian analysis were rotated using an oblique procedure; criteria to determine the number of factors rotated were the same as for Archer and Taylor (1981).[8] The resulting pattern loadings were used for interpretation.

The pattern loadings of four rotated factors are shown in Figure 11.3. What results is some degree of association between elections and referenda that occur close together in time. Hence the only election to load alone on a factor is the 1969 election, which is the most isolated in time. Otherwise the factor analysis has grouped elections and referenda in terms of "time clusters": 1954/56, 1960/60, and 1978/79/79. These are certainly not a sequence of normal votes, but neither are they mere random responses. We see them as reflecting consecutive mobilizations. In the unstable politics of peripheral states, we argue, there is a series of distinct mobilizations, each relating to a new pattern of faction building among the centralist leadership. In short, the politics of support reflect changing patterns in the politics of power. It is a process of alternative faction building and consequent mobilizations, with no transition to the stability of normal votes. Let us discuss each of the mobilizations identified in Figure 11.3 in more detail by using their factor score maps.

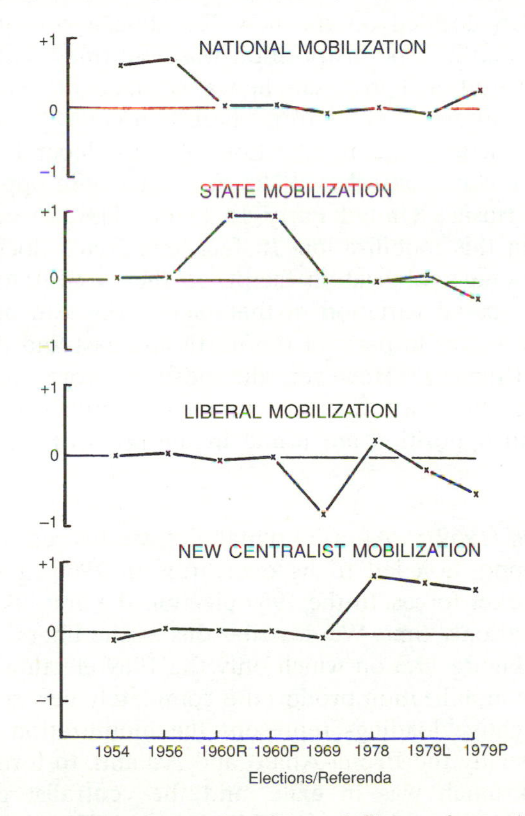


**Fig. 11.3** The changing electoral politics of Ghana: factor loading profiles for T-mode analysis

*National mobilization (1954–56).* This factor describes the pre-independence elections of 1954 and 1956, when the Convention People's Party of Nkrumah was attempting to mobilize the voters behind its leadership. This was the "national mass party" of Ghana's nationalist movement and was opposed by local independent candidates and regional parties. In 1954 its opposition came largely from independent candidates, who numbered 152 out of a total of 316 (Austin 1964, 210). In 1956 the number of independent candidates had dwindled to only 45 out of 194 (Austin 1964, 320) as regional parties emerged, notably the Ashanti-based National Liberation Party, the northern-based Muslim Association Party and Northern People's Party, and the eastern-based Togoland Congress. Hence the national mobilization described by this factor does not have an even pattern of support across the country. Figure 11.4 shows that it is weak in areas mobilized by the regional parties, notably in the north and also in parts of the central Ashanti area and parts of the east. This means that national mobilization was limited to a mass movement to the coastal area, especially in Nkrumah's home area in the southwest. This is, of course, entirely consistent with the myth of the nationalist mass party.

*State mobilization (1960).* After 1956, Busia formed the United Party out of the regional parties but was never able to fight a general election with his new party. Nkrumah decided on the new Republican constitution with a stronger executive, and in 1960 his position was confirmed with a successful Yes vote on the constitutional proposals; he was subsequently elected Ghana's first president. The pattern of votes from these two polls specify the second factor, which we define as "state mobilization." After independence Nkrumah was ensuring central state control over the new country in opposition to the federalist ideas of Busia's United Party. In Figure 11.4 we can see a new pattern of voters in this mobilization. In fact, this factor does correlate at +0.53 with the national mobilization factor, so that it overlaps with about 25 per cent of the spatial variation of that factor. This can be seen in the continuing negative scores in parts of the north and east and the one urban area in Ashanti (Kumasi).[9] However, the positive scores are much less concentrated along the coast and are more widespread if less intense. New areas of urban opposition are found in the ports of Accra and Cape Coast.

*Liberal mobilization (1969).* After Nkrumah defined his centralized state as "socialist," army opposition led to its overthrow in 1966 and the gradual liberalization of market forces. In the 1969 election the anti-Nkrumah forces won for the first and only time. We describe this as the liberal mobilization, the third factor in Figure 11.3 on which only the 1969 election loads. Figure 11.4 shows that this mobilization produced a completely new spatial pattern. The areas with negative loadings represent the mobilization of the Akan cocoa regions, especially the Brong-Ahafo and Ashanti, to form the base of Busia's victory. Nkrumah was in exile, and the centralist grouping was represented by the National Alliance of Liberals with Ewe leadership from the east. This leadership seems to be the sole reason for the new pattern of

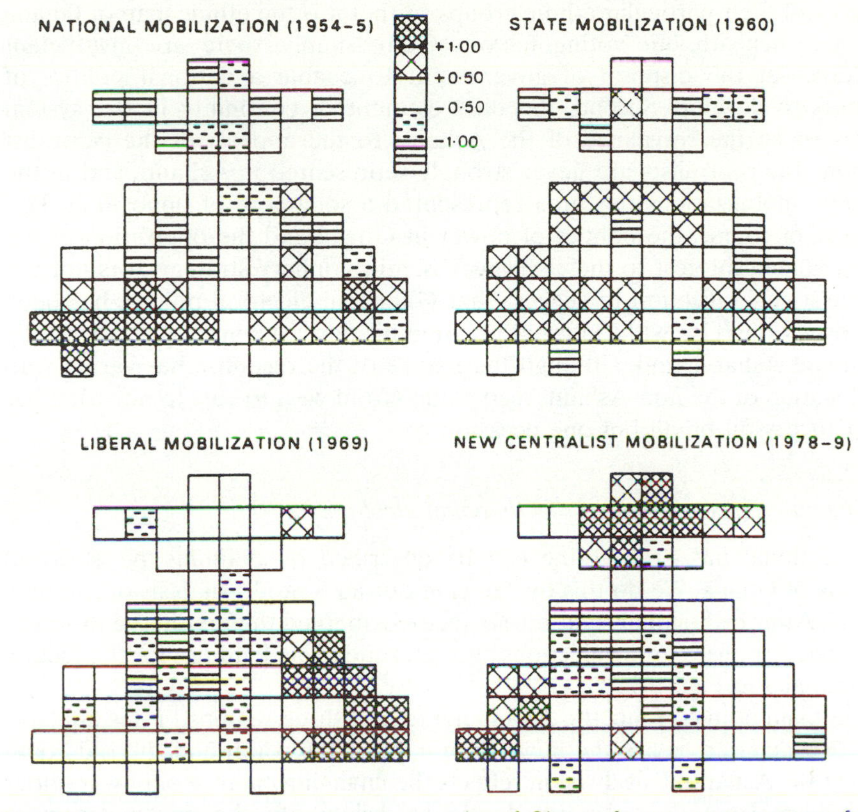


**Fig. 11.4** The changing electoral geography of Ghana: factor score patterns for T-mode analysis

positive scores concentrating in the Volta region.[10] This completely new Akan-versus-Ewe ethnic basis to the pattern reflects the transience of past mobilizations and a new temporary faction leading the centralists.

*New centralist mobilization (1978–79).* Busia's government was toppled in 1972, and the military tried to consolidate control in the "Union Government" referendum of 1978. This centralizing measure was followed by a coup that led to two civilian elections in 1979 for the legislature and the presidency. These three votes load on a single factor, which we term new centralist mobilization. Our designation reflects the return of various elements of the Nkrumah grouping in both the military and civilian governments. Figure 11.4 shows that we have another new pattern; leadership of the faction has moved to the north, and this is reflected in the map.[11] Now we have an Akan-versus-north pattern, with the latter supported by the west. The eastern concentration of 1969 is conspicuous by its absence from this factor.

Two general conclusions can be drawn from this analysis. First, each mobilization is transitory: areas of centralist support shift with the changing leadership of the pro-Nkrumah groups. New leaders bring with them the support of their particular ethnic groups so that it is the ethnic map of Ghana that lies beneath our voting factors. Quite simply, there are insufficient resources at the disposal of government to enable a national politics of support to develop. Second, the only element of continuity in the system seems to be the resistance of the Ashanti to the appeals of the centralist faction. The centralists are never strongly represented in Ashanti, and in the last two mobilizations this area represented a solid core of opposition. This reflects, of course, the politics of power in Ghana and the opposition of the cocoa landed interest to the centralists' semi-periphery strategy. It is not too simple a generalization to suggest that Ghanaian electoral politics has been concerned largely with alternative mobilizations against the potentially dominant Ashanti, and, although the nature of the response has varied with the location of the non-Ashanti leadership (from west to east to north), it has been successful on all but one occasion.

### *Delineating the sectionalism of Ghanaian elections*

The sectional basis for voting can be described by defining the electoral regions of Ghana. We do this by carrying out an S-mode analysis of the data matrix. After orthogonal rotation, six factors emerged that define six different sectional responses to the elections and referenda. The loadings for the factors are mapped in Figure 11.5.

The main ethnic group, the Akan, divide into three regions in this analysis. The two main areas are the new Akan core of the south and the old Akan core of the Ashanti. This division reflects the changing balance of power under colonialism from the formerly dominant Ashanti to the newly dominant coastal Akan. It emerged in the original electoral cleavage between Nkrumah and Busia. Areas outside these two cores, to the north of the Ashanti and to the east, form an Akan periphery. Three other peripheries are identified, each reflecting different ethnic areas: the Ewe periphery in the east, the Upper periphery in the far north, and a small northern periphery in the Northern Region.

The new Akan core, Ewe periphery, and Upper periphery each reflect alternative mobilizations of the centralist factions in the first, third, and fourth T-mode factors. The old Akan core, Northern periphery, and, to a lesser extent, Akan periphery represent areas never fully mobilized by the centralist faction. The different sectionalisms do not collapse into just two regions because the "centralist" versus "liberal" conflict has never been a nationwide contest except perhaps in 1960. Instead, we have alternative faction-building processes among both the centralists and the liberals, producing the more complicated sectionalism reported here.

## Concluding comments

We have set this study firmly within a world-economy framework because we feel that this is the only perspective within which to achieve a realistic understanding of the politics of peripheral states. Our case study has

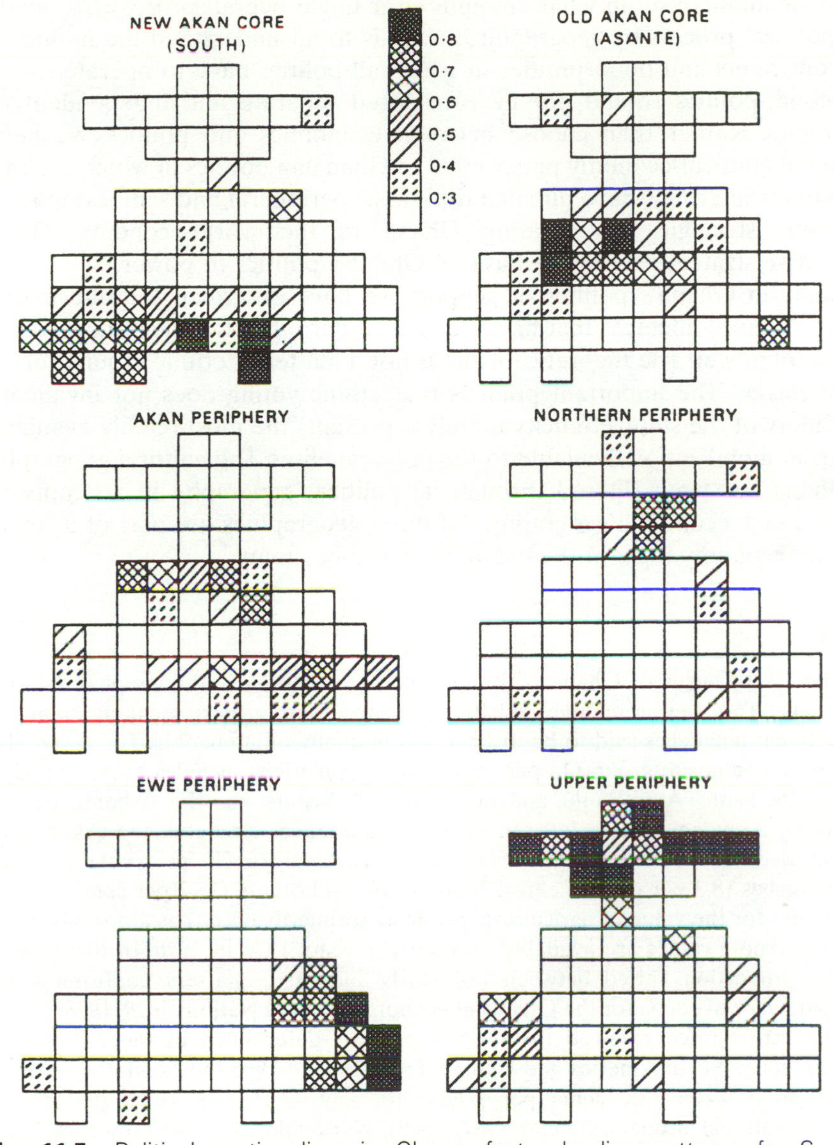


**Fig. 11.5** Political sectionalism in Ghana: factor loading patterns for S-mode analysis

illustrated the necessity to overcome three dichotomies hindering advancement in political geography.

First, there is the issue of whether explanations concerning the politics of a country should be based on factors external or internal to that country. The world-economy approach, it is often supposed, is concerned only with external or international factors at the expense of local or national political processes. Our treatment of Ghanaian politics should dispel this belief. It is not a matter of choosing a particular scale of study to emphasize; politics operates at all scales simultaneously in what amounts to a single overall process. To divide this political process by geographical scale is to misunderstand the nature of the constraints and opportunities in which all politics have to operate.

Second, politics should not be considered separate and independent of economics. Rather than choose between economics and politics, we have applied a political-economy perspective to Ghanaian politics in which the two main political groups have pursued the classic peripheral and semi-peripheral economic strategies for relating Ghana to the world-economy. These alternative strategies form the basis of Ghana's politics of power.

Finally in Ghana's politics of support we have described the alternative ethnic electoral alliances that have been used periodically to legitimate these politics of power. The theoretical issue is not a matter of ethnic groups versus social classes. The important point is that ethnic voting does not invalidate our theory of the state; ethnicity merely represents the most readily available means of mobilization available to the political elites. The cultural geography of Ghana has been filtered through its political geography in attempts to create a new economic geography. All three geographies are part of a single political-economy logic in the capitalist world-economy.

## Notes

1 The 1960 Census of Ghana is the only one that contains ethnographic data on Ghana. This census enumerated 92 disparate ethnic groups speaking some 34 different languages said to be distinct and mutually unintelligible. The Akan, the largest composite group (44 per cent of the population) includes those speaking the Twi-Fante, Anyi-Baule, and Guan dialect clusters, i.e., the Ashanti, Brong, Akim, Akuapim, Ahafo, etc. In terms of numerical strengths the Akans are followed by the Mole Dagbani (16 per cent), the Ewe (13 per cent), the Ga-Adangbes (8.3 per cent), Central Togo groups and others (18.7 per cent).

2 Figures for the value of agricultural products (primarily cocoa) as a proportion of total export values are available only for the years 1970 to 1978. In these years this proportion varied between two-thirds and three-quarters, confirming the dominance of cocoa for the Ghanaian economy (United Nations 1977, 1978, 1982). The actual price received for these exports is determined in the commodity markets of North America and Europe. From 1958 to 1965 the London spot price fell from 44.06 U.S. cents per pound to only 18.15 U.S. cents per pound (International Monetary Fund 1983, 88–9). Nkrumah fell from power in 1966. From 1969 to 1971 the London spot price fell from 45.22 U.S. cents per pound to 25.65 U.S. cents per pound (International Monetary Fund 1983, 88–9). Busia fell in 1972. All government economic planning in Ghana is ultimately dependent on the price of cocoa. Development plans are drawn up in expectation of a certain income from cocoa, but have to be periodically revised as prices fluctuate. Low prices mean harsh austerity measures and social unrest.

3 In a T-mode analysis the columns of the data matrix are elections and the rows are places; hence the factor analysis is based on correlations between elections. In an S-mode analysis the columns are places and the rows are elections; in this case the factor analysis is based on correlations between places. T-mode analysis investigates similarities between elections, S-mode analysis investigates similarities between places.

4 The centralist ideology of Nkrumah's CPP was pursued by the National Alliance of Liberals (1969) and the People's National Party (1979). The following military regimes (1972–79) have also followed the centralist tradition: The National Redemption Council, The Supreme Military Council, The Armed Forces Revolutionary Council. The present military regime of J. J. Rawlings – the Provisional National Defense Committee – is also centralist. Busia's liberalist ideology, first pursued by the UGCC, has also been resurrected several times as the National Liberation Movement, Moslem Association Party, Northern People's Party, the Togoland Congress (all of 1954–56), the United Party (1956), the Progress Party (1969–72), the Popular Front Party (1979) and the United National Convention (1979). The military regime that overthrew Nkrumah's CPP in 1966 also followed the liberalist tendency (1966–69). Thus the politics of Ghana continues to alternate between centralism (Nkrumahism) and liberalism (Busiaism).

5 The Yes vote in the 1960 Republican referendum, the Nkrumah vote in the 1960 Presidential election, and the Yes vote in the 1978 Union Government (no-party) referendum reflected the centralist position, whereas the No votes in those two referenda and the vote for Dankwa in the 1960 Presidential election reflected the liberal position.

6 We have deliberately omitted three Ghanaian elections from our analysis. The 1951 parliamentary election and 1956 Togoland plebiscite had to be deleted because they were not countrywide polls. The 1964 referendum on a one-party state for Ghana was omitted because, even though it was a countrywide poll, its results were reported only on a regional basis. As we needed constituency votes for our analysis, this election could not be used.

7 With the uneven growth of population across the country, some constituencies remained unaltered in both size and name whereas others (especially the urban ones) were divided and redivided and given new names. For instance, Kumasi North and Kumasi South constituencies of 1954–60 became Subin, Bantama, Manhyia, and Asokwa in 1969–79. The strategy we adopted in such cases was to treat Kumasi's two constituencies of 1954–60 and its four constituencies of 1969–79 as one constant unit. This aggregation was accomplished after careful consultation of various boundary demarcation documents obtained from the Ministry of Local Government and the Office of the Electoral Commissioner, Accra.

8 We used the oblique rotation procedure of SPSS called direct oblimin (Kim and Mueller 1978, 61). The default value for delta (zero) was employed in the light of experiments reported by Archer and Taylor (1981, 227). To determine the number of factors, the number being rotated, $k$, was increased until production of a factor with no loadings above 0.5. The ($k$-1) rotated solution was then chosen for interpretation. This method is justified by Archer and Taylor (1981, 228).

9 Kumasi is Constant Unit 7 in the cartogram in Figure 11.2. It is undoubtedly the nerve center of the anti-centralist forces in Ghana. Austin (1964, 1976) and Austin and Luckham (1975) have given a detailed account of the origins and operational strategies of the anti-centralist forces in Kumasi.

10 The new leader of the centralists at this time was an Ewe, K. A. Gbedemah, a one-time Finance Minister in Nkrumah's CPP cabinet. Gbedemah had been thrown out of the CPP a few years before the CPP itself was overthrown in the 1966 coup. Having lost some support from his own centralist camp, Gbedemah could not lead the resurrected centralist party (NAL) to victory. He got a good following, however, in his Ewe homeland in the Volta Region. Gbedemah's own home constituency is Keta in Constant Unit 55 (Fig. 11.2).

11 The new leader of the centralists, Hilla Limann, from Tumu district in our Constant Unit 45 (Fig. 11.2), successfully led the centralists to power again in 1979. The liberalists were defeated mainly because they split into two parties after a bitter leadership crisis following the death of their leader, K. A. Busia.

## References


Alavi, H. 1979: The state in post-colonial societies. In *Politics and state in the Third World*. ed. H. Goulbourne, pp. 40–59. London: Macmillan.

Archer, J. C. and Taylor, P. J. 1981: *Section and party: a political geography of American presidential elections*. Chichester, U.K.: Wiley.

Austin, D. 1964:. *Politics in Ghana 1946–1960*. London: Oxford University Press.

Austin, D. 1976: *Ghana observed*. Manchester: University Press.

Austin, D. and Luckham, R., eds 1975: *Politicians and soldiers in Ghana 1966–1972*. London: Cass.

Chase-Dunn, C. K. 1982: Socialist states in the capitalist world-economy. In *Socialist states in the world-system*, ed. C. K. Chase-Dunn, pp. 21–55. Beverly Hills, Calif.: Sage.

Dikshit, R. D. and Sharma, J. C. 1982: Electoral performance of the Congress Party in Punjab (1952–1977): an ecological analysis. *Transactions, Institute of Indian Geographers* 4: 1–15.

Eyre, L. A. 1984: Political violence and urban geography in Kingston, Jamaica. *Geographical Review* 74: 24–37.

Gamble, A. 1974: *The conservative nation*. London: Routledge and Kegan Paul.

Hamilton, N. 1982: *The limits of state autonomy: post-revolutionary Mexico*. Princeton, N.J.: Princeton University Press.

Heeger, C. A. 1974: *The politics of underdevelopment*. London: Macmillan.

Hettne, B. 1980: Soldiers and politics: the case of Ghana. *Journal of Peace Research* 17: 173–93.

International Monetary Fund. 1983: *International financial statistical yearbook 1983*. Washington: IMF.

Kim, J. and Mueller, C. W. 1978: *Factor analysis*. Beverly Hills, Calif.: Sage.

O'Brien, D. C. 1979: Modernization, order, and the erosion of a democratic ideal. In *Development theory: four critical studies*, ed. D. Lehmann, pp. 49–76. London: Cass.

Osei-Kwame, P. 1980: *A new conceptual model for the study of political integration in Africa*. Washington: University Press of America.

Schattschneider, E. E. 1960: *The semi-sovereign people*. Hinsdale, Ill.: Drydon.

Taylor, P. J. 1981a: Political geography and the world-economy. In *Political studies from spatial perspectives*, ed. A. D. Burnett and P. J. Taylor, pp. 157–74. Chichester, U.K.: Wiley.

Taylor, P. J. 1981b: Factor analysis in geographical research. In *European Progress in Spatial Analysis*, ed. R. J. Bennett, pp. 251–67. London: Pion.

Taylor, P. J. 1982a: A materialist framework for political geography. *Transactions, Institute of British Geographers* NS7: 15–34.

Taylor, P. J. 1982b: The changing political map. In *The changing geography of the United Kingdom*, ed. R. J. Johnston and J. C. Doornkamp, pp. 275–90. London: Methuen.

Taylor, P. J. 1984a: Accumulation, legitimation and the electoral geographies within liberal democracies. In *Political geography: recent advances and future directions*, eds. P. J. Taylor and J. W. House, pp. 117–32. London: Croom Helm.

Taylor, P. J. 1984b: The geography of elections. In *Progress in Political Geography*, ed. M. Pacione. London: Croom Helm.

Taylor, P. J. 1984c: The political geography of electoral reform. *Geographical Journal* 150.

Taylor, P. J. 1985: *Political geography: world-economy, nation-state and locality*. London: Longmans.

United Nations. 1977: *1976 Yearbook of International Trade Statistics*. New York: UN Statistical Office.

United Nations. 1978: *1977 Yearbook of International Trade Statistics*. New York: UN Statistical Office.

United Nations. 1982: *1981 Yearbook of International Trade Statistics*. New York: UN Statistical Office.

Wallerstein, I. 1961: *Africa: the politics of independence*. New York: Random House.

Wallerstein, I. 1974: *The modern world system*. New York: Academic Press.

Wallerstein, I. 1979: *The capitalist world economy*. Cambridge: University Press.

Young, C. 1982: *Ideology and development in Africa*. New Haven: Yale University Press.

# 12 Paul Routledge, 'Putting Politics in its Place. Baliapal, India, as a Terrain of Resistance'

Reprinted in full from: *Political Geography* 11(6), 588–611 (1992)

## Introduction

Contemporary research into the nature and character of social movements has been conducted within two principal approaches: resource mobilization theory (which has addressed structural and organizational processes), and identity-oriented theory (which has addressed political and cultural processes). Neither theory, however, has attempted to analyse social movement agency from the perspective of place – i.e., why particular movements emerge *where* they do. Recent research by geographers has articulated the importance of such a spatial perspective and has theorized models to analyse it. This paper explores the mediation of social movement agency by place within the cultural context of contemporary India. India provides a fascinating area of study given the emergence in the past decade of numerous place-specific social movements. Frequently these movements locate their struggles within what Gramsci (1971) has termed 'civil society'. This paper examines the experience of one contemporary social movement, the Baliapal movement, to exemplify the importance of the spatial perspective, and to explore what I term the 'terrain of resistance'. It also examines the interplay of processes of domination and resistance and the implications of movement agency within civil society. First, an overview is presented of social movement theory and then the importance of place to the study of social movements is considered. Following this, the emergence of social movements in India is discussed and a detailed case-study is presented of the Baliapal movement. This is followed by some concluding observations.

## Theoretical approaches to social movement research

While there are two principal approaches to the study of social movements – resource mobilization theory and identity-oriented theory – there has been a tendency to rely on resource mobilization theory to examine and explain movement agency. The resource mobilization approach (or structural-organizational paradigm) is associated with, among others, the work of Oberschall (1973), Zald and McCarthy (1977), Pickvance (1976a, 1976b, 1977, 1984) and Tilly (1978). It takes as the object of its analysis the collective action of groups with opposed interests, focusing upon the goals, organization and leadership of the movements that result, the resources and opportunities available to the movements, and the strategies employed by them. This perspective is concerned with movement processes over time and is keenly interested in the role of political parties (whether regional, nationalist or communist) in organizing the disaffected and the role of the state as a mechanism of repression.

Guha (1989b) argues that this perspective sees protest as instrumental and oriented towards specific economic and political goals, hence 'success' becomes the gauge by which the significance of protest is measured. As Sharp (1973) has noted, it is not always possible to conclude categorically that a particular social movement has been a clear success or failure. Indeed, elements of both may be present in certain situations. Also there may be longer-term effects of movement agency that are as yet unknown or unquantifiable (such as psychological effects) and struggles may also imbue subtle and indirect political effects, etc.

The other approach to movement agency, identity-oriented theory (or the political–cultural paradigm), is particularly associated with the work of Alain Touraine (1985) and Manuel Castells (1983), and seeks to understand how collective actors strive to create the identities and solidarities that they defend. The approach also attempts to understand how structural and cultural developments within society (such as social relations of power and domination and cultural orientations) contribute to the character and expression of a social movement. The identity-oriented perspective criticizes resource mobilization theory because the latter is viewed as studying social movement strategies as if actors are defined by their goals and not by the social and power relations in which they are situated.

In addition, the identity-oriented perspective accepts the importance of the resource mobilization theory, but argues that crucial to a fuller understanding of resistance are the systems of political legitimacy that exist and the interplay between ideologies of domination and subordination. This analysis emphasizes the expressive dimensions of social protest, especially its cultural and religious idioms, and the nature of local class relations. Hence the significance of a movement is not just in terms of what the movement accomplishes or fails to accomplish, but also in the language in which the social actors express their discontent. Recent proponents of this paradigm are Scott (1985) and Ong (1987), who have studied everyday forms of peasant resistance that stop short of outright collective defiance.

While the identity-orientated perspective addresses those cultural idioms of protest left unexamined by resource mobilization theory, both approaches to social movement agency fail to address adequately the mediation of such agency by place. As Agnew (1987) notes, research problems in political sociology have never been clearly associated with a place perspective, although most popular political movements, such as regionalist and separatist movements, have their origin in specific places or regions.

Interestingly, while much research on social movements is location-specific or recognizes the importance of territoriality in movement practice (Sorokin, 1962; Katznelson, 1981; Hannigan, 1985), questions concerning the effects of locality upon movement action and the reasons why particular movements arise in particular places frequently remain unanswered (e.g., see Della Seta, 1978; Katz and Mayer, 1985; Walton, 1979; Burgess, 1982). As Melucci (1989) has noted, social movement research tends to unify the heterogeneity of collective action by either the use of key analytical concepts such as class struggle or through empirical generalizations. Hence empirical research frequently attempts to extrapolate a movement's locationally-specific experience to a general theory of social movement practice without due consideration of particular spatial and cultural contexts (e.g., Janssen, 1978; Marcelloni, 1979; Burgess, 1982; Lagana *et al.*, 1982; Pickvance, 1985). However, social movements are affected by, and respond to, historical, economic, political, ecological and cultural processes and relations that are themselves place-specific. It is to the importance of this spatial perspective in the study of social movements that this paper will now turn.

**Social movements and place**

The spatial mediation of social structure, social relations and social power has received recent attention by several geographers including Massey (1984), Agnew (1987), Soja (1988) and Harvey (1989). Spatial practices, notes Harvey, 'take on their meanings under specific social relations of class, gender, community, ethnicity or race and get "used up" or "worked over" in the course of social action' (1989: 223). Referring to Lefebvre (1974), Harvey goes on to note (p. 254) that tension and conflict arise within society over the uses of space for individual and social purposes and the domination of space by the state and other forms of class and social power. This conflict can give rise to social movements whose aim is to liberate space by resisting the processes of domination. Such conflicts are grounded in particular places, since place is the arena where social structure and social relations intersect, giving rise to relations of power, domination and resistance.

Foucault (1980) has argued that power is diffused throughout society, no place being free from relations of dominance and subordination. These sites of power also give rise to sites of struggle which are locally based, and which represent a movement's power from within (or power to resist, transform, etc.). This is in contradistinction to the power *over*, wielded by the forces of coercion, domination, etc. Sharp (1980) denotes these sites as loci of power which can range from cultural groups, social classes, villages, etc., and which are often traditional, established formal social groups or institutions. He argues that their status as loci will be determined by their capacity to act independently, to wield effective power and to regulate the effective power of others.

These loci of power/sites of struggle therefore have a strong geographical component; they emerge as Apter and Sawa's 'mobilization space' (1984: 7) and Harvey's 'spaces of resistance' (1989: 213). The study of place, therefore, provides the context where social movement agency challenges the dominating practices of private capital and the state. Place can also determine political identity, political activity and the social context of movement agency and the territorial and cultural settings within which it occurs.

Agnew (1987) recognizes three main constituent elements in the concept of place: locale, location and sense of place. Locale refers to the settings in which everyday social interactions and relations are constituted, whether formal or informal. Location refers to the geographical area encompassing the locale as defined by social, economic and political processes operating at a wider scale – i.e., the impact of the 'macro-order' in a place (uneven development, uneven effects of government policy, etc.). Sense of place refers to the local 'structure of feeling' (Agnew, 1987: 28) or the subjective orientation that can be engendered by living in a place – i.e., geographical and social modes around which human activities take place, such as home, work, school, etc. – which create, *in toto*, a sense of place.

By the analysis of these theoretical components of place, valuable insights can be obtained into the character and nature of social movements, revealing not only why movements emerge, but also why they emerge where they do. Such analysis lends itself particularly to the identity-oriented perspective which attempts to uncover what I term the 'cultural expressions of movement resistance', that place-specific 'language of discontent' (Guha, 1989b: 3) which motivates and informs social movement agency. It also complements those insights into movement organization and strategy provided by resource mobilization theory. Therefore, an understanding of the mediation of social movement agency by place provides us with what I term the 'terrain of resistance' – the specific historical, political, economic, ecological and cultural context of movement agency.

Juxtaposing, but in dialectical relation to, these resistances are the processes of integration and repression by the state (see, for example, Castells, 1977; Burgess, 1982; Leontidou, 1985). In this analysis of Indian social movements, it is argued that the Indian State has responded through a repressive unity of power that spans coercion, mediation and seduction (after Vaneigem, 1983). Coercion refers to the use of both legitimized and subterranean violence by the central and state governments. Mediation is the process whereby the State acts as reconciler of social antagonisms and arbiter of disputes within society.

Seduction is the process whereby the State attempts to co-opt dissent via bribery and rewards such as rehabilitation plans, compensation schemes, etc. This model finds relection in Ranajit Guha's (1989c) model of coercion–collaboration–persuasion and owes much to Gramsci's (1971) model of coercion and consent.

In his theory of state domination and social hegemony, Gramsci made an analytical distinction between civil society (where the influences of ideas, culture, and institutions such as schools and trade unions, operate through consent); political society (constituting state institutions such as the army, police and bureaucracy, which operate through domination); and economic formations (constituting the social relations of production and the technical means of production). Social hegemony is attained by the many ways in which education, religion, the media and other institutions of society join, directly and indirectly, to form a common social–moral language separate from, although interlinked with, the coercive dominance of the ruling class (see Gramsci, 1971; Said, 1979). Hegemony is conceived of as being exercised in the economy (in factories and offices), in the state, in the law, legal process, and state educational institutions, and in civil society, in the mass media, in the arts and in religions (where a regime acquires popular consent through its domination of these non-coercive spheres) (Bocock, 1986).

The concept of hegemony is dynamic, a political process played out between contending ideologies, hence:

> embedded in every ideology that legitimizes domination there is a sub text, a legitimizing ideology of resistance . . . forms of domination structure forms of resistance. While protest normally arises in response to domination and attempts to resist it, most forms of domination actually enable resistance (Guha, 1989b: 98).

Within India the interplay between the processes of domination and resistance are frequently place-specific and located within what Gramsci (1971) terms civil society. In order to examine these processes and to investigate the mediation of place upon movement agency and the cultural expressions of that agency, the experience of the social movement that has emerged in Baliapal, India, will be considered. Before analysing the Baliapal movement, the context in which contemporary social movements have emerged in India will be discussed briefly.

## Social movements in India

Since Independence in 1947, the processes of modernization and development in India have contributed to the economic and political centralization of power in the hands of national and regional ruling elites. Indian society has become polarized between the middle- and upper-class elites (who control and benefit from an increasingly technocratic and militarized state), and the masses (i.e., the rural poor, women, children, *dalits, adivasis*, urban slum- and pavement-dwellers) all of whom provide cheap labor to support the rural and urban economies (R. Kothari, 1986a). Although the struggles of these masses have entered the mainstream of Indian politics, they continue to be economically, politically and culturally disadvantaged (see Selbourne, 1977; Myrdal, 1980; Omvedt, 1980, 1982; Bagchi, 1982; Nandy, 1984; Dutt, 1984; Banerjee, 1984).

Indeed, the process of economic development has led to the pauperization, dislocation and even the destruction of many low-class groups in India (Routledge, 1987). A few recent examples will suffice for elucidation. The Narmada River Valley Project – a present scheme to construct over 160 dams along the river from Gujarat to Madhya Pradesh – will entail the submergence of 375 000 hectares of forest and 80 000 hectares of agricultural land and will necessitate the displacement of over one million (predominantly *adivasi*) people. The Singrauli industrial development scheme in Madhya Pradesh – a series of dams, coal-mining projects and power stations commencing in 1961 – has displaced in excess of 200 000 people (50 000 of whom disappeared, leaving no trace of their whereabouts) and threatens another two million with displacement (Sharma, 1985). The Bhopal disaster has claimed at least 3500 lives and has left at least 300 000 permanently maimed and injured, most of whom have yet to receive any compensation (Visvanathan and Kothari, 1985).

The development process in India reflects the capitalist nature of the contemporary Indian State (Dutt, 1984). The modernizing project 'encodes a structure of domination and violence' (Visvanathan, 1985: 24), a violence that permeates Indian society and is nowhere more naked than when it is manifested as a reflex of power to challenges made against the State. This was epitomized by the Emergency of 1975–77 when virtually all constitutional safeguards (including the right to life, liberty, free speech and protection against arbitrary arrest and detention) were suspended (see Selbourne, 1977). Over the past decade, in what has been termed the 'Backdoor Emergency' (Joshi, 1984: 14) the Indian State has increasingly employed the politics of repression to uphold its control, power and dominance through the introduction of 'Black Laws' (Joshi, 1989: 3) that have institutionalized some of the worst exigencies of the Emergency period (see Joshi, 1985; Kannabiran, 1985; Balagopal, 1988; Roy, 1988).

In response to these political, economic and ecological processes, myriad 'new social movements' (Omvedt, 1985; Majumdar, 1977; Ghosh, 1989; Guha, 1989a) or 'union groups' (Roy, 1983; Sethi, 1984; Dhanagre, 1988) have arisen to challenge the social hegemony and dominating discourse of the Indian State. Historically – i.e., since the First World War – Indian social movements have frequently, but not exclusively, been associated with political parties. Until Independence, many movements were associated with the Indian National Congress – more so than left parties. Since then, competing parties (particularly those of the communist left) have attempted to win electoral support by championing people's movements (see, for example, Lynch, 1969, 1974; Gough, 1974; Wiebe, 1975; Ghose, 1975; Oommen, 1975, 1977; Omvedt, 1980, 1982; Dhanagre, 1983a; and Banerjee, 1984). However, because political parties have cynically manipulated people's movements for their own purposes, many have become wary of such support. Hence the new social movements are frequently autonomous of political parties although some have formed alliances with voluntary organizations and non-governmental organizations (Kumar, 1984; Fernandes, 1984). They are also often locally-based and single-issue-oriented and represent attempts to create political alternatives outside the purview of organized electoral politics. Examples of such movements include ecological struggles such as the Chipko movement (see Berreman, 1987; Guha, 1989b); anti-nuclear struggles such as the Kaiga movement (see Alvares, 1988); indigenous and economic rights struggles such as the Jharkhand movement (see Corbridge, 1987, 1988; Parajuli, 1990); and the anti-eviction, anti-missile-base struggle of the Baliapal movement. It is to this movement that I shall now turn.

## The Baliapal movement as a terrain of resistance

Baliapal, located in the north of the state of Orissa on India's Bay of Bengal coast, has become the site of an ongoing struggle between the resident farmers and fisherfolk and the central government and military establishment of India. The emergence and character of the Baliapal movement is a result of a variety of place-specific factors that can be elucidated by a consideration of location, locale and sense of place.

### *Location: site and militarization*

The importance of location was paramount in the government's decision, in October 1984, to choose Baliapal as the appropriate site for the National Testing Range (hereafter NTR) – a base for the testing and launching of rockets, satellites and ballistic missiles with a range of up to 5000 km. Although other sites were considered, Baliapal was chosen for several reasons:

1. it was away from major towns, ports and population centers;
2. it did not affect air and sea routes;
3. it was free from undulations, being very flat;
4. it was cyclone free with over 200 clear days a year;
5. the sea bed is shallow, hence enabling easy missile recovery; and
6. the east coast of India is out of the range of Pakistani radar surveillance (PUDR, 1988).

The final approval of the site took place on 21 May 1986; the base would cost an estimated Rs 3000 crores (US$ 2310 m); it would cover an area of 160 sq. km. and necessitate the eviction of approximately 100 000 people from 130 villages. Because of popular resistance to the decision (due to Baliapal containing some of the most fertile agricultural land in India and being the most densely populated area of Orissa), the central government announced in August 1986 that the NTR would be reduced in size to 102 sq. km. (costing US$ 840 m) and would only affect 45 000 people in 55 villages (AIFOFDR, 1986).

According to Brigadier R. S. Kannan, Area Commander for the entire Balasore district of Orissa, the NTR is part of a wider integrated military system that is being developed in Orissa (AIFOFDR, 1986). This includes:

1. Asia's largest radar observation and ground-control station at Nilgiri;
2. air-force bases at Charbatia and Rasgobindpur;
3. naval bases at Chilka Lake and Gopalpur;
4. an ammunitions industry at Saintala; and
5. a MIG fighter assembly plant at Sunabeda (see Figure 12.1).

Therefore, Baliapal was not only locationally important as a site for the NTR, it was also locationally important within the wider spatial context of the militarization of Orissa.

Increasingly, therefore, the local economy of Baliapal would be tailored to service the needs of military personnel located there, leading to price increases, an influx of hitherto unnecessary consumer products, alien military social values, and the inexorable movement from civilian to military rule (AIFOFDR, 1986). Indeed, the Defence Ministry has acknowledged that 400 sq. km. will eventually be required for the NTR, incorporating 350 km of the Orissa coast from Baliapal to Chilka Lake (Sinha, 1987).

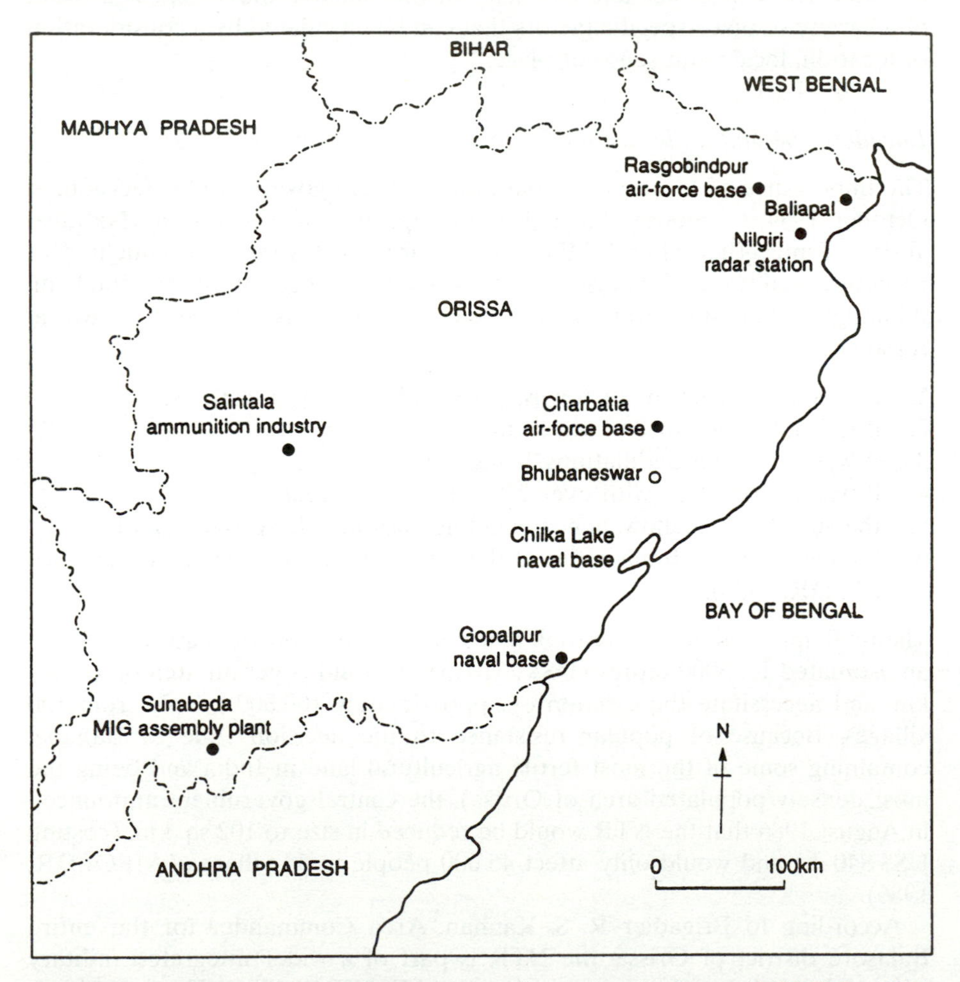


**Fig. 12.1** Location of Baliapal and other military sites in Orissa (AIFOFDR, 1986)

Faced with the threat of eviction from their homes and lands, a people's movement arose in the Baliapal area to resist the location and construction of the NTR. Crucial to understanding this movement are factors of location and locale as they pertain to the local economy, and locale and sense of place as they pertain to the character of the movement and its motivational force.

*Location and locale: the local economy*

Orissa has virtually no industrial development (except for some dams and the previously mentioned military projects) and has received little state intervention in the economy to address backwardness (Interview; Delhi, 1989). The location of Baliapal in an area of unusually high soil fertility (with access to rich river and sea resources) has led to the development of a very prosperous local economy – an 'oasis of development' in an otherwise very poor state. Indeed, because of its locational attributes, Baliapal is known as the 'Granary of Orissa' and it was in defense of these place-particular advantages that the movement was first mobilized. A consideration of the economic specifics of Baliapal's locale provides us with one of the important motivations behind the emergence of the movement.

Baliapal consists of a predominantly homestead economy. The area produces a variety of major cash crops such as *padi* (rice), groundnuts, cashews and coconuts as well as bananas, papaya, mangoes, lemons, grapefruits, onions, aubergines and other vegetables. Of the approximately 100 000 population, 60 percent are indigenous to the area and 40 percent have migrated into the area from other parts of Orissa, West Bengal and Andhra Pradesh. The majority of the population (66 percent) are poor and middle peasants (with landholdings of 1–5 acres) and landless peasants (8.5 percent). The rest are richer peasants with holdings of 5–10 acres (16.5 percent) and wealthy peasants with over 10 acres (8.5 percent) (Interviews; Baliapal, 1990).

Until the late 19th century, the main cultivating castes in this area were the Khandayats and Rajus. They were joined in the 20th century by migrant castes from Bengal, the Barajias and Golas. Today, the upper strata of Baliapal society are dominated by Rajus, Golas and some Barajias as well as a section of Khandayats (proportionately the largest of all the cultivating castes in this area). While it is rare to find a poorer class Raju or a Gola, the Khandayats are economically more differentiated (Patel, 1989a). The non-cultivating castes were divided into those who were employed in fishing (such as the Jhali and Khejali) and the artisan castes (such as the Dhobi and Gopals). Currently, 17 percent of Baliapal's population are Scheduled Castes and almost 5 percent are Scheduled Tribes, predominantly Santhal (PUDR, 1988).

There have been two developments in this area during the past 20 years that have caused major changes in the local economy; namely, the commercialization of fish and existing crops such as rice and coconut; and the introduction of new cash crops such as peanut, cashew and, most importantly, betel vine (*paan*). The commercialization of agriculture gave immediate benefits to certain sections of the cultivating peasantry who intensified the use of land for agriculture by introducing lucrative second crops such as groundnut (Patel, 1989a).

The benefits of these agricultural changes initially accrued to the upper sections of the peasantry who used their access to land to start second crops. However, the most important changes occurred with the introduction and extensive cultivation of *paan* and cashewnut cultivation. *Paan* cultivation was introduced by the Barajia caste who brought it into this area from Bengal during the early 20th century. Its production gradually expanded over the following decades as Banares *paan* traders established brokers in this area to buy the *paan*. This *paan* is known locally as the *Jaggi paan* and was marketed as *Banarasi patta* after it reached Banares.

About 20 years ago, this chain of supply was severed when *paan* producers from Baliapal established their own networks of supply of leaves to Delhi, Agra and, later, to Bombay. Simultaneously, the quality *paan* market slowly shifted from the north Indian cities to Bombay which began to demand bigger and better quality green leaves. The Baliapal producers-turned-traders, therefore, tried to keep the Bombay market supplied with this quality *paan*. They discovered that the best quality *paan* could be produced 500 m from the coast, in the sand dunes skirting the coastline of the Baliapal block. Once cultivated, this variety started to command high prices. This led to the encroachment of this tract by the cultivating castes and by the fisherpeople who started experimenting with *paan* cultivation on their homestead land or on encroached land. All were highly successful in their attempts at paan cultivation (Patel, 1989a).

Baliapal now produces 70–80 percent of the national market supply of the *Banarasi* variety of *paan* (Patel, 1989b). *Paan* is intensively cultivated on small tracts of land in bamboo enclosures called *barajas* where the betel vine grows in rows. A *baraja* can be as small as 3 decimals (1 decimal = 0.01 acre) and as large as 16 decimals. Yearly incomes from *paan* cultivation range from US$ 1200 to US$ 5300, depending upon the size of the *baraja*. Most of the families who own land, irrespective of class, indulge in some betel cultivation since it is so lucrative. The cultivators who own small amounts of land use exclusively family labor while the larger land-owners employ agricultural laborers. An estimated 30 000 *paan barajas* exist in Baliapal (Patel, 1989a, 1989b).

The other lucrative occupation is fishing – given the plentiful supply of fish from the rivers and coast, and the high demand for fish from neighboring West Bengal. Most homesteads own a fishpond where the family farms fish, while some of the landless families (many of whom are migrants) fish in the Subaranekha river or the Bay of Bengal. Those landless peasants who do not fish get yearly occupation in the *paan barajas* (7 months) and on the rice *padis* (5 months), and can earn wages of up to Rs 30 per day (US$ 2), approximately three times the national average wage of agricultural laborers (Patel, 1989a).

The wealth of the local economy, through *paan*, fishing and cash crops, has provided a degree of economic prosperity for all. All castes and classes have an (albeit unequal) economic stake in the area, in its land, sea and resources, and hence when threatened by the location of the NTR they have mobilized together to resist it, a process relatively uncommon in contemporary Indian politics. Interestingly, the participation of peasants in the Baliapal movement has been focused against the NTR because it threatens to destroy their land and displace them from their homes, rather than because the NTR represents an intensification of the militarization of Orissa. Indeed, much of the early resistance to the missile base argued for a relocation of the NTR to an area less rich in resources, less populated and less beautiful (than Baliapal). In this sense, the movement was clearly place-specific rather than issue-specific.

*Locale: social relations and movement structure*

While class, caste and gender inequalities exist within the Baliapal area and are reflected in the dominance of upper-caste and wealthy males within the social structure of the community, some of the class and caste antagonisms that exist elsewhere in India have been somewhat ameliorated by the high percentage of the population who own land and the (albeit unequal) economic prosperity enjoyed by most of the peasants. The Baliapal movement owed much of its strength to the collective sense of community and economic well-being that was shared by most of the villagers and that transcended caste, class and gender divisions.

The emergence of the resistance owed much to the work of two men, Gadhagar Giri and Gannanath Patra. Giri, a local Janata Party member and longtime resident, decided upon the use of non-violent sanctions and his mass appeal enabled him to mobilize all sections of the population and convert them into a community. Patra (a member of the Maoist Unity Committee of Communist Revolutionaries [Marxist–Leninist]) provided the organizational structure to the movement, mobilizing the lower classes into four fronts (of students, women, youth and fisherfolk). Together the two men integrated all sections of the population without disturbing the class, caste, gender and political divisions that exist within the local society. They were able to mobilize what Chatterjee, in his research on peasant mobilizations in Bengal, has termed the 'communal mode of political power' (Chatterjee, 1982: 12) whereby individual and sectional identities and rights are derived by virtue of membership of the community (Chatterjee, 1983: 317). When confronted by an external oppressive force which is perceptually distant from the everyday life of its members, the community acts as a collective whole. In the case of Baliapal, the entire community was threatened with displacement by the national security interests of the Indian State and therefore collectively resisted the threat irrespective of the class and caste differences of its members.

Chatterjee (1982) also notes that when communities act collectively there is often a correspondence between the leadership of the community and those who have risen to economically superior positions. In Baliapal, the Resistance Committee that was formed to lead the agitation comprised upper caste, male, educated *paan* traders and *panchayat* (village council) leaders (Interviews; Delhi, 1989). This reflected their class dominance within local society because:

1. as the wealthiest peasants and largest landowners, they had the most to lose by the location of the NTR;
2. unlike the poorer peasants, they had spare time and resources to attend meetings and organize;
3. being well-educated, they possessed advantages regarding movement organizing (e.g. writing press releases, liaising with government officials, etc.); and
4. they were already the primary representatives of the communities in their roles as *panchayat* leaders.

Indeed, their role in leading the agitation was accepted by the villagers since they were representing the interests of the entire community in its resistance to eviction.

This was further compounded by the revival of the *Vichar* institution whereby discussions concerning village problems (such as land disputes) find solution in a consensus process without recourse to the law-enforcement and judicial establishment (Patel, 1989a, 1989b). Giri revived the institution since land (ownership, rights, etc.) had been the one issue of contention and dispute which could have divided the movement (Patel, 1990). Eight of the Committee members conduct the *Vichar* sessions in the village *panchayats*, the other four acting as arbitrators. The *Vichar* has settled over 400 land disputes (some awaiting resolution for over 30 years), and has seen the area's crime rate fall since 1985 (Interviews; Baliapal, 1990). It also helped provide legitimacy to the leadership of the *paan* traders on the Committee and through them the upper castes. It also provided the context for binding the various caste and class interests of the area into a cohesive community (Patel, 1990).

The Resistance Committee consisted of 13 members who made the decisions concerning the agitation program, strategy and tactics. Below them was a Council of 229 members (including the landless, women, peasants, etc.) who were nominated by the villages in the Baliapal area. Below this Council of 229 were the Village Councils and the four Fronts (Interviews; Baliapal, 1990).

Another important aspect of Baliapal's social relations that contributed to the movement's structure was the position of women within the community. Women do most of the cultivation in the homestead economy as well as the household and reproductive activities. Due to the *paan* and fish economies, women have a strong stake in the land, the sea and their resources. They are intimately tied to the natural environment, their militancy being motivated by their perceived need to protect their right to the soil and to continue their lineage (Interview; Delhi, 1989). As a result, women were amongst the most militant members of the movement.

*Movement tactics*

> We want land, not missile base;
> We want peace, not war.
>
> (Slogan painted on the Kalipada barricade, Baliapal)

The movement's tactics of non-violent resistance have spanned methods of intervention, non-co-operation and protest and persuasion. The catalyst for the movement came in April 1986 when the District Collector (a senior administrative officer who coordinates all central and state government policies in the district) visited Chaumukh village in the Baliapal area to inspect work for the NTR. Upon his arrival he was *gheraoed* (surrounded) by 15 000 people and forced to walk 8 km back to Kalipada village. At Kalipada, the outside limit of the project area, barricades were erected to prevent his re-entry. Later barricades were set up on the other three approach roads to the Baliapal area (see Figure 12.2). Since then, neither the Collector

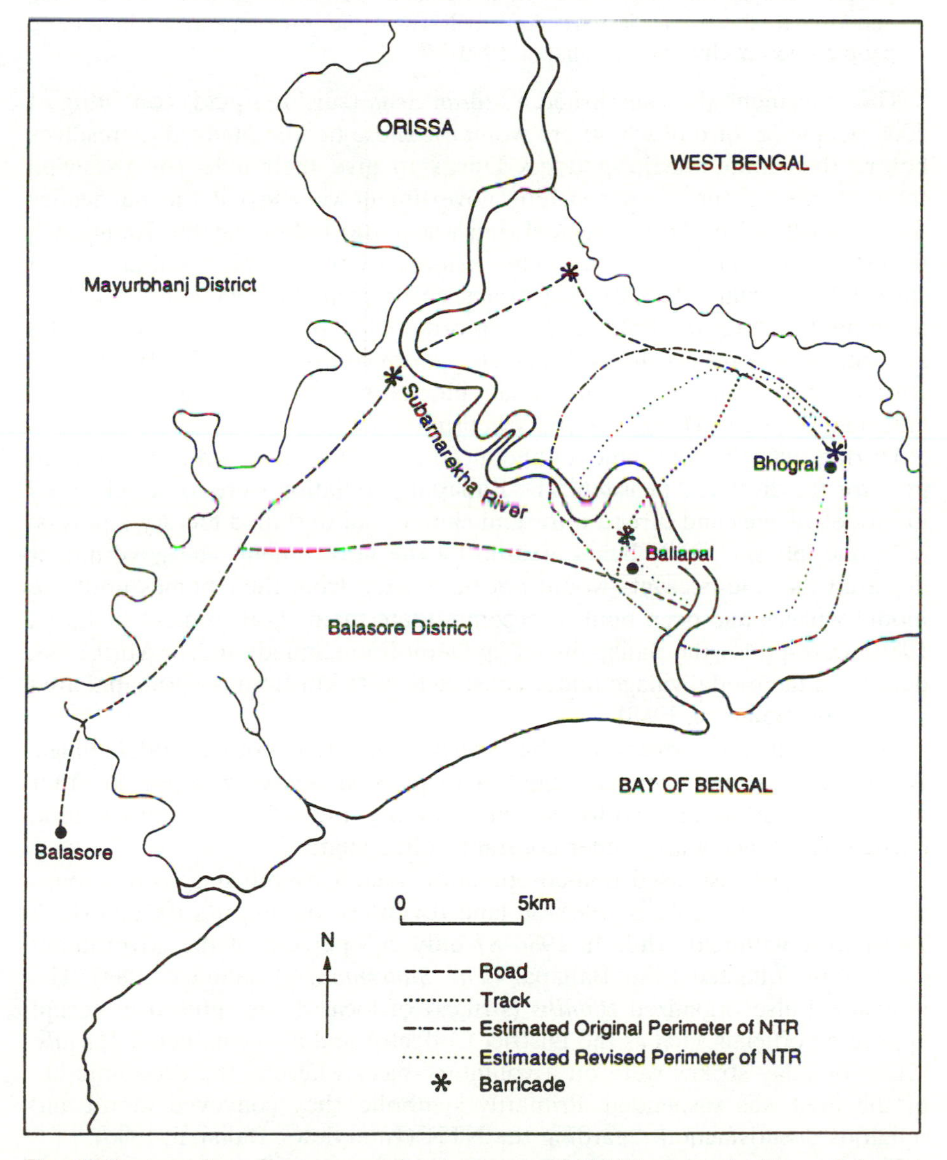


**Fig. 12.2** Location of movement barricades and NTR (Unnayan Archives, Calcutta)

nor other government officials have been allowed into the area (Interviews: Baliapal. 1990). Over the barricade at Kalipada a sign stated in Oriya:

> Land is ours; Sea is ours; Government Officials go back.

People staffed the barricades around the clock. In order to warn people of approaching government vehicles, conch shells were blown and *thalis* (metal plates) were beaten thereby summoning thousands of villagers to the barricades. Once there the villagers lay down in the road forming human road blocks. As one peasant activist told me:

> People work as usual, but if the call comes then we drop whatever we are doing and rush to the barricades. We don't believe in the government, we believe in people's power (Interview: Baliapal, 1990).

The movement also established a *Maran Sena* (suicide squad) consisting of 5000 people (a third of whom are women). These people pledged themselves before the Hindu warrior-goddess Durga to give their lives (by throwing themselves in front of approaching government vehicles) if the barricades were breached. The *Maran Sena* slogan was 'After killing me, the Range will be established on my corpse'. One example of the people's militancy will suffice. In February 1988, 24 magistrates, accompanied by 3000 armed police, attempted to enter the Baliapal area in order to 'explain' to the residents the government's plans for the villagers. They were confronted by 20 000 women, children and men forming a human blockade and were prevented from entering the area (*The Telegraph*; 4 February 1988).

During 1986, the government announced a rehabilitation and compensation plan for the displaced peasants. The Baliapal population were to be relocated in model villages and turned into semi-skilled and unskilled factory laborers; only one job per family being assured by the government. The government declared that the peasants would not be evicted from their homes until the model villages had been built to accommodate them (Government of India, 1986). In response, the Baliapalis set up Demolition Squads and, in April 1988, destroyed the model village under construction 15 km from the Baliapal area (Interview; Balasore, 1988).

Subsequently, the government has nearly completed another model village. A characterless, uniform settlement, with peasant houses grouped together under the shadow of a cotton-spinning factory, the village remains behind barbed-wire fencing and under constant police guard.

The villagers also used non-co-operation tactics by refusing to pay loans and taxes. For example, in 1985–86 land revenues totalling Rs 100 000 (US$ 7800) were withheld, while in 1986–87 only 2–3 percent of the government dues were collected from Baliapal (*The Statesman*, 21 January 1988). The movement also organized *Bandhs* (strikes) of local shops, refused to accept appointed officials such as the District Collector and also conducted *Hartals*. These one-day strikes were on a voluntary basis whereby the economic life of the area was suspended. Primarily symbolic, they conveyed moral and religious dissatisfaction regarding the NTR (Interviews; Baliapal, 1990).

The movement also utilized many methods of protest and persuasion including conducting *Dharnas* ('sitting in protest') outside of the District Collector's office and outside local police stations; holding public speeches and public meetings; and writing letters of opposition to government officials, publishing leaflets, pamphlets and books, and painting wall slogans. The movement also held numerous demonstrations (e.g., in June 1986, 10 000 demonstrated in Balasore, the district capital, demanding the relocation of the base). They wrote and performed songs, music and plays detailing the struggle (PUDR, 1988). At the peak of the struggle (1986–88) approximately 50 000 people were participating in the movement (Interview; Baliapal, 1990).

*Sense of place: motivation and expression of the movement*

The Baliapal movement was informed and motivated by a potent sense of place which refined and strengthened the economic motivation provided by the locale. This sense of place was epitomized by the movement's ideology of '*Bheeta Maati*' (our soil) articulated as 'our soil; our earth; our land'. As one activist remarked to me: 'For Baliapalis the land is our mother; our earth; our home. This is in the hearts of the people' (Interview; Baliapal, 1990). The principal carriers of this ideology were the Baliapali women, mobilized via the Women's Front. As explained earlier, their motivation came from an intimate economic and cultural connection with the land. They staffed the barricades and blew the conch shells to harken the villagers to the barricades and they confronted government officials when they attempted to enter the area. Through the articulation of *Bheeta Maati*, the Baliapal women have drawn upon the Hindu and peasant folk culture of which they are part.

In Hindu tradition, the earth is a purifying agent and remedy for disease and acts as a provider of strength and (agricultural) abundance (Crooke, 1896). Throughout much of India, the land is worshipped as 'Mother Earth'. In Orissa, every village/territory has a goddess who is a variant of Durga (or Kali), the Great goddess. As a village mother she is the particular earth within that village or territory. Hence people are born of the earth; their home land is the earth from which they are born. Those born on the same portion of earth, in the same village, share the same mother, namely the village goddess. As her children they are all one and form a kin-like community, the goddess being the earth of the community (Marglin and Marglin, 1990).

The movement's ideology drew upon the cultural and economic dimensions of the peasant's everyday reality. It converted the cultural sentiment for the land – the peasant's sense of place – into a political demand for the absolute right to its continued use through religious and mythical tools. Stories from the *puranas* (sacred Hindu texts) and the *Mahabharata* (one of the most famous Hindu epics) as well as tales about Hindu deities such as Shiva, Kali and Durga have been intertwined with political action to confront the State.

> Religion to such a community provides an ontology, an epistemology as well as a practical code of ethics, including political ethics. When this community acts politically, the symbolic meaning of particular acts – their signification – must be found in religious terms (Chatterjee, 1982: 31).

Movement ideology has been articulated in songs and dramas that have emerged as a potent expression of the Baliapal resistance. The songs and dramas have been performed locally in the Baliapal villages to empower and motivate people; to provide an ongoing oral history of the movement; to inform people about current developments in the struggle; to teach non-violent methods of resistance; and to increase awareness regarding the agitation (Interviews; Baliapal, 1990). The songs were written by a Baliapali *paan* trader (and movement leader), Purushottam Behera, to console himself after his wife died. When people appreciated his songs and drama he decided to continue writing and the songs became an integral part of the resistance. The songs celebrate the agricultural richness and diversity of Baliapal and the prosperity it has brought to the peasants, and articulate those commonly-held feelings that comprise a Baliapali's sense of place. They also evoke Hindu folklore and belief to provide religious and moral credence to the Baliapalis' claim to the land. A couple of verses from the *Maran Sena Song* will elucidate this point. In the first verse, the people's ancestral right to the land is evoked, to be defended to the death if necessary. In the second verse, the role of women in the struggle is articulated:

> The earth of our birth
> Is our demand
> For this I shall die if I have to
> Deserting this land I shall never go
> This is the path of my fate.
>
> Blow upon your conches
> Mothers, aunts and sisters
> To shake the parliament with your sound
> Destroy the evildoers and their descendants
> In mother Durga's form.
>
> (By Purushottam Behera. Translation by Jana Tathya Kendra, People's Information Centre, part of the Unnayan voluntary organization, Calcutta).

The role of the women is associated with the warrior-goddess Durga who embodies the female attribute of *Shakti* (power, energy and action) and is herself a destroyer of evil. Durga is also evoked as the spiritual guide of the *Maran Sena*. Conch shells (blown to warn the villagers of the government's approach) are found in every Hindu home in Baliapal, being used in salutation and *puja* (prayer) to the Hindu Lord of the Universe, Jagannath, whose major temple is located in Puri, Orissa.

In the following song, the threat to the people of Baliapal (i.e., the missile base) and the source of this threat (the government) are clearly articulated:

> *O Badshahi Sarkar* [O Imperial Government]
>
> O Badshahi Sarkar,
> Don't you dare tamper with our democracy.
> O war-mongers listen,
> Don't ravage this country after buying votes.

O vote-mongers listen,
You say a thousand industries in a thousand days.
In the name of industry, you set up military bases.
Are you not ashamed,
O Badshahi Sarkar?
Listen brothers and sisters
This monstrous missile base will devour Baliapal;
The government will offer hefty prices for your land
Not 'for the spinning mill' but for this inauspicious base.
If you sell your land, you're digging your own grave;
They will pay you for your death summons.
It is not a spinning mill, it is a death factory
It will spin your doom . . .
Don't dig your own graves.
(By Purushottam Behera. Translation by Jana Tathya Kendra).

In this song, the government is portrayed as a colonizer ('O Imperial Government') subverting democracy (ravaging the country after buying votes), and using tactics of subterfuge and seduction to achieve its ends (setting up military bases in the name of industry and offering hefty prices for land). The song also warns the peasants of Baliapal of the consequences of relinquishing their land ('you're digging your own grave') and, in the final line of the song, entreats them to resist.

The emergence of a people's movement in Baliapal and the adoption of a potent variety of non-violent tactics has transformed the *padi* fields and jungles of the area into a terrain of resistance. The emergence and articulation of this terrain is a result of a coalescence of factors. The emergence of the terrain has historical and political roots in the Indian government's national security policy decisions, and their designation of Baliapal as a space dispensable to their military requirements. From the perspective of the movement, the terrain has economic roots in the *paan* and cash-crop economy of Baliapal and the threat to this posed by the NTR; and it has cultural roots in the Baliapali's religious and ancestral sentiment to the land. The articulation of this terrain of resistance has been geographically localized in and around the 130 affected villages of the Baliapal area, and physically expressed most directly through the erection of the barricades, the human road blocks and the *Maran Sena*. The terrain has been articulated socio-politically within civil society where the government's legitimacy to make policy decisions which are antithetical to local community interests has been challenged by a withdrawal of peasant consent. The terrain has been articulated culturally in the form of songs, poems, the *Vichar* process, various religious idioms and the movement's non-violent tactics – those 'little tactics of the habitat' (Foucault, quoted in Soja, 1988: 74). Together, these cultural expressions of resistance provide insights into the spirit of the resistance which has motivated and inspired the peasants of Baliapal.

It is important to note that, although differentiation has been made between various 'topical' terrains, they form a complex and inter-related whole. Baliapal's terrain of resistance – as any terrain of resistance –

comprises an interwoven web of historical, political, cultural, economic, ecological, geographical, social and psychological relationships. Interestingly, the area of Baliapal does not have a history of resistance, unlike other areas of India where particular communities have had their own traditions of resistance continuing over centuries (e.g., the peasant forest movements of Uttarkhand in Uttar Pradesh and the *adivasi* movements of the Chotanagpur area of Bihar). Where this does occur, the history of protest in a particular place is vitally important in understanding the terrain of resistance.

The constituent elements of place (locale, location and sense of place), and their mediation of movement agency, provide important insights into the landscape of struggle and its place-specific character. A consideration of location provides us with the (regional) economic, political, military/strategic and geographical factors that have contributed to the emergence of the Baliapal movement. A consideration of locale provides us with the local economic, ecological and social-relational factors that have contributed to the emergence and structure of the movement. A consideration of the Baliapalis' sense of place provides us with cultural, religious and psychological factors that have shaped the character of the movement. Although we can differentiate analytically between urban and rural, or gender and ethnic terrains of resistance, the complex dynamics of human agency require research into the specifics of movement practice in particular temporal and spatial contexts.

### Government response to the movement

The government's response to the movement has spanned the spectrum of coercion, seduction and mediation. Seduction came in the form of the rehabilitation and compensation scheme designed to persuade the peasants to leave their homes and lands. Coercion took various forms:

1. the government set up an economic blockade of the area, such as preventing kerosene from entering the area, and imposing fines on bullock carts and vehicles leaving the area with produce bound for market (Interview; Cuttack, 1988);
2. notices were air-dropped from helicopters onto the villages threatening the peasants with reprisals if they did not cease the agitation (Nair, 1987);
3. activists were arrested and detained without trial and preventative arrests were made at many of the large demonstrations (PUDR, 1988); and
4. armed police were deployed around the Baliapai area reaching 8000 in number in May 1988 when the state government decided to launch a military assault on Baliapal to evict the peasants (*Indian Express*, 19 May 1988).

The movement was able to defer the attack due to several factors:

1. accounts in the press alerted the movement to the government's plans;
2. the Orissa Administrative Service refused to participate in the co-ordination of the operation;

3. the movement maintained around-the-clock vigils at the four barricades; and
4. mediation talks were initiated between the movement and the state government which ultimately defused the situation (Interviews; Calcutta, Balasore, Baliapal, 1990).

When faced with threats of government coercion the movement responded non-violently and also attempted to stress the cultural and religious ties of the community to the earth. They attempted to show how the common lineage of the villagers (where all people are brothers and sisters in relation to the village goddess) even extended to the police. An anecdote told to me by an activist in Baliapal highlights this point. In December 1986, 45 armed personnel of the Central Reserve Police Force attempted to enter the area:

> They [the police] were armed with halogen lamps, water cannons, tear gas and petrol to ignite the villages. As they attempted to enter the area they were met by hundreds of women who said 'Sons, you have mothers like us, if you are ready to kill your mother, then fire at our hearts'. The women were followed by children of 10 to 15 years old, who addressed the police: 'Father, you have sons like us, if you kill our parents, how will we live? If you kill our parents, then take us with you'. At this, the children began to cry. Incredibly, the police also began to cry and lower their guns. They embraced the children as the local people began to applaud and shout; 'The police are our brothers' (Interview; Baliapal, 1990).

While the government's response to the movement spanned the coercion–seduction–mediation spectrum it fell short of prolonged systematic violence. This can be attributed to several factors:

1. the non-violent nature of the movement, which added legitimacy to the struggle (due to the role of non-violence in the political and social memory of the Indian state) and made it more difficult for the government to rationalize violence against the movement;
2. the movement's constant public profile, aided by the media and the support of oppositional political parties and voluntary organizations;
3. the support of political parties for the movement which added legitimacy to the struggle and reduced the social distance between the movement and the government;
4. the *status quo* goals of the movement which did not pose a direct challenge to the state's legitimacy within political society; and
5. the limitations placed upon the Indian defense budget (witness the secession struggles in Punjab, Kashmir and Assam) which prevented the NTR from becoming the government's primary national security priority.

### Current situation

Despite their militancy, the poor peasants and women of the Baliapal movement have failed to participate in the leadership of the movement. It continues to reflect the class, caste and gender structure of the local area, remaining in the hands of male, upper-class cultivating castes. While an argument can be made that these wealthy castes control the movement for their own class interests (since they stand to lose the most, economically, from eviction), I have tried to show that in the unique context of Baliapal the movement has been united across class, gender and caste differences through a potent common interest. Interestingly, two processes are at work here within the context of Indian civil society. At the national and local levels, the movement represents a challenge to the government's attempts to secure social hegemony (regarding national security policies). At the local level, despite class, caste and gender inequalities, the movement has achieved hegemonic consent, due to its articulation of popular resistance to the NTR. This is fueled by the strong cultural–religious connection of the people to the land and to the locale, and the (albeit unequal) economic interests that all Baliapalis have in remaining in the area.

Since 1988, Giri has died and Patra has been ousted from the leadership of the movement (due to his desire that the movement adopt violent tactics). As a result, the movement leadership has become susceptible to external political influences, particularly from political parties. This has impeded the struggle, diluted the ideological thrust of the movement and created three factions, one owing allegiance to the Janata Party, one to the Congress (I) Party, and the third independent of both (neutral). Although the mass base of the movement has remained militantly opposed to the NTR, some communities have become polarized around party lines. This was accentuated during the November 1989 national elections and during the 1990 Orissa state elections. Indeed, during 1989 and 1990 the Resistance Committee met only at irregular intervals and limited itself to participation in the *Vichar* system. As a result, movement strategy changed from confrontation to containment (Patel, 1989b, 1990).

In November 1990, the then Prime Minister of India, V. P. Singh (of the Janata Dal Party), announced that the location of the NTR would be changed. The same day, however, he resigned and at present a Congress (I) government is in power in New Delhi determined to construct the NTR in Baliapal. However, at present (April 1992) there is no government action to construct the NTR in Baliapal; the barricades have been removed; and, in the words of one movement leader, 'the movement is sleeping' (Interview; Baliapal, 1992). The reason for this is that the state government, under the Janata Dal, has made no move to acquire the land for the NTR or issue eviction notices to the Baliapal residents. While the Baliapal peasants have returned to farming and fishing, they remain determined to reactivate the movement if the need arises in the future.

## Conclusion

Due to a coalescence of historical, political, economic and cultural factors, Baliapal's 'oasis of development' in north Orissa has been transformed into a terrain of resistance. The resistance has taken the form of a non-violent social movement comprising the peasant farmers and fisherfolk of the area.

The articulation of this resistance terrain has involved an interwoven web of relationships that pertain to politics, culture, economics, geography and psychology.

While the Baliapal movement is defensive in its goal to prevent people's eviction from the area and to maintain the *status quo*, it is offensive in its challenge to the national security decision-making process of the Indian State and the underlying premise that national security interests must always take precedence over local concerns. The people's movement has challenged this process of political and cultural subordination by the State by withdrawing their consent to the legitimacy of the State's policy decisions and enacting non-violent resistance aimed at civil society (in the Gramscian sense).

The effects of state domination and the non-violent resistance to it have been place-specific. Hence a deeper understanding of the Baliapal movement can be obtained by considering the mediation of movement agency by place. The theoretical components of place (location, locale and sense of place) provide valuable insights into the 'terrain of resistance', contributing to the understanding of the political character of the movement. The analysis of the settings where social relations are constituted (locale) and the wider socio-political processes at work (location) have contributed to an understanding of why the Baliapal movement emerged where and when it did.

Hence we can understand the importance of location – given Baliapal's role as the granary of Orissa, as the site of the NTR, and in the context of the militarization of Orissa and the national security dictates of the Indian State. The movement is essentially confined to the Baliapal area due to that area being designated as the site for the NTR. However, the struggle, albeit localized, touches upon issues that have potential implications far beyond the north Orissa area. Given the military nature of the NTR, the resistance to it challenges the notion of the primacy of national security needs (and the consequent militarization process that this engenders) and the assumption of India's ruling class that local interests must give way to national interests (however defined).

We can understand the importance of locale, with reference to the social relations of Baliapal's peasant population; Baliapal's cash crop economy; and the beauty and fertility of the Bay of Bengal coast. The lucrative *paan* and fishing economy has provided a degree of prosperity to all Baliapal peasants (including the landless), and this provided a powerful economic motivation for the villages to unite across class, caste and gender inequalities to resist the NTR. The social relations of the locale are reflected in the hierarchical structure of the movement organization, whose leadership consists of upper-caste, wealthy male village leaders and traders, and the militant involvement of women. Indeed, while challenging the national government's policy to locate the NTR in Baliapal, the movement's demands remain focused upon the maintenance of the economic and cultural relations and processes of the locale. Hence, although creating a terrain of resistance against the government within civil society, the class, caste and gender relations of the locale remain unchallenged. During interviews with activists, it has been argued that because the movement's focus has been to retain the *status quo* within the area, the possible class, caste and gender cleavages that may have divided and weakened the movement have been avoided.

Given the wealth of the Baliapal area in relation to the rest of Orissa, it has been difficult to develop a common cause with other areas and social groups in either Orissa or elsewhere. This has been compounded by the low emphasis given to the peace question, given the place-specific nature of the mobilization (against the threat of eviction) rather than an issue-specific mobilization against militarization. Stressing the threat that the NTR would pose to the rest of Orissa as well as to India may have provided the movement with a greater mobilization potential outside the Baliapal area.

We can understand the importance of the Baliapalis' sense of place through their ideology of *Bheeta Maati*, tied as it is to religious idioms and a cultural sentiment for the land and the locale. This spirit of the movement was expressed through songs and dramas; the use of conch shells and *thalis* to summon people to the barricades; the invocation of the goddess Durga as spiritual force behind the *Maran Sena*; the militant involvement of women; and the attempts of the movement to cut across gender, caste and class barriers, facilitated in part by the *Vichar* system. The Baliapalis' profound sense of place has informed, motivated and inspired the peasants to continue their resistance to the NTR – despite government seduction and coercion. An understanding of this geographical notion can provide important insights into how place-specific sites of struggle involve the affirmation of local culture, identity and ideology.

These spatial mediations provided the reasons for the government's original decision to locate the NTR in Baliapal; the emergence of the movement and the government's response to it; and hence the articulation of Baliapal as a terrain of resistance. They have also motivated the resistance and fueled the militancy of the peasants. Although the movement has remained localized in the Baliapal area, this has contributed to the movement's success in preventing the eviction of the peasant population. The effectiveness of the barricades has been enhanced (since only four dirt-track roads provide access to the area), and the agricultural abundance of the area has enabled the movement to be virtually self-sufficient in food despite the government's economic blockade.

As has been mentioned, the Baliapal struggle is located within civil society where the movement represents a popular withdrawal of consent from the social hegemony of the State. As a terrain of resistance, Baliapal exemplifies a national phenomenon. Throughout India, place-specific struggles (around issues of development, militarization, ecology, etc.) are challenging the hegemony of the State within civil society, transforming India into a 'terrain of resistances'.

> Civil society, then, is seen in action terms as the domain of struggles, public spaces, and political processes. It comprises the social realm in which the creation of norms, identities, and social relations of domination and resistance are located (Cohen, 1985: 700).

In the case of Baliapal, the failure of the State to achieve social hegemony can be attributed to several factors:

1. The potency of local religious and cultural beliefs and values, which, in affirming the Baliapalis' close ties to the land, were in sharp opposition to those values of national security articulated by the State. In a spatial sense, these values were articulated as 'defensible space' by the movement (i.e., Baliapalis' cultural sentiment for their homes and lands was to be defended, to the death if necessary) and 'dispensable space' by the State (i.e., the local economy and peasant culture of Baliapal were dispensable to the national security needs of the State).
2. The movement received support within civil society from the media and voluntary organizations – indeed, some chains of equivalence were created via district and state level support groups, and some national solidarity campaigns, etc. – which themselves challenged the government's decision to locate the NTR in Baliapal.
3. The support of opposition political parties for the movement which added legitimacy to the struggle and provided further challenges to the State's hegemony.

Regarding social movement agency within civil society, it has been shown that it is important to consider the spatial context within which movement agency takes place. The context of place, that 'terrain of resistance', and the cultural expressions that articulate discontent provide important insights into why movements emerge where and when they do. Social movement research has tended to unify the heterogeneity of collective action by either the use of key analytical concepts such as class struggle or through empirical generalizations. Hence, empirical research frequently attempts to extrapolate a movement's locationally-specific experience to a general theory of social movement practice without due consideration of particular spatial and cultural contexts (e.g. Janssen, 1978; Marcelloni, 1979; Burgess, 1982; Lagana *et al.*, 1982; Pickvance, 1985).

However, as the case of Baliapal has shown, collective action is focused upon cultural codes which are themselves spatially specific. The decision-making processes of the movement have been informed by cultural and ideological considerations. Culture and ethnicity can create 'imagined spaces' (see Harvey, 1989), reflecting a community's sense of place. The ideology which emanates from this articulates a process of positive assertion (of local values and life-styles) and resistance to intervening values of domination. Hence the use of spatial and political–cultural paradigms vitiates against the homogenization of movement practice, whereby agency is explained by purely strategic, structural and organizational concepts. Also, by analysing the cultural expressions of resistance, we can begin to understand movement agency through the voices of its participants rather than through the mediation of elite and establishment discourse.

The Baliapal movement can be located in the context of other on-going, localized non-violent struggles in India. These sites of struggle pose place-specific challenges within civil society to the dominating discourse of the State. Such movements are responses to an increasing tension between the demands and interests of local communities and the antithetical policies and agendas of a centralized and bureaucratic State apparatus. As such, they represent part of a grassroots-based democratization process under way within Indian society. Many of these struggles are attempting to:

> extend the field of political activity and to democratize new and existing public spaces at the expense of state control and its technocratic model of society (Cohen, 1985: 704).

The implications that these movements' agency may have for the further democratization of Indian society are as far-reaching as they are as yet unresolved. For, while the ethnic, linguistic and cultural diversity of India may vitiate against hegemony imposed by the national government, such diversity provides barriers to the formation of mass (i.e., regional or national) movements. Hence, it seems likely that the processes of domination and resistance will continue to be played out within Indian civil society at the local level. For it is within specific locales, which constitute specific social, cultural, economic and political relations, and which are confronted by the consequences of specific government policies, that social movements can best hope to mobilize popular support against those policies and achieve hegemonic consent.

In this paper, it has been shown that the resource mobilization theory is very useful in understanding social movement strategy, and its relation to the social structure and to political parties, but that it fails to address the expressive dimensions of resistance. Identity-oriented theory, however, provides the means of understanding the spirit of movement agency – that which inspires and empowers people, understood through a movement's ideology and cultural idioms of protest, the 'language of lived experience' (Vaneigem, 1983: 75). An analysis of the components of place provides an understanding of the context within which movement agency interpellates the social structure. Finally, while the success or failure of a social movement is of crucial concern when people are confronted with struggles for cultural survival, the very fact that they resist speaks directly to the critical elements of human transformation: participation, communication and self-realization.

## References

Agnew, J. A (1987). *Place and Politics: The Geographical Mediation of State and Society*. Boston, MA: Allen and Unwin.

AIFOFDR (1986). *People vs Baliapal Misssile Base: Report of the All-India Federation of Organizations for Democratic Rights.* Delhi: AIFOFDR.

Albert, D. H. (1985). *People Power: Applying Nonviolence Theory*. Philadelphia, PA: New Society.

Alvares, C. (1988). Showdown at Kaiga. *Illustrated Weekly of India* 29 May, 9–17.

Appen (1987). *Urgent Action Report*. Malaysia: Asia–Pacific People's Environment Network.

Apter, D. E. and Sawa, N. (1984). *Against the State: Politics and Social Protest in Japan.* Cambridge, MA: Harvard University Press.

Bagchi, A. K. (1982). *The Political Economy of Underdevelopment.* New York: Oxford University Press.

Balagopal, K. (1988). We shall have our own Mandelas. *Economic and Political Weekly* October, 2039–2043.
Banerjee, S. (1984). *India's Simmering Revolution*. London: Zed Press.
Banks, J. ( 1972). *The Sociology of Social Movements*. London: Macmillan.
Bardhan, P. (1985). *The Political Economy of Development in India*. Delhi: Oxford University Press.
Baru, S. (1988). State and industrialization in a post-colonial capitalist economy. *Economic and Political Weekly* January, 143–150.
Bayley, D. (1970). Public protest and the political process in India. In *Protest Reform and Revolt: A Reader in Social Movements* (J. Gusfield ed.) pp. 298–308. London: Wiley.
Belden-Fields, A. (1988). In defense of political economy and systemic analysis: A critique of prevailing theoretical approaches to the new social movements. In *Marxism and the Interpretation of Culture* (C. Nelson and L. Grossberg eds) pp. 141–156. Chicago, IL: University of Illinois Press.
Berreman, G. (1987). The Chipko movement in the Indian Himalayas. In *Dimensions of Social Life: Essays in Honor of David G. Mandelbaum* (P. Hockings ed.) pp. 345–368. Berlin: Mouton de Gruyter.
Bocock, R. (1986). *Hegemony*. Sussex: Ellis Horwood Ltd.
Boggs, C. (1986). *Social Movements and Political Power*. Philadelphia, PA: Temple University Press.
Bonner, A. (1990). *Averting the Apocalypse*. Durham, NC: Duke University Press.
Borja, J. (1977). Popular movements and urban alternatives in post-Franco Spain. *International Journal of Urban and Regional Research* 7, 157–173.
Burgess, R. (1982). The politics of urban response in Latin America. *International Journal of Urban and Regional Research* 6, 465–479.
Castells, M. (1977). *The Urban Question*. London: Edward Arnold.
Castells, M. (1983). *The City and the Grassroots*. Berkeley, CA: University of California Press.
Castells, M. (1985). Commenting on C. G. Pickvance's 'The rise and fall of urban movements . . .' *Society and Space* 3, 55–61.
Chatterjee, P. (1982). Agrarian relations and communalism in Bengal, 1926–1935. In *Subaltern Studies*, Vol. 1 (R. Guha ed.) pp. 9–38.
Chatterjee, P. (1983). More on modes of power and the peasantry. In *Subaltern Studies*, Vol. 2 (R. Guha ed.) pp. 311–349.
Cohen, J. L (1985). Strategy and identity: New theoretical paradigms and contemporary social movements. *Social Research* 52, 4, 663–716.
Collier, D. (1979). *The New Authoritarianism in Latin America*. Princeton, NJ: Princeton University Press.
Committee to Assist the Struggle of Bhograi and Baliapal (1988). *News and Views on Missile Testing Range at Baliapal and Bhograi*. Calcutta: CASBB.
Corbridge, S. (1987). Industrialization, internal colonialism and ethnoregionalism: the Jharkhand, India, 1880–1980. *Journal of Historical Geography* 13, 3, 249–266.
Corbridge, S. (1988). The ideology of tribal economy and society: politics in the Jharkhand, 1950–1980. *Modern Asian Studies* 22, 1, 1–42.
Crooke, W. (1896). *The Popular Religion and Folklore of Northern India, Vols 1 and 2*. London: Archibald Constable and Co.
Della Seta, P. (1978). Notes on urban struggles in Italy. *International Journal of Urban and Regional Research* 2, 303–329.
De Souza, A. (1983). *Urban Growth and Urban Planning*. New Delhi: Indian Social Institute.

Dhanagre, D. N. (1975). *Agrarian Movements and Gandhian Politics*. Agra: Agra University.

Dhanagre, D. N. (1983a). *Peasant Movements in India 1920–1950* Delhi: Oxford University Press.

Dhanagre, D. N. (1983b). The Green Revolution and social inequalities in rural India. *Bulletin of Concerned Asian Scholars* 15, 3, 2–13.

Dhanagre, D. N. (1988). Action groups and social transformation in India. *Lokayan Bulletin* 6, 5, 37–59.

Draaisma, J. and van Hoogstraten, P. (1983). The squatter movement in Amsterdam. *International Journal of Urban and Regional Research* 7, 406–415.

Dutt, S. (1984). *India and the Third World*. London: Zed.

Fernandes, G. (1986). The battle for Baliapal. *The Other Side* August, 41–51.

Fernandes, W. (1984). External support for action groups: The role of macro organizations. *Social Action* 34, 145–155.

Foss, D. A. and Larkin, R. (1986). *Beyond Revolution: A New Theory of Social Movements*. Massachusetts: Bergin and Garvey.

Foucault, M. (1977). *Language, Counter Memory, Practice*. Ithaca: Cornell University Press.

Foucault, M. (1980). *Power/Knowledge*. New York: Pantheon Books.

Gamson, W. (1975). *The Strategy of Social Protest*. Homewood, IL: Dorsey Press.

Ghose, A. (1984). Impact of Delhi Anti-encroachment Bills. *Economic and Political Weekly* September, 1564–1566.

Ghose, S. (1975). *Political Ideas and Movements in India*. New Delhi: Allied.

Ghosh, A. (1989). Civil liberties, uncivil State. *Seminar* March, 34–37.

Gough, K. (1968). Peasant resistance and revolt in South India. *Pacific Affairs* 41, 4, 526–544.

Gough, K. (1974). Peasant uprisings in India. *Economic and Political Weekly* 9, 32–4.

Government of India (1986). *National Testing Range–Baliapal: A Comprehensive Report on the Project and Rehabilitation*. Cuttack: Orissa Government Press.

Gramsci, A. (1971). *Prison Notebooks*, ed. by Q. Hoare and G. M. Smith. New York: International Publishers.

Guha, R. (1983). *Elementary Aspects of Peasant Insurgency in Colonial India*. Delhi: Oxford University Press.

Guha, R. (1989a). The Problem. *Seminar* March, 12–15.

Guha, R. (1989b). *The Unquiet Woods*. Delhi: Oxford University Press.

Guha, R. (1989c). Dominance without hegemony and its historiography. In *Subaltern Stuctures*, Vol. 6 (R. Guha ed.) pp. 210–309.

Gurr, T. (1970). *Why Men Rebel*. Princeton, NJ: Princeton University Press.

Gusfield, J., ed. (1970). *Protest, Reform and Revolt: A Reader in Social Movements*. London: John Wiley and Sons.

Habermas, J. (1981). New social movements. *Telos* 49, 33–37.

Hannigan, J. A. (1985). Alain Touraine, Manuel Castells and social movement theory: a critical appraisal. *Sociological Quarterly* 26, 4, 435–454.

Harvey, D. (1989). *The Condition of Postmodernity*. Oxford: Basil Blackwell.

Hasson, S. (1983). The emergence of an urban social movement in Israeli society: an integrated approach. *International Journal of Urban and Regional Research* 7, 157–173.

Heberle, R. (1986). Types and functions of social movements. In *International Encyclopedia of the Social Sciences* pp. 438–444. New York: Macmillan.

Janssen, R. (1978). Class practices of dwellers in Barrios Populares. *International Journal of Urban and Regional Research* 2, 147–159.

Joshi, B. (1984). India and the Backdoor Emergency. *South Asia Bulletin* 5, 2, 14–24.

Joshi, B. (1986). *Untouchable: Voices from the Dalit Movement*. London: Zed Press.

Josh, B. (1989). National identity and development: India's continuing conflict. *Cultural Survival Quarterly* 13, 2, 3–7.

Kannabiran, K. (1985). Erosion of constitutional safeguards. *Economic and Political Weekly* May, 786–788

Katz, S. and Mayer, M. (1985). 'Gimme shelter': self-help housing struggles within and against the State in New York City and West Berlin. *International Journal of Urban and Regional Research* 9, 15–45.

Katznelson, I. (1981). *City Trenches*. New York: Pantheon.

Klandermans, B., ed. (1989). *International Social Movement Research*. Greenwich, CT: Jai Press.

Kodama, K. and Vesa, U., eds (1990) *Towards a Comparative Analysis of Peace Movements*. Hants: Dartmouth.

Kothari, R. (1986a). Masses, classes and the State. *Economic and Political Weekly* February, 210–216.

Kothari, R. (1986b). NGOs, the State and world capitalism. *Economic and Political Weekly* December, 2177–2182.

Kothari, R. (1988). Integration and exclusion in Indian politics. *Economic and Political Weekly* October, 2223–2227.

Kothari, S. (1985). Ecology vs development: the struggle for survival. *Lokayan Bulletin* 3, 4, 7–22.

Kriesberg, L., ed. (1978–88). *Research in Social Movements, Conflict and Change*. Greenwich, CT: JAI Press.

Kumar, K. (1984). People's Science and development theory. *Economic and Political Weekly* 14 July, 1082–1084

Lagana, G., Pianta, M. and Segre, A. (1982). Urban social movements and urban restructuring in Turin, 1969–76. *International Journal of Urban and Regional Research* 6, 223–245.

Lefebvre, H. (1974). *La Production de l'espace*. Paris: Editions Anthropos.

Leontidou, L. (1985). Urban land rights and working class consciousness in peripheral societies. *International Journal of Urban and Regional Research* 4, 533–555.

Lofland, J. ( 1991). *Protest: Studies of Collective Behaviour and Social Movements.* New Brunswick, NJ: Transaction Publishers.

Lynch, O. (1969). *The Politics of Untouchability*. New York: Columbia University Press.

Lynch, O. (1974). Political mobilization and ethnicity among Adi-Dravidas in a Bombay slum. *Economic and Political Weekly* July–September, 1657–1668.

Lynd, S. (1966). *Nonviolence in America: A Documentary History*. New York: Bobbs-Merrill.

Majumdar, T. (1977). The urban poor and social change: a study of squatter settlements in Delhi. *Social Action* 27, 3, 216–240.

Malik, S. C. (1977). *Dissent, Protest and Reform in Indian Civilization*. Delhi: Delhi University Press.

Marcelloni, M. (1979). Urban movements and political struggles in Italy. *International Journal of Urban and Regional Research* 3, 251–267.

Marglin, S. and Marglin, F., eds ( 1990). *Dominating Knowledge: Development, Culture and Resistance*. Oxford: Clarendon Press.

Massey, D. (1984). *Spatial Divisions of Labour: Social Structures and the Geography of Production*. London: Methuen.

McAdam, D. (1982). *Political Process and the Development of Black Insurgency*. Chicago, IL: Chicago University Press.

McManus, P. and Schlabach, G., eds (1991). *Relentless Persistence: Nonviolent Action in Latin America*. Philadelphia, PA: New Society.

Melucci, A. (1989). *Nomads of the Present*. London: Radius.

Mouffe, C. (1988). Hegemony and new political subjects: toward a new concept of democracy. In *Marxism and the Interpretation of Culture* (C. Nelson and L. Grossberg eds) pp. 89–104. Chicago, IL: University of Illinois Press.

Myrdal, J. (1980). *India Waits*. Madras: Sangam Books.

Nair, V. (1987). Resistance unto death against the establishment of Baliapal missile test range. *Ananda Bazaar Patrika* 9 May, 11–15.

Nandy, A. (1984). Culture, State and rediscovery of Indian politics. *Economic and Political Weekly* 19, 49, 2078–2083.

Nandy, A. (1988). *Science, Hegemony and Violence*. Delhi: Oxford University Press.

Oberschall, A. (1973). *Social Conflicts and Social Movements*. New Jersey: Prentice Hall.

Offe, C. (1985). New social movements: challenging the boundaries of institutional politics. *Social Research* 52, 4, 817–868.

Olives, P. (1976). The struggle against urban renewal in the Cité d'Alicarte (Paris). In *Urban Sociology: Critical Essays* (C. Pickvance ed.) pp. 174–197. New York: St Martin's.

Omvedt, G. (1980). *We Will Smash This Prison*. London: Zed.

Omvedt, G. (1982). *Land, Caste and Politics in Indian States*. New Delhi: Authors Guild Press.

Omvedt, G. (1983a). The new peasant movement in India. *Bulletin of Concerned Asian Scholars* 15, 3, 14–23.

Omvedt, G. (1983b). Capitalist agriculture and rural classes in India. *Bulletin of Concerned Asian Scholars* 15, 3, 2–13.

Omvedt, G. (1985). After the failures, a 'new-style communism'. *The Guardian* 27 March, 17.

Ong, A. (1987). *Spirits of Resistance and Capitalist Discipline*. Albany, NY: SUNY Press.

Oommen, T. K. (1975) Agrarian legislations and movements as sources of change. *Economic and Political Weekly* 19, 49, 2078–2083.

Oommen, T. K. (1977). From mobilization to institutionalization: the life cycle of an agrarian labour movement in Kerala. In *Dissent, Protest and Reform in Indian Civilization* (S. C. Malik ed.) pp. 286–302. Delhi: Delhi University Press.

Parajuli, P. (1990). *Grassroots movements and popular education in Jharkhand, India*. Unpublished PhD thesis, Stanford University.

Parajuli, P. (1991). Power and knowledge in development discourse: new social movements and the State in India. *International Conflict Research* 127 February, 173–190.

Patel, S. (1989a). Baliapal agitation: socio-economic background. *Economic and Political Weekly* March, 604–605.

Patel, S. (1989b). Agni, cyclone and Baliapal agitation. *Economic and Political Weekly* June, 1381–1382.

Patel, S. (1990). Baliapal agitation: leadership crisis. *Economic and Political Weekly* 9 July, 1238–1240.

Pickvance, C. G. (1976a). *Urban Sociology: Critical Essays*. New York: St. Martin's.

Pickvance, C. G. (1976b). On the study of urban social movements. In *Urban Sociology: Critical Essays* (C. Pickvance ed.) pp. 198–218. New York: St Martin's.

Pickvance, C. G. (1977). From social base to social force: some analytical issues in the study of urban protest. In *Captive Cities* (M. Harloe ed.) pp. 175–183. New York: Wiley and Sons.

Pickvance, C. G. (1984). Book review. *International Journal of Urban and Regional Research* 8, 588–591.
Pickvance, C. G. (1985). The rise and fall of urban movements and the role of comparative analysis. *Society and Space* 3, 31–53.
PUDR (1988). *Bheeta Maati*. Delhi: People's Union of Democratic Rights.
Routledge, P. (1987). Modernity as a vision of conquest: development and culture in India. *Cultural Survival Quarterly* 11, 3, 63–66.
Roy, A. (1988). Quickening pace of authoritarianism and the Left. *Economic and Political Weekly* September, 1561–1562.
Roy, S. (1983). Non-aligned activists. *Seminar* November.
Said, E. (1979). *Orientalism*. New York: Vintage Books.
Scott, J. C. (1985). *Weapons of the Weak*. New Haven, CT: Yale University Press.
Selbourne, D. (1977). *An Eye to India*. New York: Penguin.
Sengupta, N., ed. (1982). *Fourth World Dynamics: Jharkhand*. New Delhi: Authors Guild.
Sethi, H. (1984). Groups in a new politics of transformation. *Economic and Political Weekly* February, 305–316.
Sharma, S. (1985). Development and diminishing livelihood. *Lokayan Bulletin* 3, 5, 77-84.
Sharp, G. (1973). *The Politics of Nonviolent Action*, 3 vols. Boston, MA: Porter Sargent.
Sharp, G. (1980). *Social Power and Political Freedom*. Boston, MA: Porter Sargent.
Sinha, A. (1987). People's war on missile range. *Surya India* 11, 4, 8–17.
Smelser, N. J. (1962). *The Theory of Collective Behaviour*. London: Routledge and Kegan Paul.
Soja, E. (1988). *Postmodern Geographies: The Reassertion of Space in Critical Social Theory*. London: Verso.
Sorokin, P. (1962). *Social and Cultural Dynamics*. New Jersey: Free Press.
Tilly, C. (1978). *From Mobilization to Revolution*. Reading, MA: Addison–Wesley.
Touraine, A. (1981). *The Voice and the Eye*. New York: Cambridge University Press.
Touraine, A. (1985). An introduction to the study of social movements. *Social Research* 52, 4, 749–787.
Unnayan (1988). *Baliapal File*. Calcutta: Unnayan Archives.
Unseem, M. (1975). *Protest Movements in America*. Indianapolis: Bobbs-Merrill.
Vaneigem, R. (1983). *The Revolution of Everyday Life*. London: Rebel Press.
Visvanathan, S. (1985). From the annals of the laboratory state. *Lokayan Bulletin* 3, 4, 23–47.
Visvanathan, S. and Kothari, R. (1985). Bhopal: the imagination of a disaster. *Lokayan Bulletin* 3, 4, 48–76.
Walton, J. (1979). Urban political movements and revolutionary change in the Third World. *Urban Affairs Quarterly* 15, 1, 3–22.
Wiebe, P. (1975). *Social Life in an Indian Slum*. New Delhi: Vikas.
Wolf, E. (1969). *Peasant Wars of the Twentieth Century*. New York: Columbia University Press.
Zald, M. N. and McCarthy, J. D., eds (1977). Resource mobilization and social movements: a partial theory. *American Journal of Sociology* 82, May, 33–47.

# SECTION FIVE

# *PLACES AND THE POLITICS OF IDENTITIES*

## Editor's introduction

The 'new' social movements for racial civil rights, ecology, feminism and gay rights that began to spring up in Europe and North America in the 1960s are not simply attempts to obtain material changes but also and crucially 'struggles over signification; they are attempts simultaneously to make a nonstandard identity acceptable and to make that identity livable' (Calhoun 1991, 51). Much contemporary politics is then a politics of identity, involving struggles over the recognition and legitimacy of different social identities within the wider society. The novelty of this can be overplayed. The instrumental character of some older movements, such as the labour movement, has obscured the extent to which labour struggles also usually involve questions of recognition and self-respect. Nevertheless, the concern for identity is distinctively modern if not just recent (Calhoun 1994). When all-encompassing 'identity schemes' such as kinship prevail within a society there is little or no problem. It is when we must live in a world where social networks are diffuse and there is limited cultural consensus that persons face the difficulty of identifying who they are in relative isolation. Identity politics is about struggling to establish the recognition of collective differences in identity within a society in which those differences are either not acknowledged or involve negative evaluations and sanctions.

Until recently, social science, including political geography, has paid only scant attention to issues of identity and identity politics (Anderson and Gale 1992; Carter *et al.* 1993). Politics have been viewed as being about the pursuit of shared material interests. Only the advent of movements explicitly pushing the case for 'new identities' has alerted us to the extent to which the 'identity question' informs all political activities to one degree or another. Active involvement in any kind of interest group or political party involves a sense of political identity without which solidarity or loyalty are impossible to achieve. Shared symbols, myths and rituals are central to the process of forming political identity (e.g. Kertzer 1993; Duncan 1990). 'Identity politics' is only a more explicit and stronger version of the same thing. As Mackenzie (1976, 133) has suggested, those 'powerful abstractions' –

nation, race, religion, class – have long been the stuff of politics. They are about identity as much as they are about interests.

Identities are established through the stories that people tell themselves and others about their experiences. In this way individual persons come to see themselves as parts of larger collectivities with common experiences and histories. People present themselves to one another in terms of stories, and tell stories about one another (Dennett 1991). These narratives are attempts to create a unified self, one that makes the self intelligible. Stories, with their linear forms and ascriptions of subjective powers, are the organizing devices that express the self. Lives and stories become intertwined to define identities (see, e.g., MacIntyre 1981; Kirby 1991; Widdershoven 1993). From this perspective, identity is about the connection drawn between a 'self' and a community of communicators or storytellers (Mackenzie 1976).

One feature of modern societies is the extent to which communication takes place over ever-widening distances. Indeed, the struggle for stable identities in an unstable world is a result of the breakdown of unreflexive and totalistic identity schemes in which 'everyone knows their place'. Yet, it is remarkable that one can still say with some confidence that 'Those who share a place share an identity' (Mackenzie 1976, 130). This is so for several reasons. First, because even as people strive politically to establish identities that are not necessarily place-specific they do so within a geographical 'field' of shared relevance, such as a state (Calhoun 1994, 25). Second, as they struggle for one identity people usually share other identities of which the most important can be that associated with those around them. People have multiple identities and loyalties that derive from the overlapping social worlds in which they live their lives (Calhoun 1994, 26). Third, communication, social interaction and reactions to distant events are all filtered through the routines and experiences of everyday life. For most people these are still geographically constrained. Even if not always localized, 'shared social spaces' still define the limits for the social appropriateness of given identities (Mackenzie 1976, 131). Fourth, and finally, 'imagined geographies' are important within many identities, such as that of 'African-American' with its roots in African diaspora, slavery and the southern US and contemporary expression in the 'black ghetto', and the 'migrant identities' of diasporic groups caught between different social worlds (Mohanty 1991; Davies 1994; Morley and Robins 1995). Places are shared, even if only in the imagination.

The concept of identity has come in for some criticism. The term itself implies a certain natural solidity and permanence to 'identities' that the politics of identities are all about establishing. There would be no point to the politics if the identities were not either in question or subject to denigration by more powerful 'others'. To Handler (1994, 30) the intellectual problem lies in treating groups as 'bounded objects in the natural world. Rather, "they" are symbolic processes that emerge or dissolve in particular contexts of action. Groups do not have essential identities; indeed, they ought not to be defined as things at all.' The danger of 'essentializing' identities does not mean, however, that identities are socially constructed 'out of thin air' without any meaningful relationship to 'natural differences'. Given that much racial discrimination is based on reactions to biological differences such as skin colour, for example, it is not surprising that an 'African-American' identity would involve reference to skin colour (Calhoun 1994, 17). However much intellectuals may be uncomfortable with this, it is an intrinsic part of stories of identity among African-Americans. Finally, the relationship of identities to interests remains ambiguous in much of the literature on the politics of identities (Calhoun 1991). Perhaps the theoretical problem here lies in drawing too neat a line between the two, as if identities and shared interests must always stand in opposition to one another. This judgement may well be the result of an overreaction to the prevalence of 'rational choice' and other interest-based accounts of politics in the social sciences (see the Introduction to Section 4).

Interestingly, given the generally postmodern tenor of the concept of identity, the relationship of places to the politics of identities has been addressed from within the three broad streams of political-geographical thinking. Each type of perspective, however, tends to pick up on a distinctive aspect of the relationship. From a spatial-analytic point of view the focus is on the *boundaries* (both social and jurisdictional) that help define identities. Political-economic approaches are more concerned with the processes of *spatial exclusion/inclusion* which help to create the circumstances in which groups can acquire identities. Postmodern perspectives, broadly construed, privilege the way in which identities are expressed through attempts at *associating identities with places*. The three articles and extracts I have selected for this section are representative of these emphases within this growing area of political geography.

Early in the twentieth century the sociologist George Simmel (1971 edn) claimed that processes of socialization often take spatial configurations. By this he meant that spatial proximity or distance serves to define the limits of social 'defense' and 'offense' between social forms. The 'life-principles' of some social forms, however, such as ethnic ones, are strongly spatial whereas that of a church, to use Simmel's example, tends not to be. But he thought that most social forms have a spatial fixity to them such that 'the proximity or distance, the uniqueness or the plurality which characterize the relations of social groups to their territory are therefore often the root and symbol of their structure' (Spykman 1925, 147). Political boundaries, both within and between states, often justified as *representing* the existing social boundaries between groups, give an extra boost to the spatial differentiation of groups and their identities. Indeed, some groups push to have territorial boundaries formally delimited in order to reinforce their differences with others. This is the theme of the article by **Alexander B. Murphy (Reading 13)** in his discussion of the 'social construction of space' in Belgium by Flemish (Dutch-speaking) and Walloon (French-speaking) groups.

The process of defining regions for the different groups is seen as a crucial part of their social differentiation. Identities have been pursued through formal attempts at dividing the national space into mutually exclusive enclaves for the two linguistic groups. The impact of the regionalization process is further illustrated by the case of Brussels, the capital, whose populace is seen as acquiring something of a distinctive identity (neither fully Flemish nor Walloon) by virtue of its position as a separate jurisdiction. Regional boundaries, therefore, create or reinforce social identities that in cases such as Belgium are the leitmotif of the national political process itself. Though the formal division of national space through competitive identity construction is certainly not unique to this one example (Paasi 1986; Murphy 1991), there is some question as to whether it must always lead to a totally regionalist solution. In the Italian case, for example, the advent of political regionalism in Northern Italy has not been based on administrative boundaries nor has it generated an equivalent exposition of regional identities elsewhere. Social boundaries have been enough (Agnew 1995).

Boundaries can be thought of within a different frame of reference: in terms of processes of spatial exclusion and inclusion. This is the approach taken by **Loïc J. D. Wacquant (Reading 14)** in his account of contemporary (postFordist) Black ghettos in American cities. These ghettos are different from the ones of 30 years ago; they have much greater internal segregation by social class and changed relations with the wider American society. Wacquant, true to a political-economic perspective, chooses to emphasize the *external* factors that have 'reshaped the social and symbolic territory within which ghetto residents (re)define themselves and the collectivity they form' and the '*internal* production of its specific social order and consciousness only indirectly' (p. 272). He distances himself from both those perspectives that see the ghetto as a wasteland occupied by a homogeneous 'underclass' and those which substitute a 'populist celebration of the "value of blackness" ' for a 'rigorous assessment of the state and the fate of the ghetto at the close of the Fordist era' (p. 272). (Fordism refers to the era of mass production/consumption associated with relatively strong trade unions, government intervention and a tacit 'wage bargain' between labour and capital guaranteed by government.) Using Chicago as his illustrative case, Wacquant shows that the 'new' ghetto has two distinctive features: (1) a decaying inner core with satellite working-class and middle-class neighbourhoods and (2) a massive amount of physical decay and social collapse in the ghetto core. His marshalling of empirical evidence, however, suggests that the bleak prognoses of underclass theorists (such as William Julius Wilson, much abused by the celebrants of contemporary ghetto culture, see Remnick 1996) may not be so far off the mark. The 'organizational infrastructure' that gave the 'classic ghetto' its communal strength and identity (churches, lodges, the black press, etc.) has largely withered away. In this setting a ghetto identity is a stigma from which 'nearly everybody is desperately trying to escape' (p. 281).

What Wacquant wants most of all is to avoid 'blaming the victims'. It is the abandonment of the ghetto by both government and society that must be indicted for the creation of the conditions that have produced the new one, not the people who live there. His own analysis, however, reveals that fear of living or being in the decaying inner ghetto is not hysterical. Life there *is* dangerous. This is not just another fabrication of 'the man'. The underclass thesis may be off the mark but neither is there much to celebrate. The identities of the classic ghetto appear beyond retrieval in the context of a political economy that writes off the inner city as a 'reservation' or 'Bantustan', a wild zone beyond ordinary society.

This narrative of the creation of a 'spoiled' identity – from classic ghetto to hyperghetto – stands in stark contrast to those accounts which find continuing strength for identity politics in the contemporary ghetto. An author such as bell hooks (1992, 17), for example, sees continuing evidence of the richness of 'oppositional black culture' emanating from within the ghetto. The tension between the two positions is expressed most clearly by David Harvey (1993, 64) when he writes:

> The identity of the homeless person (or the racially oppressed) is vital to their sense of selfhood. Perpetuation of that sense of self and of identity may depend on perpetuation of the processes which gave rise to it. . . . [T]he mere pursuit of identity politics as an end in itself (rather than as a fundamental struggle to break with an identity which internalizes oppression) may serve to perpetuate rather than to challenge the persistence of those processes which gave rise to those identities in the first place.

This is the dilemma of identity politics from a political-economic perspective: that perpetuating an identity may mean sacrificing interests.

A happier coincidence between identity and interests is mapped by **Benjamin Forest** (**Reading 15**) in his account of the territorial strategy employed by a group of gay men in West Hollywood, California to declare and underpin their identity by creating a new local government jurisdiction in which they would have a major stake. Forest surveys the coverage given to the incorporation of the new municipality by the 'gay press' and identifies a set of themes relating to 'gayness' which were associated with the place itself. As a 'gay place' West Hollywood became a concrete referent for a more abstract identity. Identity politics and place were intimately connected in this case, suggesting to Forest that this was a sophisticated way of stabilizing and advertising a particular gay identity. The place is crucial because it allowed a set of usually separate features of gayness (creativity, aesthetic sensibility, maturity, etc.) to be associated together with West Hollywood. A 'moral narrative' was constructed that allowed gays to be presented both as good citizens and as inhabitants of a gay place. This is taken to illustrate a more general claim about how representations of a place figure in the making and remaking of identities, a major theme of poststructuralist and postcolonial thinking in political geography (see, e.g., Duncan and Ley 1993; most chapters in Keith and Pile 1993; Thrift and Pile 1995; Duncan 1996; Nuttall *et al.* 1996).

Whether the identity politics involves ethnicity, race or sexual orientation, therefore, there are important links to the places within which the identities in question are defined and pursued. What one makes of identity politics and how one construes its precise geographical character, however, will depend on the theoretical perspective you adopt. Like the identities themselves, the meanings ascribed to the role of place in identity politics are highly contestable.

## References

Agnew, J. A. 1995: The rhetoric of regionalism: the Northern League and Italian Politics, 1983–1994. *Transactions of the Institute of British Geographers*, 20: 156–72.

Anderson, K. and Gale, F. (eds) 1992: *Inventing places: studies in cultural geography*. Melbourne, Australia: Longman Cheshire.

Calhoun, C. 1991: The problem of identity in collective action. In J. Huber (ed.) *Macro-micro linkages in sociology*. Thousand Oaks CA: Sage.

Calhoun, C. 1994: Social theory and the politics of identity. In C. Calhoun (ed.) *Social theory and the politics of identity*. Oxford: Blackwell.

Carter, E., Donald, J. and Squires, J. (eds) 1993: *Space and place: theories of identity and location*. London: Lawrence and Wishart.

Davies, C. B. 1994: *Black women, writing and identity: migrations of the subject*. London: Routledge.

Dennett, D. 1991: *Consciousness explained*. Boston MA: Little, Brown.

Duncan, J. S. 1990: *The city as text: the politics of landscape interpretation in the Kandyan kingdom*. Cambridge: Cambridge University Press.

Duncan, J.S. and Ley, D. (eds) 1993: *Place/culture/representation*. London: Routledge.

Duncan, N. (ed.) 1996: *BodySpace: Destabilising geographies of gender and sexuality*. London: Routledge.

Handler, R. 1994: Is 'identity' a useful cross-cultural concept? In J. R. Gillis (ed.) *Commemorations: the politics of national identity*. Princeton NJ: Princeton University Press.

Harvey, D. 1993: Class relations, social justice and the politics of difference. In

M. Keith and S. Pile (eds) *Place and the politics of identity.* London: Routledge.

hooks, b. 1992: Loving blackness as political resistance. In *Black looks, race and representation.* Boston MA: South End Press.

Keith, M. and Pile, S. (eds) 1993: *Place and the politics of identity.* London: Routledge.

Kertzer, D.I. 1993: *Ritual, politics and power.* New Haven CT: Yale University Press.

Kirby, A. P. 1991: *Narrative and the self.* Bloomington IN: Indiana University Press.

MacIntyre, A. 1981: *After virtue: a study in moral theory.* Notre Dame IN: University of Notre Dame Press.

Mackenzie, W. J. M. 1976: *Political identity.* London: Penguin.

Mohanty, C. T. 1991: Cartographies of struggle. In C. T. Mohanty, A. Russo and L. Torres (eds) *Third world women and the politics of feminism.* Bloomington IN: Indiana University Press.

Morley, D. and Robins, K. 1995: *Spaces of identity: global media, electronic landscapes and cultural boundaries.* London: Routledge.

Murphy, A. B. 1991: Regions as social constructs: the gap between theory and practice. *Progress in Human Geography*, 15: 23–35.

Nuttall, S., Gunner, L. and Smith-Darian, K. (eds) 1996: *Text, theory, space: post-colonial representations and identity.* London: Routledge.

Paasi, A. 1986: The institutionalization of regions: a theoretical framework for understanding the emergence of regions and the constitution of regional identity. *Fennia*, 164: 105–46.

Remnick, D. 1996: Dr. Wilson's neighborhood. *The New Yorker*, 29 April/6 May: 96–107.

Simmel, G. 1971: *On individuality and social forms.* Chicago: University of Chicago Press.

Spykman, N. J. 1925: *The social theory of George Simmel.* New York: Russell and Russell.

Thrift, N. and Pile, S. (eds) 1995: *Mapping the subject: geographies of cultural transformation.* London: Routledge.

Widdershoven, G. 1993: The story of life: hermeneutic perspectives on the relation between narrative and life history. In R. Josselson and A. Lieblich (eds) *The narrative study of lives.* Thousand Oaks CA: Sage.

# 13 Alexander B. Murphy, 'Linguistic Regionalism and the Social Construction of Space in Belgium'

Reprinted in full from: *International Journal of the Sociology of Language* 104, 49–64 (1993)

The three major language regions of Belgium, Flanders, Wallonia, and Brussels, have become indelible features of the Belgian political, social, and cultural landscape. Their importance is not just administrative; they have assumed a high degree of functional significance as well. It is commonplace to consider major fiscal, economic, and demographic developments in terms of differences among and between Flanders, Wallonia, and Brussels. Separate institutions have been established to deal with matters ranging from education to water use in each of the regions. And by most indicators, the language region is a significant basis for social and political identification.

Although the present language regions do not have deep historical roots as political or social units, they now provide the spatial backdrop for many of the questions that are asked about Belgium. Articles are written about the economy in Wallonia, language use in Brussels, and education in Flanders. One of the first questions that is posed about any new policy is whether it will benefit one or another language region disproportionately. Framing questions in these terms makes considerable sense, of course; with the constitutional revisions of the past 20 years, significant powers have devolved to the language regions. The regional councils of Flanders, Brussels, and Wallonia make independent decisions on issues ranging from environmental protection to foreign trade. At the same time, too frequently the questions that we ask about Belgium's language regions seem to take their existence for granted, almost as if they were natural and untransmutable spatial givens.

In this article I seek to shed some light on language and society in contemporary Belgium by looking at the social significance of Belgium's language regions. My analysis focuses on Flanders, Wallonia, and Brussels. There is also a small German-language region in eastern Belgium, but its lack of prominence at the national level leads me to concentrate attention elsewhere. I argue that Western social science has largely failed to acknowledge regionalism as a force in industrialized societies; by extension, the language regions of Belgium are generally treated simply as backdrops to ethnolinguistic processes. Recent challenges to the ways we have thought about regionalism in the "developed" world point the way to a view of regions as socially constructed places that reflect and shape social, political, economic, and cultural processes. Treating Wallonia, Brussels, and Flanders in these terms can shed light on a number of different facets of the Belgian situation, including the degree of interaction between Flemings and Walloons, the growth of a Brussels identity, the differential status of bilingualism in the three major language regions, and the continuing volatility of the Fourons/Voeren issue.

**Conceptual background**

One of the most notable features of Western social science between 1945 and the early 1970s was the tendency to treat the state as the only territorial unit of great significance in industrialized societies (Agnew 1987, 1989). This can be seen in the way in which the term "nation-state" came to be used. Literally a "nation-state" is a sovereign political territory composed of a single group of people who see themselves as one (Connor 1978). Few states actually conform to this model, of course (Mikesell 1983). Yet most Western social scientists came to use the term to refer to any sovereign state, no matter how ethnically heterogeneous that state might be.

Given this preoccupation with the state, it is not surprising that questions of substate ethnicity and nationalism received scant attention. Not fitting well with state-based models of society, they could only be ignored. In the early 1970s, however, Walker Connor (1972) drew attention to the extent to which scholars had failed to engage even basic issues of ethnicity and substate nationalism. His call for research on the subject, combined with the growing salience of substate nationalist movements in many parts of the world, led to a surge of interest in ethnicity and nationalism.

There is now a voluminous literature documenting ethnic and national movements all over the world. But a deeper understanding of the dynamics of ethnonationalism is arguably hindered by the failure to probe the nature and significance of regional structures and arrangements. We pay lip service to the fact that Flemings live in Flanders, Bretons in Brittany, Welsh in Wales, and Québécois in Québec without considering how or why Flanders, Brittany, Wales, or Québec came to be significant perceptual or functional units, and the implications of that process.

The importance of considering such matters is suggested by a number of questions that one might ask about the dynamics of substate nationalism. How has the development of functional language regions in Belgium affected the growth of Walloon or Flemish identity? How has the lack of a clearly defined conception of the spatial extent of Brittany affected French policies toward the region? How has the creation of strong links between southern Wales and England changed Welsh nationalist ideas about what is Wales? How has the decision to locate the capital of Canada in Ottawa affected the rise of substate nationalism in Québec? All of these questions offer potentially valuable insights, yet most of them have been slighted because the regional context is so frequently taken for granted.

The tendency to ignore the social nature and significance of territorial processes is coming under increasing attack in the theoretical literature on society (see Thrift 1983; Giddens 1985; Markusen 1987; Soja 1989). Commentators are being challenged to view regions not simply as spatial surrogates for social groups, but as part of the fabric of society itself (e.g. the literature on human territoriality; see Soja 1971; Sack 1981, 1986). Yet to date few empirical studies of ethnonationalism have explicitly addressed how or why an ethnic region acquires perceptual or functional significance and the implications of that process. This omission can be seen in the literature on the recent conflict between the Armenians and the Azerbaidzhanis in the Soviet Union. Most commentaries on the subject begin with a discussion of the ethnic characteristics and relative location of the Armenian and Azerbaidzhan Soviet Socialist Republics. What is missing is a consideration of the ways in which the creation of those two republics with their particular boundaries fed Armenian–Azerbaidzhani animosity, an issue that is at the heart of the current conflict (Wixman 1986).

Treating regions as social constructions is of particular importance when the focus of study is language, ethnicity, or nationalism, since ethnic and cultural differences are intimately bound up with territorial arrangements and understandings (Williams and Smith 1983). This is at least implicitly recognized in a number of recent studies of culture and nationality (see Rudolph and Thompson 1989). Yet even studies focusing on regional change or the devolution of power to regional administrations often limit themselves to a consideration of the types of new political and economic arrangements that are emerging and their likely impact on the institutional framework for intergroup relations in the state. The challenge is to acquire a better understanding of the ways in which regionalism and regional change shape the context in which identities and understandings are formed, and in which individuals and groups interact.

**Language and regionalism in Belgium**

The three major official language regions of Belgium are products of the long and sometimes volatile history of relations between the so-called language communities (see Verdoodt 1977). What we now call Flanders and Wallonia did not emerge as significant perceptual units until the late nineteenth century at the earliest, and during the course of the twentieth century all three major language regions gradually acquired greater functional significance. Their boundaries were not firmly established until the early 1960s, and they did not emerge as administrative regions until after the constitutional revisions of 1970 (see Coudenberg Group 1989).

The changing nature and significance of linguistic regionalism in Belgium is deeply implicated in the evolution of relations among and between Flemings, Walloons, and residents of Brussels. As I have sought to point out elsewhere (Murphy 1988), the all too frequent practice of projecting the present regional structure into the past has obfuscated important aspects of the history of language and regionalism in Belgium. Indeed, it represents a complete abstraction of regional formations from society. Yet the importance of viewing regionalism as a social construction is not just needed for historical understanding. As the following four examples reveal, it can shed light on important aspects of contemporary Belgium as well.

*Interaction between Walloons and Flemings*

Ever since the 1930s, laws have been in place mandating the use of French in public affairs in Walloon communes and Dutch in Flemish communes (Maroy 1969). Outside of Brussels the only exceptions to a policy of strict regional unilingualism can be found in communes along the language boundary and around Brussels where so-called language facilities are provided for speakers of the extraregional language (see McRae 1986). With the constitutional reforms of the past 20 years, the linguistic character of public institutions in Flanders and Wallonia is now well established.

The administrative partitioning of Belgium along language lines institutionalized the division between language communities in a way that may well be encouraging the growth of a regional ideology that sees Flanders as Flemish and Wallonia as Walloon (see Kabugubugu and Nuttin 1970–1971; Murphy 1988). With the rise of separate economic, social, cultural, and political institutions, people in northern and southern Belgium alike are constantly being confronted with images and arrangements that focus attention on the regional, as opposed to the national, level (Murphy 1988). It is unlikely that this process is benign with respect to feelings of regional identification. Moreover, it is reasonable to assume that regional thinking is being fueled by the concrete and symbolic legacies of the movements that sought to bring about the administrative division of Belgium along language lines. These include regional flags, regional holidays, and an extensive literature celebrating local history and culture.

Regional administrative structures and regionalist sentiment are affecting interaction between the Flemings of Flanders and the Walloons of Wallonia. This is most clearly seen in the institutional realm, where the partitioning of Belgium along language lines has resulted in the division of political parties, government ministries, universities, the state-run media, and some labor unions. Since these institutions were once loci of at least limited interaction between Flemings and Walloons, their division creates increased separation between the residents of Flanders and Wallonia. It is more difficult to ascertain the extent to which more general patterns of interregional interaction have been affected by the rise of linguistic regionalism. Indeed, there continue to be many forces promoting interaction, including the large number of workers from all over the country who commute to Brussels each day and the general popularity of tourist destinations in both Flanders and Wallonia. Nonetheless, there is some indication that the language boundary represents an increasingly significant barrier to interaction in Belgium.

One of these indications comes from a study by the Belgian geographer Jacques Charlier (1985) of patterns of communication by telephone in Belgium in 1982. His study revealed remarkably little contact between Flanders and Wallonia in comparison with contacts within those regions or between either of those regions and Brussels. Another striking indication of the role of the language boundary as an interaction barrier is the paucity of migration between Flanders and Wallonia.

Table 13.1, which sets forth data for interregional migration during 1987,

**Table 13.1** Migration between Belgian districts, 1987

| *Place of emigration* | *Place of immigration* | | | | | | | | | | | |
|---|---|---|---|---|---|---|---|---|---|---|---|---|
| | *Flemish districts* | | *Walloon districts* | | *Brussels* | | *Flemish districts outside Brabant* | | *Walloon districts outside Brabant* | | *Brabant* | |
| | *N* | *%* | *N* | *%* | *N* | *%* | *N* | *%* | *N* | *%* | *N* | *%* |
| Flemish districts | 68,053 | 79.3 | 6,322 | 7.4 | 11,427 | 13.3 | 57,662 | 67.2 | 4,423 | 5.2 | 23,717 | 27.6 |
| Walloon districts | 5,374 | 8.1 | 50,171 | 75.7 | 10,757 | 16.2 | 3,150 | 4.8 | 47,109 | 71.1 | 16,043 | 23.2 |
| Flemish districts outside Brabant | 58,182 | 89.7 | 3,433 | 5.3 | 3,231 | 5.0 | 50,259 | 77.5 | 3,183 | 4.9 | 11,404 | 17.6 |
| Walloon districts outside Brabant | 4,048 | 7.0 | 46,739 | 80.9 | 6,996 | 12.1 | 2,929 | 5.1 | 43,677 | 75.6 | 11,177 | 19.3 |

Source: Calculated from *Annuaire Statistique* (1988: 34–35).

reveals the extent to which the language boundary functions as a barrier to movement between Flanders and Wallonia. Taking all officially Flemish and Walloon districts into account, only 7.4 percent of those emigrating from a Flemish district to another district in Belgium moved from Flanders to Wallonia: almost 80 percent remained in Flanders and the rest went to Brussels. The figures are almost as skewed in the Walloon case: 8.1 percent moved to Flanders, as opposed to more than 75.7 percent who remained in Wallonia. Since the districts of Brabant are closely linked to the capital and encompass a number of communes with language facilities, an even better picture of the role of the language boundary as a barrier to interregional interaction can be seen by excluding the Brabant districts from consideration. Not taking those districts into account, the figures for migration from Flanders to Wallonia and from Wallonia to Flanders drop to around 5 per cent.

Given that some of those who are moving between Flanders and Wallonia are almost certainly foreigners (see Van der Haegen 1990), the regionalization of Belgium seems to be having a significant impact on interaction patterns. Part of this can be explained by simple language differences, but such differences were not enough to stem a significant migration of Flemish or Dutch speakers to southern Belgium during the nineteenth century when the Sambre-Meuse valley was a significant early center of industrialization (see Ministère de l'Industrie et du Travail 1896). It seems likely, then, that linguistic regionalism, with its associated arrangements and ideologies, is having a negative impact on interregional interaction.

Although the regionalization process in Belgium was arguably both necessary and warranted, it is important to recognize that resultant barriers to interregional interaction can foster misunderstanding and can lead to the growing apart of peoples. Regional isolation certainly helps to explain why surveys show that residents of Flanders and Wallonia have a much higher level of sympathy and understanding for those living in their own language region than for those living in other language regions (see McRae 1986: 109). More generally, the interaction issue suggests that one of the primary challenges facing Belgium in the years ahead is to sustain mutual cooperation and understanding in a system that is increasingly separated and disjointed. To make this point is not to argue for strong integrationist policies; such policies usually backfire under any circumstances (Murphy 1989a). Rather, it is to encourage thinking about how institutions can be structured to encourage cooperation and interaction on issues that do not threaten language community rights. This, in turn, will require a departure from the long-dominant idea that a nation is a group of people sharing a single language and a single culture (see Pattanayak and Bayer 1987).

### *The growth of a Brussels identity*

With the recent establishment of a formal regional government for Brussels, the Belgian capital now has a fully empowered regional administration on a par with Flanders and Wallonia (see Couttenier 1989). The status of Brussels has long been a central issue in the debate over language and regionalism in

Belgium (see Witte and Baetens Beardsmore 1987). Historically Brussels was a Brabantic city in which the vast majority of inhabitants spoke a Germanic regional language (see Verdoodt 1989: 115–120). During the past century and a half, however, the city has undergone considerable "frenchification," and Dutch has become the standardized tongue of Germanic speakers in the Belgian capital. Both French and Dutch are now official languages, but French is the mother tongue of a substantial majority of the current residents. As an officially bilingual region with significant competence over such matters as economic policy, public works, environmental protection, and scientific research, Brussels has acquired a territorial distinctiveness that transcends its role as national capital. It is now a distinct political unit and a focus for regional identity, particularly for Francophone residents of the capital (Lefèvre 1979).

The tendency to ignore the nature and implications of the regionalization process is clearly revealed in many recent studies of Brussels. A point of entry for many commentators is either the comparative status of French and Dutch speakers in Brussels or the nature and extent of contact between Francophone Bruxellois and their Dutch-speaking counterparts. Although such matters are of considerable interest, they suggest that the most significant manifestation of the communities problem in the Belgian capital is the distinction between French and Dutch speakers. Not only does this ignore the difficulty of assigning residents of the capital to one language group or the other: it fails to recognize that the rise of Brussels as a distinct regional unit within the Belgian state may be fostering a Brussels identity among French and Dutch speakers alike.

It would certainly be wrong to suggest that the people of Brussels constitute a single, cohesive community. At the same time, the degree to which the present regional structure is fostering a distinct set of shared concerns among residents of the capital is arguably at least as important as the distinctions that exist between Francophone and Dutch-speaking residents of the capital (see Murphy 1989b). Indeed, the regionalization process has introduced functional discontinuities between Brussels and the other language regions and has led to the frequent juxtaposition of the interests of Brussels against those of Flanders and Wallonia. This in turn is arguably fostering a growing sense of regional distinctiveness in the capital that will increasingly set its residents apart from the inhabitants of surrounding areas.

The controversy that often surrounds the allocation of funds to regional administrations provides a case in point. For residents of the capital, feelings of solidarity with one language group or another can easily be overshadowed by concerns that a smaller per capita share of national subsidies goes to Brussels than to Flanders or Wallonia. The feelings of regional distinctiveness that arise from such concerns are exacerbated by another feature of the regionalization process: the tendency to break down information about economic, demographic, political, and social matters by language regions. Residents of the capital are constantly being confronted with statistics showing how they differ collectively from their Flemish and Walloon counterparts. They regularly read the results of surveys that take as their starting point the division of Belgium into the three distinct language regions. It is almost inconceivable that the presentation of information in this manner has no effect on feelings of regional identity.

Once a sense of regional distinctiveness begins to emerge, it can be fed by the obvious distinctions that exist between regions. For residents of the Belgian capital, there is much to set Brussels apart. It is the home of many of the most important national cultural, economic, and political institutions. In addition, the regionalization process has led to the establishment of a variety of institutions that are strictly devoted to matters concerning the capital. Brussels is also the seat of important international bodies including the Commission of the European Communities. For many residents, these institutions are a source of both employment and pride, and they encourage an international outlook that is less prevalent in other Belgian cities.

None of this is meant to suggest that language is no longer a socially divisive force in Brussels. Yet despite continued controversy over the linguistic makeup of governmental agencies, conflicts over language use in public establishments are almost certainly less frequent than they were 30 years ago (Murphy 1989b). At the same time, Brussels is the one part of Belgium where institutional arrangements continue to encourage contact between Dutch and French speakers. Hence, the continued growth of a sense of distinctiveness is likely to be one of the more important aspects of the Belgian political scene in the years ahead.

### *Linguistic capabilities*

There are specific linguistic consequences to the regional developments discussed above. By most assessments, the establishment of Flanders and Wallonia as strictly unilingual regions has led to the decline of unilingual Francophones in the communes of Flanders that lack language facilities. Moreover, it may be discouraging use and/or knowledge of the "other" national language among Dutch-speaking Flemings and French-speaking Walloons (see De Vriendt and Van de Craen 1990). By contrast, the establishment of Brussels as an officially bilingual region and the growth of a sense of distinct regional identity in the capital have arguably encouraged bilingualism there.

The lack of available census information on language since 1947 and the uncertainty of much of the prior linguistic census data make it difficult to verify linguistic trends with any certainty (Baetens Beardsmore 1980). Nonetheless, there is considerable indirect evidence to substantiate the trends noted above, as revealed in a recent study of bilingualism in Belgium by De Vriendt and Van de Craen (1990). In the case of Flanders, census figures for language use in nine major Flemish cities in 1930 and in 1947 show an average drop in the percentage of residents using French all or most of the time from 5.95 percent to 4.11 percent of the population (calculated from figures in De Vriendt and Van de Craen 1990: 36). This corresponds to a period during which a policy of territorial unilingualism first acquired legal expression (Sonntag 1991). The heightened emphasis on territorial unilingualism in recent years has, by all indications, led to a further decline in the use of French among those living in Flemish communes without language facilities (De Vriendt and Van de Craen 1990: 40–41). At the same time, a recent survey by Radio-Télé-Luxembourg confirms that there are virtually no unilingual Francophones in Flanders outside of the communes with language facilities (De Vriendt and Van de Craen 1990: 39–40).

In the case of Wallonia, the effects of the regionalization process are more ambiguous. Although Walloon knowledge of Dutch has always been limited, a significant majority of Walloon students choose Dutch as their second language of formal study, and any Walloon child wishing to begin second language study as early as the fifth year of schooling must take Dutch. At the same time, a small but not insignificant minority of Walloon students who begin language study later in their schooling opt for English or German over Dutch (Hamers 1981). Whether this can be attributed to the partitioning of Belgium along language lines is, of course, an open question. But the current administrative structure, with its emphasis on the language regions as semiautonomous territorial units, may help to explain why some Walloon students express the view that learning English or German is more important than learning Dutch.

In the Brussels case, census data from the nineteenth and early twentieth centuries show a fairly steady increase in bilingualism in the capital from under 30 percent of the population in 1866 to almost 44 percent in 1947 (McRae 1986: 295). Since the 1947 census, bilingualism has further increased in Brussels (De Vriendt and Van de Craen 1990: 43–44), as has intermarriage between French and Dutch speakers (Louckx 1975). These developments reflect the emergence of a Brussels identity that is not based on a particular language. If anything, there is a growing sense that one of the important and attractive features of the capital is its bilingual character (Van de Craen and De Vriendt 1987). An indication of this is found in a survey conducted by the French-language newspaper *Le Soir*, which showed that around 70 percent of the Brussels population believes that people dealing with the public, whether in the private or the public sector, should be bilingual (reported in De Vriendt and Van de Craen 1990: 46). The commitment to bilingualism is also suggested by the increasing willingness of parents speaking one of the national languages to send their children to schools operating in the other (Gielen and Louckx 1984).

All of this suggests that an important link exists between language and regionalism. Changes in language use cannot be understood adequately as responses to abstract cultural or political stimuli. Rather, language use is shaped by the territorial understandings and arrangements that hold sway at any time. Hence, the future of bilingualism in Belgium will necessarily be closely tied to the evolution of regional structures and territorial identities.

### *The Fourons/Voeren case*

To the outside observer, one of the curiosities of the Belgian situation is the extent to which the linguistic situation in the Fourons/Voeren has attracted such attention and has so often been the catalyst for interregional tension. As recently as 1987, the Belgian government was brought down by an inability to resolve the Fourons/Voeren controversy (see Couttenier 1988), and aspects of the situation remain unsettled. Throughout the past 25 years, the Fourons/Voeren has been the staging ground for major demonstrations arising from a controversy over the commune's incorporation into Flanders in the early 1960s and, more recently, the actions of the mayor of the commune, José Happart.

Despite a long association with the province of Liège and a special bilingual status that dated from 1947, when the present official language boundary was established, in the early 1960s the commune was transferred to the Flemish province of Limburg. The commune was given special French language facilities, but the transfer was rejected by a majority of the local residents. Even though the mother tongue of most of the commune's residents is a Germanic dialect, a substantial majority did not wish to have their close economic and cultural ties with Liège severed (CRISP 1979; Hermans and Verjans 1983). Consequently, many resisted the incorporation of the commune into Flanders; their views found expression in a new political party named "Retour à Liège." That party has consistently captured a majority of local council seats.

The controversy has never been resolved, and the election of José Happart as mayor in 1983 further complicated the situation. As a commune that is officially part of Flanders. the Fourons/Voeren must by law be governed in Dutch. Happart, a member of the "Retour à Liège" party, claimed that he only knew French and refused to use Dutch in the execution of his duties. The resulting crisis has been a consistent source of tension not only within the commune, but between Flemings and Walloons generally.

Although the details of the Fourons/Voeren case are relatively well known, it is remarkable that so few commentators have questioned how or why a controversy in a small, economically peripheral region comprising less than 0.05 percent of the Belgian population can attract so much attention. The Fourons/Voeren is certainly not critical to either Flemish or Walloon political/cultural ambitions, and the fate of the commune scarcely attracted notice prior to the discussions leading up to the early 1960s language laws. Since the cultural and economic characteristics of Flanders and Wallonia shed no light on the matter, consideration should be given to the role of the regionalization process itself.

The language issue in Belgium during the nineteenth century primarily centered on the right of individual Flemings to use their language in the schools, the courts, the military, and the government (see Clough 1930). With the adoption of a territorial approach to language rights in the early decades of the twentieth century, the focus shifted from individual rights to territorial rights (McRae 1975). The goal was to establish language regions with particular legal requirements regarding the use of language that would apply to everyone living within the region. The process culminated in the constitutional revisions of the past 20 years, which institutionalize a largely territorial conception of language rights. An inevitable outgrowth of this development is that the types of issues that are now seen as important are primarily of a territorial nature. They include the territorial extent of the language regions, the linguistic status of communes along the language border, and the regional status of Brussels (see Murphy 1988: 181–182).

Against this backdrop the situation in the Fourons/Voeren begins to make more sense. To many Flemings, the actions of Happart and other supporters of the "Retour à Liège" agenda represent a direct blow to the integrity of the territorial structure that sustains and protects Flemish interests in the Belgian state. In their view, giving in on the issue would only precipitate arguments to readjust the language border in other places, most notably around Brussels. This would undermine the integrity of the current regional structure and open the way to a defeat of Flemish rights through a gradual "frenchification" of Flanders.

By contrast, supporters of "Retour à Liège" see the incorporation of the commune into Flanders as an example of Flemish territorial hegemony. They argue that the present regional structure has no legitimacy unless it is based on the territorial aspirations of local populations. Since a majority of the country's population is already located in Flanders, they see the desire to keep the Fourons/Voeren in Flanders as a Flemish attempt to use their majority status to deny the rights of individuals who wish to align themselves with Wallonia.

Many Belgians reject both of the positions outlined above, of course. They see the Fourons/Voeren situation as little more than an annoying problem that should be resolved as quickly and painlessly as possible through compromise. Nonetheless, it remains an important issue for some precisely because the regionalization process has made it important. When rights are defined territorially, the nature and extent of the resulting territorial structures are of critical importance. The Fourons/Voeren case is sufficiently peripheral that a reasonable compromise could result in an easing of tensions. A more serious long-term issue is the status of the officially Dutch-speaking communes around Brussels, a number of which contain significant Francophone minorities or even majorities. Given the current emphasis on territorial issues and the economic and political significance of these communes, they are likely to be foci of considerable interregional tension in the years ahead.

## Conclusion

The recent history of linguistic regionalism in Belgium should remind us that Flanders, Wallonia, and Brussels, like all regions, are socially constructed places. They are not just convenient or politically significant compartments for analysis; they are part of what is being analyzed. Studies of the economy of Wallonia, language use in Brussels, and education in Flanders will necessarily be limited if the language regions of Belgium are simply treated as containers. Wallonia, Brussels, and Flanders reflect and influence arrangements and ideas that are integral to the existence and subsequent development of the social, economic, cultural, and political characteristics of those regions.

Treating the language regions of Belgium as social constructs can also contribute to the project of moving beyond a social science dominated by state-centered analyses and by occluded theories of regionalism. For all the surge of interest in the concepts of place and region in the recent literature on society, we are only beginning to understand the nature and significance of perceptual and functional compartmentalizations of space. The ideological dimensions of regionalism have barely been addressed (Paasi 1986; Murphy 1991). Exploring the social significance of regions and regional change in Belgium, as well as in other contexts, can provide insights into the creation and development of group identities and intergroup relations. It is an endeavor that goes to the heart of our efforts to understand and explain the relationship between language and society.

## References

Agnew, John A. (1987). *Place and Politics: The Geographical Mediation of State and Society*. Boston: Allen and Unwin.

—(1989). The devaluation of place in social science. In *The Power of Place: Bringing Together Geographical and Sociological Imaginations*, J. A. Agnew and J. S. Duncan (eds.), 9–29. Boston: Unwin Hyman.

*Annuaire Statistique* (1988). *Annuaire Statistique de la Belgique*. Brussels: Ministère des Affaires Economiques.

Baetens Beardsmore, Hugo (1980). Bilingualism in Belgium. *Journal of Multilingual and Multicultural Development* 1(2), 145–154.

Charlier, Jacques (1985). Les flux téléphoniques interzonaux en Belgique en 1982: une approache multivariée. Unpublished paper presented to the Study Group on the Geography of Communication, International Geographical Union, Montpelier, France, November 18–19.

Clough, Shepard B. (1930). *A History of the Flemish Movement in Belgium: A Study in Nationalism*. New York: R. H. Smith.

Connor, Walker (1972). Nation-building or nation-destroying? *World Politics* 24(3), 319–355.

—(1978). A nation is a nation, is a state, is an ethnic group, is a . . . *Ethnic and Racial Studies* 1(4), 377–400.

Coudenberg Group (1989). *The New Belgian Institutional Framework*. Brussels: Coudenberg Group.

Couttenier, Ivan (1988). Belgian politics in 1987. *Res Publica* 30, 201–231.

—(1989). Belgian politics in 1988. *Res Publica* 31, 302–328.

CRISP (Centre de Recherche et d'Information Socio-Politiques) (1979). Le problème des Fourons de 1962 à nos jours. *Courrier Hebdomadaire du C.R.I.S.P.* 859 (November 23).

De Vriendt, Sera and Van de Craen, Piet (1990). *Bilingualism and Belgium: A History and an Appraisal*. Centre for Language and Communication Studies Occasional Paper No. 23. Trinity College, Dublin.

Giddens, Anthony (1985). *The Constitution of Society*. Berkeley: University of California Press.

Gielen, Gerda and Louckx, Freddy (1984). Sociologisch onderzoek naar de herkomst, het taalgedrag en het schoolkeuzegedrag van ouders met kinderen in het nederlandstalig basisonderwijs van de Brussels agglomeratie. *Taal en Sociale Integratie* 7, 161–208.

Hamers, Josiane (1981). The language question in Belgium. *Language and Society* 5, 17–20.

Hermans, Michel and Verjans, Pierre (1983). Les origines de la querelle fouronaise. *Courrier Hebdomadaire du C.R.I.S.P.* 1019 (December 2).

Kabugubugu, Amédée and Nuttin, Joseph R. (1970–1971). Changement d'attitude envers la Belgique chez les étudiants flamands. *Psychologica Belgica* 11, 23–44.

Lefèvre, Jacques (1979). Nationalisme linguistique et identification linguistique: le cas de Belgique. *International Journal of the Sociology of Language* 20, 37–58.

Louckx, Freddy (1975). *Wetenschappelijk Onderzoek van de Taaltoestanden in de Brusselse Agglomeratie: Autochtone Brusselaars op de Tweesprong*. Brussels: Nederlandse Commissie voor de Cultuur van de Brusselse Agglomeratie.

McRae, Kenneth D. (1975). The principle of territoriality and the principle of personality in multilingual states. *Linguistics* 158, 33–54.

—(1986). *Conflict and Compromise in Multilingual Societies: Belgium*. Waterloo, Ontario: Wilfrid Laurier University Press.

Markusen, Ann R. (1987). *Regions: The Economics and Politics of Territory*. Totowa, NJ: Rowman and Littlefield.

Maroy, Pierre (1969). L'évolution de la législation linguistique belge. *Revue du Droit Public et de la Science Politique en France* 82, 449–501.

Massey, Doreen (1978). Regionalism: some current issues. *Capital and Class* 6, 106–125.

Mikesell, Marvin W. (1983). The myth of the nation state. *Journal of Geography* 82(6), 257–260.

Ministère de l'Industrie et du Travail (1896). *Revue du Travail*. Brussels: Ministère de l'Industrie et du Travail.

Murphy, Alexander B. (1988). *The Regional Dynamics of Language Differentiation in Belgium: A Study in Cultural-Political Geography*. Geography Research Series No. 227. Chicago: University of Chicago.

—(1989a). Territorial policies in multiethnic states. *Geographical Review* 79(4), 410–421.

—(1989b). The territorial dimension of sociolinguistic patterns and processes in Brussels. In *Taal en Sociale Integratie* 13, P. Van de Craen (ed.), 117–128. Brussels: VUB Press.

—(1991). Regions as social constructs: the gap between theory and practice. *Progress in Human Geography* 15(1), 23–35.

Paasi, Anssi (1986). The institutionalization of regions: a theoretical framework for understanding the emergence of regions and the constitution of regional identity. *Fennia* 164, 105–146.

Pattanayak, D. B. and Bayer, J. M. (1987). Laponce's "the French language in Canada: tensions between geography and politics" – a rejoinder. *Political Geography Quarterly* 6(3), 261–263.

Rudolph, Joseph R., Jr. and Thompson, Robert J. (eds.) (1989). *Ethnoterritorial Politics, Policy, and the Western World*. Boulder, CO: Lynne Rienner.

Sack, Robert D. (1981). Territorial bases of power. In *Political Studies from Spatial Perspectives*. A. D. Burnett and P. J. Taylor (eds.), 53–71. New York: Wiley.

—(1986). *Human Territoriality: Its Theory and History*. Cambridge Studies in Historical Geography. Cambridge: Cambridge University Press.

Soja, Edward W. (1971). *The Political Organization of Space*. Commission on College Geography Resource Paper No. 8. Washington, D.C.: Association of American Geographers.

—(1989) *Postmodern Geographies: The Reassertion of Space in Critical Social Theory*. London: Verso.

Sonntag, Selma K. (1991). *Competition and Compromise Amongst Elites in Belgian Language Politics*. Plurilingua XII. Brussels: Research Centre on Multilingualism; Bonn: Dümmler.

Thrift, Nigel J. (1983). On the determination of social action in space and time. *Environment and Planning D: Society and Space* 1, 23–57.

Van de Craen, Piet and De Vriendt, Sera (1987). Réalités et politiques linguistiques: le cas de Bruxelles. *Etudes de Linguistique Appliquée* janv.–mars, 110–116.

Van der Haegen, Herman (1990). De bevolking van vreemde nationaliteit en haar impact op de Belgische demografie. *Bevolking en Gezin* (1), 7–36.

Verdoodt, Alben (1977). *Les Problèmes des Groupes Linguistiques en Belgique*. Bibliothèque des Cahiers de l'Institut de Linguistique de Louvain, 10. Louvain: Peeters.

—(1989). *Regional and Minority Languages of the Member Countries of the Council of Europe*. Québec: Presses Laval.

Williams, Colin H. and Smith, Anthony D. (1983). The national construction of social space. *Progress in Human Geography* 7, 502–518.

Witte, Els and Baetens Beardsmore. Hugo (eds) (1987). *The Interdisciplinary Study of Urban Bilingualism in Brussels*. Clevedon, Avon: Multilingual Matters.

Wixman, Ronald (1986). Applied Soviet nationality policy: a suggested rationale. In *Passé Turco-Tatar – Présent Soviétique*, C. Lemercier-Quelquejay, G. Veinstein, and S. E. Wimbush (eds), 449–468. Louvain: Peeters.

# 14 Loïc J. D. Wacquant, 'The New Urban Color Line: the State and the Fate of the Ghetto in PostFordist America'

Excerpt from: Craig Calhoun (ed.), *Social Theory and the Politics of Identity*, Chapter 9. Oxford: Blackwell (1994)

Tryin' to survive, tryin' to stay alive
The ghetto, talkin' 'bout the ghetto
Even though the streets are bumpy, lights burnt out
Dope fiends die with a pipe in their mouth
Old school buddies not doin' it right
Every day it's the same and it's the same every night
I wouldn't shoot you bro' but I'd shoot that fool
If he played me close and tried to test my cool
Every day I wonder just how I'll die
The only thing I know is how to survive.

Too Short, "The Ghetto."[1]

## From race riots to silent riots: changing visions of the ghetto

Twenty years after the uprisings that lighted fires of frustration and rage in the black slums of the American metropolis, the ghetto has returned to the frontline of national issues. Only, this time, the open racial uprisings that tore through the Afro-American communities of northern cities in defiant revolt against white authority have given way to the "slow rioting" (Curtis 1985) of black-on-black crime, mass school rejection, drug trafficking and internal social decay.[2] On the nightly news, scenes of white policemen unleashing state violence on peaceful black demonstrators demanding mere recognition of their elemental constitutional rights have been replaced by reports on drive-by shootings, homelessness, and teenage pregnancy. Black ministers, local politicians, and concerned mothers still agitate and demonstrate, but their pleas and marches are less often directed at the government than at the drug dealers and gangs who have turned so many inner-city neighborhoods into theaters of dread and death. The vision of "Negro" looters and black power activists reclaiming forceful control over their community's fate (Boskin 1970) and riding the crest of a wave of racial pride and self-assertion has given way to the loathsome imagery of the "underclass," a term that purports to denote a new segment of the minority poor allegedly characterized by behavioral deficiency and cultural deviance (Auletta 1982; Sawhill 1989), a menacing urban hydra personified by the defiant and aggressive gang member and the dissolute if passive teenage "welfare mother," twin emblematic figures whose (self-)destructive behavior is said to represent the one a physical threat and the other a moral attack on the integrity of American values and national life.

The wave of social movements that energized the black community and helped lift collective hopes through the 1960s (Morris 1984; McAdam 1981) has subsided and, with it, the country's commitment to combating racial inequality. This is well reflected in the changing idiom of public debate on the ghetto. As the "War on Poverty" of Lyndon B. Johnson was replaced by the "War on Welfare" of Ronald Reagan (Katz 1989), the issue of the societal connection between race, class, and poverty was reformulated in terms of the personal motivations, family norms, and group values of the residents of the inner city, with welfare playing the part of the villain. The goals of government policy, too, were downgraded accordingly: rather than pursue the eradication of poverty – the optimistic target that the Great Society program was set to reach by 1976 as a tribute to the nation's bicentennial – and the diminution of racial disparities, the state is now content to oversee the containment of the first in crumbling minority enclaves (and in the jails that have been built at an astounding pace in the past decade to absorb the most disruptive of their occupants) and "benign neglect" of the second. Accordingly, the focus of social research has shifted from the urban color line to the individual defects of the black poor, from the ghetto as a mechanism of racial domination and economic oppression (Clark 1965; Liebow 1967; Rainwater 1970), and the structural political and economic impediments that block the full participation of poor urban blacks to the national collectivity, to the "pathologies" of the so-called underclass said to inhabit it and to the punitive measures that may be employed to minimize their claim upon collective resources and to force them into the peripheral segments of an expanding low-wage labor market (e.g., Ricketts and Sawhill 1988; Mead 1989).[3]

These shifts in the symbolic representation and political treatment of the ghetto, however, can hardly efface the fact that the ominous forewarning of the 1968 National Advisory Commission on Civil Disorders (Kerner Commission 1989: 396, 389) has come true: "The country [has moved] toward two societies, separate and unequal" as a consequence of "the accelerating segregation of low-income, disadvantaged Negroes within the ghettos of the largest American cities." While the black middle class has experienced real, if tenuous, progress and expansion thanks largely to governmental efforts and (secondarily) to increased legal pressure upon corporate employers (Collins 1983; Landry 1987; Son *et al.* 1989), urban black poverty is more intense, more tenacious, and more concentrated today than it was in the 1960s (Wilson 1987). And the economic, social, and cultural distance between inner-city minorities and the rest of society has reached levels that are unprecedented in modern American history as well as unknown in other advanced societies.

*Not the same old ghetto*

Is this to say, borrowing the words of historian Gilbert Osofsky (1971: 189), that there is an "unending and tragic sameness about black life in the metropolis," that of the "enduring ghetto," which perpetuates itself through time unaffected by societal trends and political forces as momentous as the onset of a postindustrial economy, the enactment of broad civil rights and affirmative action legislation, and the reorganization of urban space under the twin pressures of suburban deconcentration and central-city gentrification? Quite the contrary. For underneath the persistence of economic subordination and racial entrapment, the ghetto of the 1980s is quite different from that of the 1950s. The *communal* ghetto of the immediate post-war era, compact, sharply bounded, and comprising a full complement of black classes bound together by a unified collective consciousness, a near-complete social division of labor, and broad-based communitarian agencies of mobilization and representation, has been replaced by what we may call the *hyperghetto* of the 1980s and 1990s (Wacquant 1989, 1991), whose spatial configuration, institutional and demographic makeup, structural position and function in urban society are quite novel. Furthermore, the separation of the ghetto from the rest of American society is only apparent: it is one of "lifeworld," not "system," to use a conceptual distinction elaborated by Habermas (1984). It refers to the concrete experiences and relations of its occupants, not to the underlying ties that firmly anchor them in the metropolitan ensemble – if in exclusionary fashion. For, as I shall argue in this chapter, there are deep-seated causal and functional linkages between the transformation of the ghetto and changes in the structure of the US economy, society, and polity over the past three decades.

Analysis of the economic and political factors that have combined to turn them into veritable domestic "Bantustans" reveals that ghettos are not autonomous social entities that contain within themselves the principle of their own reproduction and change. It demonstrates also that the parlous state of America's historic "Black Belts" is not the simple mechanical result of deindustrialization, demographic movements, or of a skills or spatial "mismatch" rooted in ecological processes, and still much less the product of the rise of a "new" underclass, *in statu nascendi* or already "crystallized" into a "permanent" fixture of the American urban landscape (Loewenstein 1985; *Chicago Tribune* 1986; Nathan 1987), whether defined by its behavior, income, culture, or isolation. It is the product, rather, of a transformation of the *political* articulation of race, class, and urban space in both discourse and objective reality.

The ghetto is still with us but it is a different "kind" of ghetto: its internal makeup has changed along with its environment and with the institutional processes that simultaneously chain it to the rest of American society and ensure its dependent and marginal location within it. To understand these differences, what the ghetto is and means to both insiders and outsiders, one must sweep aside the discourse of the "underclass" that has crowded the stage of the resurging debate on race and poverty in the city (Fainstein 1993) and reconstruct instead the linked relations between the transformation of everyday life and social relations inside the urban core, on the one hand, and the restructuring of the system of forces, economic, racial, and political, that account for the particular configuration of caste and class it materializes. Accordingly, the main focus of this analysis will be on the *external* factors that have reshaped the social and symbolic territory within which ghetto residents (re)define themselves and the collectivity they form, and it addresses the *internal* production of its specific social order and consciousness only indirectly. This emphasis is not born of the belief that structural determination constitutes the *alpha* and *omega* of identity formation, far from it. It rests, rather, on two premises, one theoretical and the other empirical.

The first premise is that elucidation of the objective conditions under which identity comes to be constructed, asserted, and disputed in the inner city constitutes a sociological prerequisite to the analysis of the experiential *Lebenswelt* of the ghetto and its embedded forms of practice and signification. It is in this objective space of material and symbolic positions and resources that are rooted the strategies deployed by ghetto residents to figure out who they are and who they can be. While I have no doubt that such an analysis remains unfinished absent the complement of an "indigenous perspective" (*à la* Aldon Morris) throwing light on the complexities of identity formation "from below" (or, to be more precise, from within), I also believe that populist celebration of "the value of blackness" and of the richness of "oppositional black culture" (hooks 1992: 17) offers neither a substitute, nor an adequate starting point for a rigorous assessment of the state and fate of the ghetto at the close of the Fordist era.

The second premise of this inquiry is that the reality of the ghetto as a physical, social, and symbolic place in American society is, whether one likes it or not, largely being decided – indeed imposed – from outside, as its residents are increasingly dispossessed of the means to produce their own collective and individual identities. A brief contrast of the opposed provenance, uses, and semantic charge of the vocabularies of "soul" and "underclass" is instructive in this respect. The notion of soul, which gained wide appeal during the racial turmoil of the 1960s, was a "folk conception of the lower-class urban Negro's own 'national character' " (Hannerz 1968: 54). Produced from within for in-group consumption, it served as a symbol of solidarity and a badge of personal and group pride. By contrast, "underclass status" is established wholly from the outside (and from above) and forced upon its putative "members" by specialists in symbolic production – journalists, politicians, academics, and governmental experts – for purposes of control and disciplining (in Foucault's sense of the term) and without the slightest concern for the self-understanding of those who are arbitrarily lumped into this analytical fiction. Whereas the folk concept of soul, as part of an "internal ghetto dialogue" toward an indigenous reassessment of black identity (Keil 1966), was appraisive, the idiom of underclass is a derogatory label, an identity that nobody claims except to pin it on an Other. That even "insurgent" black intellectuals such as Cornel West should embrace the idiom of underclass is revealing of the degree to which the ghetto has become an *alien object* on the landscape of American society.

*Three preliminary caveats*

Three caveats are in order before drawing a portrait of social conditions and living in the contemporary inner city, using Chicago as an illustrative case. First, it must be emphasized that the ghetto is not simply a topographic entity or an aggregation of poor families and individuals but an *institutional form*, that is, a particular, spatially based, concatenation of mechanisms of *ethnoracial closure and control*. Briefly put, a ghetto may be ideal-typically characterized as a bounded, racially and/or culturally uniform sociospatial formation based on the forcible relegation of a negatively typed population – such as Jews in medieval Europe and African-Americans in the modern United States – to a reserved territory in which this population develops a set of specific institutions that operate both as a functional substitute for, and as a protective buffer from, the dominant institutions of the encompassing society (Wacquant 1991). The fact that most ghettos have *historically* been places of widespread and sometimes acute material misery does not mean that a ghetto necessarily has to be poor – certainly, the "Bronzeville" of the 1940s was more prosperous than most Southern black communities – nor that it has to be uniformly deprived.[4] This implies that the ghetto is not a social monolith. Notwithstanding their extreme dilapidation, many inner-city neighborhoods still contain a modicum of occupational, cultural, and family variety. Neither is the ghetto entirely barren: amidst its desolation, scattered islets of (relative) economic and social stability persist, which offer fragile but crucial launching pads for the strategies of coping and escape of its residents, and new forms of sociability continually develop in the cracks of the crumbling system.

Second, one must resist the tendency to treat the ghetto as an alien space, to see only what is different in it, in short to *exoticize it*, as proponents of the scholarly myth of the "underclass" have been wont to do in their grisly tales of "antisocial" behavior that resonate so well with journalistic reports (from which they are often drawn in the first place) and with common class and racial prejudice against the black poor. Indeed, a cursory sociology of sociology would show that most descriptions of the "underclass" reveal more about the *relation* of the analyst to the object, and about his or her racial and class preconceptions, fears, and fantasies, than they do about their putative object; and that representations of "underclass areas" bear the distinctive mark of the ostensibly "neutral" (that is, dominant) gaze set upon them from a distance by analysts who, all too often, have rarely set foot in one.[5] Ghetto dwellers are not a distinctive breed of men and women in need of a special denomination; they are ordinary people trying to make a living and to improve their lot as best they can under the unusually oppressive and depressed circumstances imposed upon them. Though their cultural codes and patterns of conduct may, from the standpoint of a secure outside observer, appear peculiar, quixotic or even "aberrant" (a word so often reiterated when talking about the ghetto that it has become virtually oxymoronic with it), on closer examination, they turn out to obey a social rationality that takes stock of past experiences and is well suited to their immediate socioeconomic context and possibilities (Wacquant 1992a).

The third caveat stresses, against the central premise of American poverty research, that the ghetto does not suffer from "social disorganization" – another moralizing concept that had by now best be banned from the social sciences. Rather, it is *organized differently* in response to the relentless press of economic necessity, social insecurity, racial enmity, and political stigmatization. The ghetto comprises a particular type of social order, premised on the racial marking and dualization of space, "organized around an intense competition for, and conflict over, the scarce resources" that suffuses an environment replete with "social predators" (Sanchez-Jankowski 1991: 22, 183–92) and politically constituted as inferior. Finally and relatedly, one must keep in mind that ghetto dwellers are not part of a separate group somehow severed from the rest of society, as many advocates of the "underclass" thesis would have us believe. They belong, rather, to unskilled and socially disqualified fractions of the black working class, if only by virtue of the multifarious kinship and marital links, social ties, cultural connections, and institutional processes that cut across the alleged divide between them and the rest of the Afro-American community (Aschenbrenner 1975; Collins 1983: 370; Pétonnet 1985).[6]

## From the "communal" ghetto of the 1950s to the "hyperghetto" of the 1980s

The process of black ghettoization – from initial piling up and expansion to sudden white flight and disinvestment, followed by abrupt increases in joblessness, crime, educational retardation, and other social dislocations – is old and well known: it goes back to the initial formation of the ghetto as an institution of *racial exclusion* in the early decades of the century.[7] It must be emphasized at the outset that blacks are the only group to have experienced ghettoization in American society. White immigrants of various peripheral provenance (Italians, Irish, Polish, Jews, etc.) initially lived in heterogeneous *ethnic neighborhoods* which, though they may have been slums, were temporary and, for the most part, voluntary way-stations on the road to integration in a composite white society; they were not, *pace* Wirth (1927), ghettos in any sense other than an impressionistic, journalistic one. Segregation in them was only partial, and based on a mixture of class, nationality, and citizenship. The residential confinement of blacks, on the other hand, was (and still is) unique in that only Afro-Americans have had to live in areas where "segregation was practically total, essentially unvoluntary, and also perpetual" (Philpott 1978: xvi).[8] Moreover, the forced separation of blacks went beyond housing to encompass other basic institutional arenas, from schooling and employment to public services and political representation, leading to the development of a parallel social structure without counterpart among whites.

What is distinctive about black ghettoization today is, first, that it has become spatially as well as institutionally differentiated and *decentered*, split, as it were, between a decaying, if expanding, urban core on the one hand, and satellite working-class and middle-class neighborhoods located on the periphery of cities and, increasingly, in segregated suburbs often adjacent to the historic Black Belt. The second novel feature of black ghettoization in postFordist America is its sheer scale and "the intensity of the collapse at the center of the ghetto," as well as the fact that "the cycle still operates two decades after fair housing laws have been in effect" (Orfield 1985: 163). Indeed, in that very period when legal changes were presumed to bring about its amelioration, the inner city has been plagued by accelerating physical degradation, rampant insecurity and violence, and degrees of economic exclusion and social hardship comparable only to those of the worst years of the Great Depression.

*Physical decay and danger in the urban core*

Walk along 63rd Street on the South Side of Chicago, within a stone's throw of the University of Chicago campus, along what used to be one of the city's most vibrant commercial strips, and you will witness a grim spectacle repeated over and over across the black ghettos of America – in Harlem or in the Brownsville district of Brooklyn in New York City, in Camden, New Jersey, on the East Side of Cleveland, or in Boston's Roxbury.[9] Abandoned buildings, vacant lots strewn with debris and garbage, broken sidewalks, boarded-up store-front churches, and the charred remains of shops line up miles and miles of decaying neighborhoods left to rot since the 1960s.

Forty years ago, 63rd Street was called the "Miracle Mile" by local merchants vying for space and a piece of the pie. There were nearly 800 businesses, and not a single vacant lot in an eighteen-by-four block area. The neighborhood was lively as people streamed in from other parts of town, joining in crowds so dense at rush hour that one was literally swept off one's feet upon getting out of the elevated train station. Large restaurants were open around the clock; no fewer than five banks and six hotels were present; and movie houses, taverns, and ballrooms never seemed to empty. Here is the description of the street by the only white shopkeeper left from that era:

> It looks like Berlin after the war and that's sad. The street is bombed out, decaying. Seventy-five per cent of it is vacant. It's very unfortunate but it seems that all that really grows here is liquor stores. And they're not contributing anything to the community: it's all "*take, take, take!*" Very depressing. [Sighs heavily] It's an area devoid of hope, it's an area devoid of investments. People don't come into Woodlawn.

Now the street's nickname has taken an ironic twist: it is a miracle for a business to survive on it. Not a single theater, bank, jazz club, or repair shop outlived the 1970s. The lumber yards, print shops, garages, and light manufacturing enterprises have disappeared as well. Fewer than 90 commercial establishments remain, most of them tiny eating places, beauty parlors, and barber shops, apparel, food, and liquor outlets which each employs at best a handful of workers.

Perhaps the most significant fact of daily life in today's ghetto, however, is the extraordinary *prevalence of physical danger and the acute sense of insecurity* that pervade its streets.[10] Between 1980 and 1984 alone, serious crimes in Chicago multiplied fourfold to reach the astonishing rate of 1,254 per 1,000 residents. Most of them were committed by and upon residents of the ghetto. A plurality of the 849 homicide victims officially recorded in Chicago in 1990 were young Afro-American men, most of them shot to death in poor all-black neighborhoods. With the wide diffusion of drugs and firearms, mortality in major inner-city areas has reached "rates that justify special consideration analogous to that given to 'natural disaster areas' "; today males in Bangladesh have a higher probability of survival after age 35 than their counterparts of Harlem (McCord and Freeman 1990). No wonder some analysts of the urban scene openly talk of young black males as "an endangered species" (Gibbs 1989). The combined availability of guns, durable exclusion from wage-labor, and pervasiveness of the drug trade have modified the rules of masculine confrontation on the street in ways that fuel the escalation of deadly assault. A former leader of the Black Gangster Disciples muses:

> See, back then, if two gang guys wanna fight, they let 'em two guys fight *one-on-one*. But it's not like that now: if you wanna fight me, I'mma git me a gun an' shoot you, you see what I'm sayin'? Whenever you got a gun, tha's the first thin' you think about – not about *peace treaties* an' let dese two guys fight and settle their disagreement as real grown men. It's *scary now* because dese guys, they don't have – [his voice rising in shock] I mean they don't have *no value for life – no value!*

Residences are scarcely safer than the streets. The windows and doors of apartments or houses are commonly barricaded behind heavy metal gates and burglar bars. Public facilities are not spared. Elderly ghetto dwellers nostalgically evoke a time when they used to sleep in municipal parks in the summer, rolled in mosquito nets, or on rooftops and balconies in search of relief from the summer heat. Nowadays, parks are considered "no-go" areas, especially after nightfall; some are even off-limits to the youths who live in their immediate vicinity because they fall within the territory of a rival gang. Buses of the Chicago Transit Authority running routes from the downtown

Loop through the South Side are escorted by special police squad cars to deter assaults and still register several hundred violent incidents per month. Several CTA train stations on the Jackson Park line have been closed to entry in an attempt to limit crime, at the cost of denying local residents access to public transportation. Insecurity is so deep that simply maneuvering one's way through public space has become a major dilemma in the daily life of inner-city residents, as averred by this comment from an elderly South Sider on a sunny day of late June: "Oh, I hate to see this hot weather back. I mean I do like warm weather, *its the people it bring out I don't like*: punks and dope fiends, you're beginnin' to see them outa d'buildings now, on d'streets. This ain't no good."

Schools are no exception to this pattern. Many public establishments in Chicago's inner city organize parents' militias that patrol the school grounds armed with baseball bats while classes are in session. Others hire off-duty policemen to supplement security and use metal detectors to try to limit the number of guns and other weapons circulating on school grounds. A South Side elementary school on 55th Street briefly made the headlines after five youths were gunned down and killed within a few blocks in the course of a single year. Its students were found to be living in "numbing fear" of the gang violence that awaits them outside the school. Children "say they are afraid for their lives to go to school" confessed one teacher. "It seems like every year somebody's child loses their life and can't get out of 8th grade," added a mother. And the principal could only regret that the school security guards are unable to provide protection once pupils leave the premises (*Chicago Tribune* 1990).

Today's ghetto truly is "no place to be a child," as goes the title of a recent book comparing Chicago's inner city to refugee camps in war-torn Cambodia (Garbarino *et al.* 1991). Youngsters raised in this environment of pandemic violence experience enormous emotional damage and display posttraumatic stress disorders very similar to those suffered by veterans. A tenant of a South Side high-rise complex (cited in Brune and Camacho 1983: 13) concurs that Chicago "is no place to raise a family. It's like a three-ring circus around here during the hot weather. There's constantly fighting. They've been times when we had to take all the kids and put 'em in the hallway on the floor, so much gunfire around here." By age five, virtually all children living in large public housing projects have encountered shooting or death firsthand. Many mothers opt to send their offspring away to stay in the suburbs or with family in the South to shelter them from the neighborhood's brutality.

The incidence of crime in the ghetto is exacerbated by the racial closure of space in American cities. If so much violence is of the "black-on-black" variety, it is not only because inner-city Afro-Americans suffer extreme economic redundancy and social alienation. It is also that anonymous black males have become widely recognized symbols of danger (Anderson 1991: chapter 6) so that, unless they display the trappings of middle-class culture, they are routinely barred from bordering white areas where their skin color causes them to be immediately viewed as potential criminals or troublemakers: "You can't go over to the white community to do anything, because when you're seen over there, you're already stopped on suspicion. So you got to prey in your den, because you're less noticeable over there. You got to burglarize your own people" (cited in Blauner 1989: 223).

### *Depopulation, economic exclusion, and the organizational collapse of the ghetto*

Yet the continued physical and commercial decline, rising street violence, and ubiquitous insecurity of the ghetto are themselves but surface manifestations of a more profound transformation of its socioeconomic and institutional fabric. First, whereas the ghetto of the 1950s was overpopulated as the result of the swelling influx of black migrants from the South triggered by the wartime boom and the mechanization of Southern agriculture, the contemporary ghetto has been undergoing steady depopulation as better-off families moved out in search of more congenial surroundings. For instance, the core of Chicago's South Side lost close to half of its inhabitants, as the residents of Oakland, Grand Boulevard, and Washington Park decreased from about 200,000 in 1950 to 102,000 in 1980 – slipping even further down to an estimated 63,500 in 1990 according to early returns of the census. During those years, moreover, despite the construction of massive public housing high-rises, the number of housing units decreased by one third through arson (often perpetrated by absentee landlords seeking to collect insurance money) and the abandonment and destruction brought about by urban renewal programs that razed more dwellings than they built, so that overcrowding and inadequate housing are still widespread in the urban core.

But the most dramatic change in the demography of the ghetto has been the precipitous decline of the employed population caused by two mutually reinforcing factors: the continuing exodus of upwardly mobile black families and the rising joblessness of those left behind. In 1950 over half of the adults living at the heart of the South Side Black Belt were gainfully employed, a rate equal to that of the city as a whole. Chicago was still a dominant national industrial center then and half those employed blacks held blue-collar jobs. By 1980, the number of working residents had dropped by a staggering 77 per cent so that nearly three of every four persons over the age of 16 were jobless. In 30 years, the number of operatives and laborers crumbled from 35,808 to 4,963; that for craftsmen plummeted from 6,564 to 1,338, while the corresponding figure for private household and service workers fell from 25,181 to 5,203. And, whereas the black middle class multiplied fivefold citywide between 1950 and 1980, the number of white-collar employees, managers, and professionals living in the urban core was cut by half, from 15,341 to 7,394. A long-time resident of Woodlawn (who, ironically, recently moved to the city's North Side to shelter his children from the violence of the streets) complains about the disappearance of better-off families from his old South Side neighborhood:

> It [used to] be tonsa teachers livin' in d'neighbo'hood, but now they movin', *everybody move up* . . . If you look at d'community, Louie, *it's decayin': ain't nobody*

> *here*. Ain't no teachers on 63rd Street, over here on Maryland, *ain't none*, know what I'm sayin'? Ev'rybody that's got a lil' knowledge, they leavin' it. If these people would stay in an' help reshape it, *they can reshape it*. Like teachers, policemen, firemen, business leaders, all o'em *responsible: everybody leavin' out*. An' they takin' the money with 'em.

How did this happen? At the close of the war, *all* blacks, irrespective of their social status, were forcibly relegated to the same compressed spatial enclave, and they had no choice but to coexist in it. As whites fled *en masse* to the suburbs with the blessing and help of the federal government, they opened up adjacent areas in which black families from the middle class and from the upper fractions of the working class could move to create new, soon to be solidly black neighborhoods. The deconcentration of the Afro-American community, in turn, dispersed the institutions of the ghetto and increased their class differentiation.[11] Simultaneously, in a systematic and self-conscious effort to maintain the prevailing pattern of racial segregation, the city was making sure that all of its new public housing was built exclusively in existing ghetto areas (Hirsch 1983), where only the poorest would soon tolerate dwelling. By the 1970s, then, *the urban color line had effectively been redrawn along class lines* at the behest of government, with the historic core of the Black Belt containing inordinate concentrations of the jobless and dependent while the brunt of the black middle and stabler working classes resided in segregated peripheral city neighborhoods.

The consequence of this threefold movement – the out-migration of stably employed Afro-American families made possible by state-sponsored white flight to the suburbs, the crowding of public housing in black slum areas, the expulsion of the remaining ghetto residents from the wage-labor market – has been soaring and endemic poverty. In Grand Boulevard, a section of the South Side containing some 50,000 people, half the population lived under the poverty line in 1980, up from 37 per cent ten years earlier, and three of every four households was headed by a single mother. With a median family income below 7,000 dollars per annum (less than a third of the citywide figure), many families in fact did not even reach half the poverty line. Six residents in ten had to rely on one or another form of public assistance in order to subsist.

The social and economic desolation of today's ghetto is distinctly perceived by its inhabitants, as data from the Urban Family Life Survey show.[12] Asked how many men are working steadily in their neighborhood, 55 per cent of the residents of Chicago's traditional Black Belt (South Side and West Side together) answer "very few or none at all," compared to 21 per cent in peripheral black areas harboring a mix of poor, working-, and middle-class families. A full half also declare that the proportion of employed males in their area diminished over the preceding years. One adult in four belongs to a household without a working telephone (only one in ten in outlying black areas) and 86 per cent to a household that rents its living quarters (as against about half among blacks in low-poverty areas); nearly a third reside in buildings managed by the Chicago Housing Authority (CHA), though the latter oversees only 4 per cent of the city's housing supply.

It is abundantly clear that the urban core today contains mainly those dispossessed fractions of the black (sub)proletariat who are unable to escape its blighted conditions. Given a choice, fewer than one in four residents of Chicago's ghetto would stay in their neighborhood, as opposed to four in ten in low-poverty black tracts. Only 18 per cent rate their neighborhood as a "good or very good" place to live, contrasted with 42 per cent in peripheral black areas, and nearly half report that the state of their surroundings has worsened in the past few years. Not surprisingly, gang activity is more prevalent at the heart of the ghetto: half of its inhabitants consider gangs as a "big problem" in their area, compared to fewer than a third in low-poverty black precincts. As for its future, nearly one-third foresee no improvement in their neighborhood, while another 30 per cent expect it to continue to deteriorate.

Today's ghetto dwellers are thus not only *individually poorer* than their counterparts of three decades ago in the sense that they have borne an absolute reduction in their standards of living and that the distance between them and the rest of society has widened – the federal poverty line represented half the median national family income in 1960 but only a third by 1980 (Beeghley 1984: 355). They are also considerably *poorer collectively* in several respects. First, they reside among an overwhelmingly deprived and downwardly mobile or immobile population, and therefore tend to be isolated from other components of the Afro-American community: as we saw above, the black middle class has both fled the urban core and grown outside of it.[13] Second, and as a consequence, they can no longer count on the nexus of institutions that used to give the ghetto its internal coherence and cohesion. The "Black Metropolis" of the mid-century so admirably dissected by Drake and Cayton (1962: 17) was a "distinctive city within a city" containing an extended division of labor and the full gamut of black social classes. The "proliferation of institutions" that made "Bronzeville," as its residents used to call it, the capital of black America enabled it to duplicate (though at a markedly inferior level) the organizational structure of the wider white society and to provide limited but real avenues of mobility within its own internal order.

By contrast, the hyperghetto of the late century has weathered such organizational decline that it contains neither an extended division of labor nor a representative cross-section of black classes, nor functioning duplicates of the central institutions of the broader urban society. The organizational infrastructure – the black press and the Church, the lodges and the social clubs, the political groups, the businesses and professional services, and the policy racket (or "numbers game") – that gave the classical ghetto of the 1950s its communal character and strength, and that served as an instrument of collective solidarity and mobilization, has by and large withered away, weakening the citywide networks of solidarity and cooperation typical of the communal ghetto (Mithun 1973). And, whereas, in the context of the full employment and industrial prosperity brought about by the Korean war, "the entire institutional structure of Bronzeville [was] providing basic satisfactions for the 'reasonable expectations' shared by people at various class levels"

(Drake and Cayton 1962: vol. 2, p. xi), today the prevalence of joblessness and the organizational void of the contemporary hyperghetto prevent it from satisfying even the basic needs of its residents.

Oppressive as it was, the traditional ghetto formed "a milieu for Negro Americans in which they [could] imbue their lives with meaning" (Drake and Cayton 1962: vol. 2, p. xiv) and which elicited attachment and pride. By contrast, today's ghetto is a despised and stigmatizing locale from which nearly everybody is desperately trying to escape, "a place of stunted hopes and blighted aspirations, a city of limits in which the reach of realistic ambition is to survive" (Monroe and Goldman 1988: 251).

## Concluding notes

Alejandro Portes (1972: 286, emphasis added) remarks in a famous article on the shanty towns of Latin America that "*the grave mistake of theories on the urban slum has been to transform sociological conditions into psychological traits* and to impute to the victims the distorted characteristics of their victimizers." This is an apt characterization of the recent scholarly and public policy debate on the ghetto in the United States. By focusing narrowly on the presumed behavioral and cultural deficiencies of inner-city residents or on the aggregate impact of the consolidation of a postindustrial economic order without paying due notice of the historical structures of racial and class inequality, spatial separation, and governmental (in)action that filter or amplify it, recent discussions on the so-called underclass have hidden the political roots of the predicament of the ghetto and contributed to the further stigmatization and political isolation of its residents.

There is no room here to address the numerous analytical inconsistencies, grave empirical flaws, and policy dangers of the *demi-savant* concept of "underclass,"[14] including its internal instability and heterogeneity, which make it possible to redraw its boundaries at will to fit the ideological interests at hand; its essentialism, which permits a slippage from substantive to substance, from measurement to reality, leading to mistake a *statistical artifact* for an actual social group; its far-flung negative moral connotations and its falsely "deracialized" ring allowing those who use it to speak about race without appearing to do so. Suffice it, by way of conclusion, to highlight its built-in propensity to sever the ghetto from the broader sociopolitical structures of caste and class domination of which the latter is both a product and a central mechanism.

By reviving and modernizing the century-old notion that urban poverty is the result of the personal vices and collective pathologies of the poor, the rhetoric of the "underclass" has given a veneer of scientific legitimacy to middle-class fears of the black subproletariat and blocked an accurate, historically grounded analysis of the changing political articulation of racial segregation, class inequality, and state abandonment in the American city. It has diverted attention away from the institutional arrangements in education, housing, welfare, transportation, and health and human services that perpetuate the concentration of unemployed and underemployed blacks in the urban core. By omitting to relate the state of the ghetto to the breakdown of the public sector, it has absolved the urban, housing, and educational choices made by the federal and local governments of both Democratic and Republican stripes since the mid-1970s.

Yet it is this policy of abandonment and punitive containment of the black poor that explains that, one century after its creation and two decades after the country's aborted and ill-named "War on Poverty," the American ghetto remains, to borrow a line from the preface to the Kerner Commission Report (1989: xx) of 1968, "the personification of that nation's shame, of its deepest failure, and its greatest challenge."

## Notes

1 "The Ghetto," by Leroy Hutson, Donna Hathaway, Al Eaton, and Todd Shaw, copyright © 1990, Don Pow Music; administered by Peer International Music Corporation, all rights reserved; used by permission (from the album *Short Dog's in the House*, 1990; Zomba Recording Corp.).

2 These lines were written before the South-Central Los Angeles events of April 1992, but the near-complete disappearance of the latter from public debate only weeks after their onset does not encourage me to revise this introductory statement. Indeed, most remarkable about this partially race-based outbreak of urban violence is how thoroughly it was assimilated to preexisting images and discourses on the ghetto (to the point of disfigurement since this erased its multiethnic composition as well as its class dimension) and how little impact it has had on policy and scholarly discussion on the nexus of race, class, and state in the city – as if it had been no more than a "reality show," if a particularly lurid and frightful one (Wacquant 1993).

3 Thus research on "urban poverty" over the past decade has been fixated on issues of family, welfare, and deviance (in the realms of sexuality and crime in particular), at the cost of neglecting, if not obfuscating, both the deepening class disparities and racial division of American society and the political power shifts that have allowed a range of public policies (in education, housing, health, urban development, justice, etc.) to curtail life chances in the inner city. The issues of family structure, race, and poverty have become virtually confounded (Zinn 1989), as if some necessary causal relation obtained among them. Likewise, the questions of urban decline and race have become thoroughly embroiled, so much so that the term "urban" has become a euphemism for poor blacks and other dominated ethnoracial categories (Franklin 1991: ch. 4).

4 Conversely, not all low-income areas are ghettos, however extreme their destitution: think of declining white industrial cities of the deindustrializing Midwest such as Pontiac, Michigan, rural counties of the Mississippi delta, Native American reservations, or entire portions of the United States in the 1930s. To call any area exhibiting a high rate or concentration of poverty a ghetto is not only arbitrary (what is the appropriate cut-off point and for what unit of measurement?); it robs the term of its historical meaning and empties it of its sociological contents, thereby thwarting investigation of the precise mechanisms and criteria whereby exclusion operates (discussions with Martin Sanchez-Jankowski helped me clarify this point).

5 Perhaps it was necessary, to produce this odd discursive formation, composed largely of empirically dressed moralizations and policy invocations, whose primary

function is to insulate and shelter "mainstream" society from the threat and taint of poor blacks by symbolically removing them from it, for proponents of the underclass mythology to first studiously remove themselves from the ghetto in order to "theorize" it from afar and above, and only through the reassuring buffer of their bureaucratic research apparatus. One example: it is remarkable (and unfortunately rather typical) that, of the 27 authors who contributed to the lavishly financed and publicized collection of conference papers pithily entitled *The Urban Underclass* (Jencks and Peterson 1991), only *one* has carried out extensive first-hand observation inside the ghetto.

6 In an original yet regrettably often overlooked network analytic study, Melvin Oliver (1988) provides a suggestive portrait of the urban Afro-American community as clusters of interpersonal ties that directly belies its common representation as a hotbed of social disaffiliation and pathologies. He finds in particular that the residents of Los Angeles's historic ghetto of Watts and of the newer segregated middle-class area of Crenshaw–Baldwin Hills have quite similar networks (as characterized by their size, relational context, spatial distribution, density, strandedness, and reciprocity) and that extralocal ties with kin are equally prevalent in both areas.

7 See Spear (1967), Philpott (1978), and Drake and Cayton (1962, vol. 1) in the case of Chicago's ghetto, and Kusmer (1986) and Franklin (1980) for a broader historical overview of black urbanization. It is not possible here to give an adequate treatment of the historical roots of the trajectory of the dark ghetto in the *longue durée* of its lifespan. Suffice it to point out that, even though its motor causes are situated outside of it, the transformation of the ghetto is, as with every social form, mediated in part by its internal structure so that a full resolution of its recent evolution must start a century ago, in the decades of its incubation.

8 For instance, in 1930, at a time when the all-black South Side ghetto already grouped over 90 per cent of the city's Afro-American population, Chicago's "Little Ireland" was a hodge-podge of 25 "nationalities" composed of only one-third Irish persons and containing a bare 3 per cent of the city's residents of Irish descent (Philpott 1978: 141–2).

9 Unless otherwise indicated, quotes from interviews and first-hand observations come from fieldwork I conducted on Chicago's South Side in 1988–91 in the course of an ethnographic study of the culture and economy of professional boxing in the ghetto.

10 Violence is an aspect of ghetto life that is difficult to discuss without immediately calling forth the willfully gory – and often grossly misleading – images of stereotypical media descriptions of crime and lawlessness that have become a staple of political and intellectual discourse on the "underclass." Yet, based on my ethnographic fieldwork on Chicago's South Side, I feel that any account of the ghetto must start with this violence because of its experiential acuity and enormously disruptive ramifications in the lives of those trapped in it. At the same time, I want to insist, if only by way of prolepsis, first that inner-city violence is, in its forms and organization, quite different from what journalistic accounts reveal, in some ways not as horrific and in other ways much worse, owing in particular to its routine and socially entropic character. Second, this internecine violence "from below" must be analyzed not as an expression of "pathology" but as a function of the degree of penetration and mode of regulation of this territory by the state – a response to various kinds of violence "from above" and a by-product of the political abandonment of public institutions in the urban core (Wacquant 1993b). I have tried elsewhere (see Wacquant 1992a) to offer a more

nuanced account, from the inside, of the impact of systemic insecurity on the texture of everyday life in the ghetto, as seen through the eyes and survival strategies of a professional hustler who works the streets of Chicago's South Side.

11 To be sure, this class differentiation has existed in more or less attenuated forms since the origins of the Black Belt: the latter was never the *gemeinschaftliche* compact that analysts nostalgic of a "golden age" of the ghetto that never existed sometimes invoke. However brutal, the caste division imposed by whites never obliterated internal cleavages along class lines (partly convergent with persistent skin color differences) among Afro-Americans, as can be seen, for instance, in the spread of "store-front churches" in the face of old-line Baptist and Methodist churches in the 1920s (Spear 1967: ch. 9) or in the bifurcation of the "jook continuum" and the "urban-commercial complex" in the realm of dance and entertainment (Hazzard-Gordon 1990).

12 This survey was conducted as part of the Urban Poverty and Family Structure Project (directed by William Julius Wilson) at the University of Chicago. It consists of a multi-stage, random probability sample of residents of Chicago's poor neighborhoods (defined as census tracts with at least 20 per cent poor persons in them in 1980) conducted in 1986–7. The survey covered 1,184 blacks, with a completion rate of about 80 per cent, of whom a third lived on the city's South Side and West Side. The financial support of the Ford Foundation, the Carnegie Corporation, the US Department of Health and Human Services, the Institute for Research on Poverty, the Joyce Foundation, the Lloyd A. Fry Foundation, the Rockefeller Foundation, the Spencer Foundation, the William T. Grant Foundation, the Woods Charitable Fund, and the Chicago Community Trust for this research is gratefully acknowledged.

13 The fact that an increasing number of urban middle-class blacks have never experienced ghetto life firsthand (though, having generally lived in sharply segregated, all-black areas, they are fully acquainted with discriminatory and other racist practices) is bound to affect processes of black identity formation, individual and collective. The meaning that middle-class blacks attach to a range of ghetto idioms and expressive symbols (e.g., musical genres, hairdos and dress codes, linguistic demeanor) is likely to change when exposure to them comes from family lore or from secondary sources such as formal education and the popular media rather than through native immersion.

14 See Wacquant (1992b) for an analysis of the functions of the scholarly myth of the "underclass" in the intellectual and political-journalistic fields and of the sources of its social success. For a cogent discussion of its policy liabilities, see Gans (1991).

## Selected references

Anderson, Elijah 1991: *Streetwise: Race, Class, and Change in an Urban Community*. Chicago: University of Chicago Press.

Aschenbrenner, Joyce 1975: *Lifelines: Black Families in Chicago*. Prospect Heights: Waveland Press.

Auletta, Ken 1982: *The Underclass*. New York: Vintage.

Beeghley, Leonard 1984: "Illusion and Reality in the Measurement of Poverty," *Social Problems*, vol. 31 (February): 322–33.

Blauner, Robert 1989: *Black Lives, White Lives: Three Decades of Race Relations in America*. Berkeley: University of California Press.

Boskin, Joseph 1970: "The Revolt of the Urban Ghettos, 1964–1967," in *Roots of*

*Rebellion: The Evolution of Black Politics and Protest Since World War 11*, ed. Richard P. Young. New York: Harper and Row, pp. 309–27.

Brune, Tom and Eduardo Camacho 1983: *A Special Report: Race and Poverty in Chicago*. Chicago: The Chicago Reporter and the Center for Community Research and Assistance.

*Chicago Tribune* (Staff of the) 1986: *The American Millstone: An Examination of the Nation's Permanent Underclass*. Chicago: Contemporary Books.

*Chicago Tribune* 1990: "School Lets Out, Fear Rushes In: Gangs Terrorize Area after Classes," January 24.

Clark, Kenneth B. 1965: *Dark Ghetto: Dilemmas of Social Power*. New York: Harper.

Collins, Sharon M. 1983: "The Making of the Black Middle Class," *Social Problems*, vol. 3, no. 4 (April): 369 82.

Curtis, Lynn A. 1985: *American Violence and Public Policy*. New Haven: Yale University Press.

Drake, St Clair and Horace R. Cayton 1962 [1945]: *Black Metropolis: A Study of Negro Life in a Northern City*, 2 vols, rev. and enlarged edn. New York: Harper and Row.

Fainstein, Norman 1993: "Race, Class, and Segregation: Discourses About African-Americans," *International Journal of Urban and Regional Research*, vol. 17, no. 3 (September): 384–403.

Franklin, John Hope 1980: *From Slavery to Freedom: A History of Negro Americans*, 5th edn. New York: Knopf.

Franklin, Raymond S. 1991: *Shadows of Race and Class*. Minneapolis: University of Minnesota Press.

Gans, Herbert H. 1991: "The Dangers of the Underclass: Its Harmfulness as a Planning Concept," in *People, Plans and Policies: Essays on Poverty, Racism, and Other National Urban Problems*. New York: Columbia University Press, pp. 328–43.

Garbarino, James, Kathleen Kostelny and Nancy Dubrow 1991: *No Place to be a Child*. Lexington: Lexington Books.

Gibbs, Jewelle Taylor, ed. 1989: *Young, Black and Male in America: An Endangered Species*. New York: Auburn House Publishing Co.

Habermas, Jürgen 1984 [1981]: *The Theory of Communicative Action*, vol. 1: *Reason and the Rationalization of Society*. Boston: Beacon Press.

Hannerz, Ulf 1968: "The Rhetoric of Soul: Identification in Negro Society," *Race*, vol. 9, no. 4: 453–65.

Hazzard-Gordon, Katrina 1990: *Jookin': The Rise of Social Dance Formations in African Culture*. Philadelphia: Temple University Press.

Hirsch, Arnold 1983: *Making the Second Ghetto: Race and Housing in Chicago, 1940–1960*. Cambridge: Cambridge University Press.

hooks, bell 1992: "Loving Blackness as Political Resistance," in *Black Looks: Race and Representation*. Boston: South End Press, pp. 9–20.

Jencks, Christopher and Paul E. Peterson, eds 1991: *The Urban Underclass*. Washington, DC: The Brookings Institution.

Katz, Michael B. 1989: *The Undeserving Poor. From the War on Poverty to the War on Welfare*. New York: Pantheon.

Keil, Charles 1966: *Urban Blues*. Chicago: Chicago University Press.

Kerner Commission 1989 [1968]: *The Kerner Report. The 1968 Report of the National Advisory Commission on Civil Disorders*. New York: Pantheon.

Kusmer, Kenneth L. 1986: "The Black Urban Experience in American History," in *The State of Afro-American History: Past, Present, and Future*, ed. Darlene Clark Hine. Baton Rouge and London: Louisiana State University Press, pp. 91–135.

Landry, Bart 1987: *The New Black Middle Class*. Berkeley: Univ. of California Press.

Lash, Scott and John Urry 1988: *The End of Organized Capitalism*. Madison: University of Wisconsin Press.
Liebow, Elliot 1967: *Tally's Corner: A Study of Negro Streetcorner Men*. Boston: Little, Brown and Co.
Lipsitz, George 1989: *A Life in the Struggle: Ivory Perry and the Culture of Opposition*. Philadelphia: Temple University Press.
Loewenstein, Gaither 1985: "The New Underclass: A Contemporary Sociological Dilemma," *The Sociological Quarterly*, vol. 26, no. 1 (Spring): 35–48.
McAdam, Doug 1981: *Political Process and the Development of Black Insurgency*. Chicago: University of Chicago Press.
McCord, C. and H. Freeman 1990: "Excess Mortality in Harlem," *New England Journal of Medicine*, vol. 323, no. 3: 173–7.
Mead, Lawrence 1989: "The Logic of Workfare: The Underclass and Work Policy," *Annals of the American Academy of Political and Social Science* 501: 156–69.
Mithun, Jacqueline S. 1973: "Cooperation and Solidarity as Survival Necessities in a Black Urban Community," *Urban Anthropology*, vol. 2, no. 1 (Spring): 25–34.
Monroe, Sylvester and Peter Goldman 1988: *Brothers: Black and Poor. A True Story of Courage and Survival*. New York: William Morrow.
Morris, Aldon 1984: *The Origins of the Civil Rights Movement: Black Communities Organizing for Change*. New York: Free Press.
Nathan, Richard P. 1987: "Will the Underclass Always be with Us?" *Society*, vol. 24, no. 3 (March–April): 57–62.
Oliver, Melvin 1988: "The Urban Black Community As Network: Toward a Social Network Perspective," *The Sociological Quarterly*, vol. 29, no. 4: 623–45.
Orfield, Gary 1985: "Ghettoization and Its Alternatives," in *The New Urban Reality*, ed. Paul Peterson: Washington, DC: The Brookings Institution, pp. 161–93.
Osofsky, Gilbert 1971: "The Enduring Ghetto," in *Harlem: The Making of a Ghetto. Negro New York, 1890–1930*, 2nd edn. New York: Harper, pp. 189–201.
Pétonnet, Colette 1985: "La Pâleur noire. Couleur et culture aux États-Unis," *L'Homme* 97–8 (January–June): 171–87.
Philpott, Thomas Lee 1978: *The Slum and the Ghetto: Neighborhood Deterioration and Middle-Class Reform, Chicago 1880–1930*. New York: Oxford University Press.
Portes, Alejandro 1972: "The Rationality of the Slum: An Essay On Interpretive Sociology," *Comparative Studies in Society and History* 14.
Rainwater, Lee 1970: *Behind Ghetto Walls*. Chicago: Aldine.
Ricketts, Erol 1989: "A Broader Understanding Required (Reply to Steinberg)," *New Politics*, vol. 2, no. 4.
Ricketts, Erol R. and Isabell V. Sawhill 1988: "Defining and Measuring the Underclass," *Journal of Policy Analysis and Management*, vol. 7 (Winter): 316–25.
Sanchez-Jankowski, Martin 1991: *Islands in the Street: Gangs in Urban American Society*. Berkeley: University of California Press.
Sawhill, Isabel V. 1989: "The Underclass: An Overview," *The Public Interest* 96: 3–15.
Son, In Soo, Suzanne W. Model and Gene A. Fisher 1989: "Polarization and Progress in the Black Community: Earnings and Status Gains for Young Black Males in the Era of Affirmative Action," *Sociological Forum*, vol. 4, no. 3 (Summer): 309–27.
Spear, Allan H. 1967: *Black Chicago: The Making of a Negro Ghetto, 1890–1920*. Chicago: University of Chicago Press.
Wacquant, Loïc J. D. 1989: "The Ghetto, the State, and the New Capitalist Economy," *Dissent* (Fall): 508–20.
Wacquant, Loïc J. D. 1991: "What Makes a Ghetto? Notes Toward a Comparative Analysis of Modes of Urban Exclusion," paper presented at the MSH/Russell Sage

Conference on "Poverty, Immigration and Urban Marginality in Advanced Societies," Paris, Maison Suger, May 10–11, 1991.

Wacquant, Loïc J. D. 1992a: " 'The Zone': le métier de 'hustler' dans le ghetto noir américain," *Actes de la recherche en sciences sociales*, vol. 92 (June) 38–58.

Wacquant, Loïc J. D. 1992b: "Décivilisation et démonisation: la mutation du ghetto noir américain," in *L'Amérique des français*, ed. Christine Fauré and Tom Bishop. Paris: Éditions François Bourin, pp. 103–25.

Wacquant, Loïc J. D. 1993: "Morning in America, Dusk in the Dark Ghetto: The New 'Civil War' in the American City," in *Metropolis*, ed. Phil Kasinitz. New York: New York University Press.

Wilson, William Julius 1987: *The Truly Disadvantaged: The Inner City, the Underclass and Public Policy*. Chicago: University of Chicago Press.

Wirth, Louis 1927: "The Ghetto," *American Journal of Sociology*, vol. 33 (July): 57–71.

Zinn, Maxine Baca 1989: "Family, Race, and Poverty in the Eighties," *Signs: Journal of Women in Culture and Society*, vol. 14, no. 4: 856–74.

# 15 Benjamin Forest, 'West Hollywood as Symbol: the Significance of Place in the Construction of a Gay Identity'

Reprinted in full from: *Environment and Planning D: Society and Space* 13, 133–57 (1995)

> November 6, 1984: In California, West Hollywood becomes the first 'gay city' in the United States, after voters there decide to turn the previously unincorporated area into a self-governing municipality and elect a largely gay city council to run it.
>
> *The Gay Decades* Rutledge (1992, page 231)

In this study I trace the definition of a gay identity in West Hollywood, California as expressed in a series of articles appearing in the gay press during the campaign for municipal incorporation between mid-1983 and late 1984. These articles tied the physical and social characteristics of the new city to the physical, mental, and moral character of gays living in West Hollywood. I identify seven elements of this new gay male identity – creativity, aesthetic sensibility, an affinity with entertainment and consumption, progressiveness, responsibility, maturity, and centrality – but do not claim that this is an exhaustive or exclusive list. Incorporation of the city was portrayed as a way to consolidate and legitimize this identity, so that the 'cityhood' movement acted to reduce the 'marginal' status of gays, and to draw them closer to the 'center' of US society (Shils, 1975, pages 3–16). As presented in the gay press, the incorporation of West Hollywood was less a radical project than an attempt to achieve recognition of gays as members of civil society. Indeed, several authors argue that community debates over citizenship within 'identity groups' are one of the necessary conditions for the creation of a civil society

(Shklar, 1991; Shotter, 1993). As such, the effort for incorporation resembles an ethnic group strategy (Epstein, 1987). It differs sharply from more radical strategies for oppressed groups, such as those suggested by hooks (1990), that use marginal status as a strategy of empowerment.

The conflation of place attributes with the personal qualities of gay men contributes to what Knopp (1992, page 652) describes as "sexual identity formation", and in turn, "symbolic and representational struggles over the sexual meanings associated with particular places". Using place as a symbol tends to mask the socially constructed quality of gay identity, so that it takes on a 'natural' existence.[1] The narrative construction of a 'gay city', and thus the attempt to create an identity based on more than sexual acts, suggests that the gay press sought to portray gayness as akin to ethnicity, in contrast to homophobic characterizations of gayness as a perversion, sickness, or moral failure.

The 'geographies' of gays and lesbians have received considerable attention in the last four years, although studies by geographers date back to Weightman (1980). For the most part these recent studies have been of the political economy of gay neighborhoods, or have been ethnographic studies of gays and lesbians. Generally, those working within the framework of political economy note that symbolic struggles over the meaning of places often coincide with economic and political struggles, but concentrate on these latter issues (Knopp, 1990a; 1990b; 1992). Ethnographic studies document how different places, such as the home, workplace, etc are experienced by gays and lesbians, with particular attention to the constraints imposed by '(hetero)sexed' space (Valentine, 1993a). Both of these approaches share a view that places take on importance primarily as sites of routine activities, so that the important issues are how the daily lives of gays and lesbians are constrained or empowered in particular localities. More generally, studies of the relationship between place and identity have focused on political, rather than symbolic–cultural, issues (Agnew, 1992). Humanistic geography has contributed little to this growing body of research – an unfortunate situation because the humanistic perspective offers unique insights to this relationship. [See, however, Crow (1994) and Till (1993) for two attempts to bring a more humanistic perspective to the question of place and identity.] In particular, humanistic geographers have discussed the importance of place as a center of meaning, as well as simply a site of routine activities (Tuan, 1977). The symbolic element is especially important to the normative importance of place because morally valued ways of life are often created, shaped, and reinforced through the construction of real and imagined places. Hence, in this study I do not focus on the social history of West Hollywood, nor do I make any claims about the experience or perceptions of gays and lesbians. Rather, I examine the symbolic representation of West Hollywood in the gay press during the 'cityhood' campaign. This emphasis on the creation of place through language rather than through material transformation follows from both Tuan (1991) and Barnes and Duncan (1992). I argue that two characteristics of place – its capacity to 'concretize' an idea or culture, and its holistic quality – make it a particularly effective means to create social identities.

I discuss the holistic quality of place to argue that the capacity to experience place as a whole helps to resolve the internal contradiction of identity. As Entrikin (1991a, page 19), notes, however, place is perspectival. Consequently to one degree or another, one must view this holistic quality as the result of a political decision by the gay press. Massey's (1991, page 276) description of place identity as "frequently riven with internal tensions and conflicts" suits the actual city of West Hollywood, and even (to some degree) the symbolic city described by the gay press. I argue, however, that the use of a holistic symbol was an attempt to resolve these contradictions. Thus the use of place in this fashion was a political decision, one which exploited the unique capacity of place to be experienced holistically. This is emphatically not the only possible or actual experience of West Hollywood.

Except where noted, 'gay' refers specifically to gay men because most of the material on gays used in this research came from weekly newspapers or 'newsmagazines' targeted towards this group. Several commentators, notably Bell (1991, page 323), have observed that geographical studies have largely ignored lesbians – Valentine (1993a; 1993b; 1993c), Peake (1993), and Adler and Brenner (1992) are welcome exceptions. This deficiency is particularly important because, as Bell notes, there are plural 'homosexual geographies'. In terms of the social history of West Hollywood, lesbians were at least as important as gay men in the activities of the incorporation campaign. There is a substantial lesbian population in the city and the first mayor was a lesbian.

The 'geographies' of gay men in West Hollywood certainly differ from the 'geographies' of West Hollywood lesbians. The highly visible public expression of gay men in West Hollywood in contrast to lesbians is a case in point. Authors of some studies argue that lesbians are generally less visible in urban areas because they have a relatively private orientation (Castells, 1983). More compelling are studies that have focused on the fact that lesbians, like women in general, have less access to capital, earn lower incomes, and have a higher risk of violence in public places (Adler and Brenner, 1992; Valentine, 1993a; 1993b). The symbols and meanings used in the gay press are also almost certainly class specific, an issue which has been addressed by relatively few geographers (Geltmaker, 1992, page 633; Knopp, 1987; Lynch, 1987). The gay papers used for the study targeted relatively affluent gays, which no doubt largely determined the generally nonradical nature of West Hollywood gay identity.

### Place and identity

Work by Tuan (1974; 1977; 1980; 1984) and Entrikin (199lb) suggests that place continues to have normative importance, that is, it continues to play a role in the moral evaluation of ways of life. Although place in general has long been a concern of humanistic geographers (Relph, 1976, for example), Agnew (1989) argues that place as a concept has been devalued because contemporary social science has conflated it with 'community', and concern with community has been supplanted by concern for the social. Agnew believes that place can be 'rehabilitated' through the study of social relations in place, a project which has been taken up by locality studies (Massey, 1991;

1993). Entrikin (1991b, page 61) contends that the focus on social relations forecloses examination of the normative significance of place by "overlooking the role that cultural interpretations of a morally valued way of life play in the constitution of social relations". One attempt to integrate locality, social and political relations, and moral evaluations focuses on the idea of citizenship. Smith (1989) argues that locality is the basis on which to advance the normative dimension of citizenship. It is not inconsequential then, that (1) the attempt to form a new identity was centered around a political incorporation, and that (2) this new identity included characteristics of 'good citizens'. It is not clear from Smith (1989), however, why locality is an effective vehicle for realizing the normative values of citizenship. The answer, I believe, lies in the role that place can play in moral narratives.

Morally valued ways of life are evaluated in the narratives of individuals and groups, and form, in part, the basis of self-identity. These narratives, however, have increasingly been seen as constructed and reflective, so much so that Giddens (1991, page 31) describes the modernistic conception of the self as a 'reflexive project'. Place is particularly significant when it is used to construct the normative conventions of a group. Thus "the cohesiveness of social groups is related to the constitution of individual and communal identity, which cannot be removed from the question of valued ways of life" (Entrikin, 1991b, page 82). This quality of place is recognized by Tuan (1977, page 178), who writes that place can embody a culture, and achieves an identity "by dramatizing the aspirations, needs, and functional rhythms of personal and group life". It is the symbolic value of place that makes it an effective organizer of identity.

When groups deploy a territorial strategy in creating an identity, it accentuates the importance of place. Sack (1986, page 19) defines territoriality as "the attempt by an individual or group to affect, influence, or control people, phenomena, and relationships, by delimiting and asserting control over a geographic area". The symbolic nature of gays' efforts does not, however, make their strategy any less territorial, because the primary symbolic result of territoriality is a reification of power (Sack, 1981, page 66). Tuan (1974, page 240) argues more generally that physical locations become places because they make "an entire functional realm . . . visible, tangible, and sensible: it is the embodiment of a culture". In this study I am concerned with the 'power' to create symbols and new identities, rather than with control over actions or material goods. I do not make any empirical claims about the effectiveness of the gay press in shaping or determining individual identities, but rather seek to analyze the role played by place as an organizing principle of an ideal identity. Indeed, the first project presents a different set of questions relating to the relationship between culture and personality (Bourdieu, 1977, pages 84–85).

Authors of recent works (Duggan, 1992; Stein, 1992, for example) draw on the literary critic Sedgwick (1990) to question the value and legitimacy of fixed categories of identity, particularly those related to gender and sexuality. Sedgwick (1990, page 25) argues that conceptions of sexual identity can be disrupted, even among "people of identical gender, race, nationality, class, and 'sexual orientation' ". While addressing the question of gender identity, Butler (1990, page 16) makes a related point when she suggests that "identity" is more "a normative ideal rather than a descriptive feature of experience". Butler also argues that the ways in which gender and sexual identities have been constructed are not politically neutral, but are intimately connected to heterosexual domination: "The univocity of sex, the internal coherence of gender, and the binary framework for both sex and gender are . . . regulatory fictions that consolidate and naturalize the convergent power regimes of masculine and heterosexist oppression" (page 33). These views contrast with somewhat earlier social constructionist frameworks which did not problematize the category of identity (Altman, 1982; Plummer, 1981; Weinberg, 1983).

The questions raised by Sedgwick and Butler present significant problems to claims of homogeneous, hegemonic identities, and would seem to undermine attempts to demonstrate the ontological status of group identities. Such a demonstration would require an ethnographic study to document empirically the 'success' of the gay identity advanced in the gay press, that is, the degree to which gay men in West Hollywood internalized this identity. As noted above, my ambitions for this study are more modest. I seek only to evaluate a particular 'normative ideal' of gay identity, and not to make a claim about the influence of that ideal.

It also seems misguided to abandon the concept of identity just at a time when scholars are beginning to address the emancipatory and empowering potential of place (Rodman, 1992). This concern with the loss of identity and 'the subject' has been voiced in both anthropology (Smith, 1992, page 525) and feminist studies (hooks, 1990, page 28). A perspective concerned with the use of place to resist domination seems particularly well suited to the study of gays, given the critical role of 'gay territories' in the history of gay liberation. In general, geography has neglected instances of self-identification and self-empowerment by marginalized groups, and has concentrated attention on the way in which landscape and place serve to reproduce elites, to impose racial identities, or to perpetuate the exclusion of women, nonwhites, and other marginalized groups (Anderson, 1987; 1988; Anderson and Gale, 1992; Bondi, 1992; Till, 1993). In West Hollywood there was a similar attempt to link the physical characteristics of the place with the moral, spiritual, and physical characteristics of the people living in that place, but in this instance gays exercised a relatively high degree of control over the new identity.

## Gays and public space

Geographers have tied the construction of gay identity to material and symbolic transformations of public spaces. Knopp (1987; 1990a; 1990b; 1992) and Valentine (1993a; 1993b; 1993c) have been among the most active contributors to this literature, but not the only ones (Bell, 1991). Knopp's ultimate ambition is "to identify specific ways in which sexuality is implicated in the spatial constitution of society, and, simultaneously, specific ways in which space and place are implicated in the constitution of sexual practices and sexual identity" (Knopp, 1992, page 652). Valentine (1993b; 1993c) argues that the focus on the transformation of public spaces stems largely from the concentration by geographers on gay men in urban neighborhoods. Her studies concern the social networks of lesbians, the strategies used by lesbians to negotiate and develop 'multiple sexual identities', and the availability of sites for gays and lesbians to meet. Hence most work to date has concerned the effect of gays on particular places, or has treated place largely as a location for social activities. The focus in this study differs in that my primary concern is not on the forces that resulted in the historical concentration of gays in West Hollywood, nor on the dynamics that led to an economic and political power base, nor on the social networks of gays and lesbians in West Hollywood. Rather, I analyze how conceptions of place served to organize a model identity in the gay press.

In addition to providing a gathering place, public spaces created by gays provide for relative safety, for the perpetuation of gay subcultures, and, most important to this study, provide symbols around which gay identity is centered (Castells, 1983; Godfrey, 1988; Jackson, 1989; Knopp, 1992). In the case of West Hollywood, the designation of a gay area as an independent city was important symbolically because it lent legal legitimation to gay identity, and, for a time at least, raised gays' level of political activity in West Hollywood (Moos, 1989, page 357). The coverage of the cityhood movement in national gay newsmagazines, such as *The Advocate*, suggests that the symbolic influence of West Hollywood extended to the national level.

A consequence of the struggle for public expression is the close tie between gay identity and themes of repression and confrontation. The degree of contrast with heterosexual norms, however, may vary between gay groups. Groups such as ACT-UP (AIDS Coalition to Unleash Power) and Queer Nation are part of the 'liberationist' sector of the gay liberation movement which seeks to challenge the heterosexual assumption in public and semipublic spaces in a direct and confrontational fashion (Davis, 1991; Geltmaker, 1992). Because the incorporation movement worked within, and indeed glorified, existing political, economic, and social systems, the model gay identity associated with the effort is much closer to what Davis (1991) characterizes as the 'assimilationist' tradition (which seeks acceptance by heterosexual society). Additionally, public confrontation is not the only strategy used by gays and lesbians. Rather they may 'negotiate' multiple identities which depend in part on place and location, because most gays and lesbians spend most of their time in heterosexual environments (Valentine, 1993c, page 246). Even in Valentine's study of lesbians in the United Kingdom, however, occasional (if routine) informal gatherings of lesbians in 'marginal' spaces helped foster a sense of collective identity. For gay men in the United States, however, gay social space has been particularly critical to 'coming out', so that it is not surprising that there is a close tie between gay identity and the redefinition of public and social space.

**West Hollywood: the historical context**[2]

Covering less than two square miles, West Hollywood is an irregularly shaped entity bordering the city of Los Angeles on the north, east, and south, and Beverly Hills on the west (see Figure 15.1). According to Moos (1989, page 352), it has the greatest population density of any area west of the Mississippi River. During most of the 19th century, the area that is now West Hollywood was devoted to agricultural uses. Industrial development began in the early 20th century, associated with a rail line and residential settlement sparked by

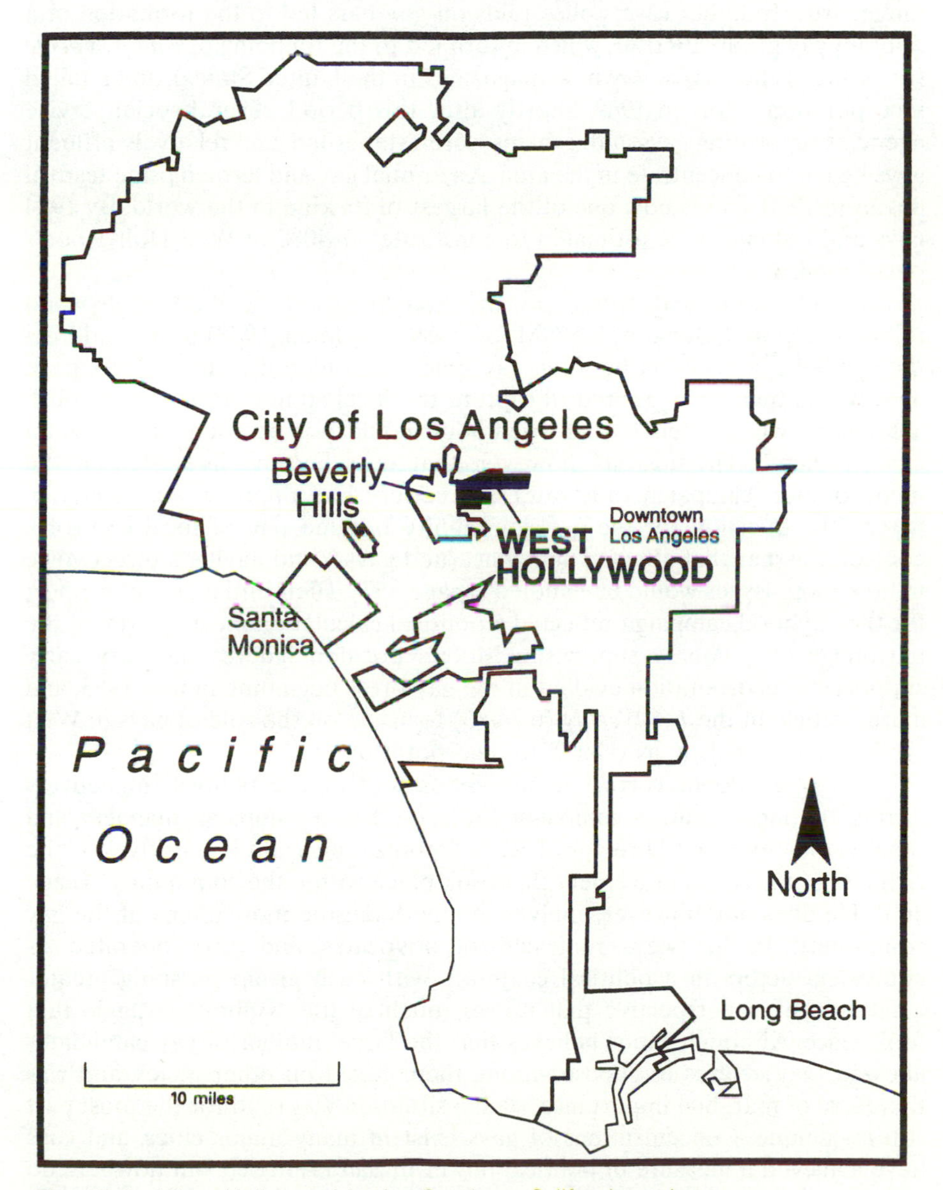


**Fig. 15.1** West Hollywood in the Southern California region

real estate development for workers in the movie industry. In 1924 the area voted against annexation to the city of Los Angeles, apparently because Los Angeles County had fewer restrictions on night-clubs. By the 1960s West Hollywood had become known as an area tolerant of 'alternative' life-styles – particularly for those associated with the music industry. As an unincorporated area, West Hollywood was policed by the county sheriff office, which (at the time) was thought to take a less oppressive stance toward gays. Geltmaker's (1992, pages 640–642) description of the current relationship between gays and the county sheriffs in Los Angeles suggests that this is no longer true. In either case, police raids on gay bars led to the formation of a militant gay group, PRIDE, which in turn led to the founding of *The Advocate* (now one of the largest gay newsmagazines in the United States), and a failed incorporation effort in 1969. Shortly after this period, several social service agencies supporting gays and lesbians were established, and relatively affluent gays began to concentrate in the area. An annual gay and lesbian pride festival began in 1970, and is now one of the largest of its kind in the world. By 1984 gays and lesbians were estimated to constitute 30–40% of West Hollywood's population.

The only published works on the incorporation of West Hollywood (Christensen and Gerston, 1987; Moos, 1989; Waldman, 1988) do not address the symbolic connection between gay spaces and identity. Moos (1989, page 366) writes that gays "desired to capture the local state in order to establish a territory where their sexual orientation could not be used as a weapon against them". He suggests, however, that gays became interested in the incorporation campaign only after the county had approved the election, when "the question for the gay community became one of local control – control of issues that affected and concerned gays – and having a direct voice in how those issues would be handled" (page 357). Their initial lack of support for the cityhood campaign reflected a political calculation meant to retain the patronage of a county supervisor. Moos's position ignores the very early support for incorporation evident in the gay press beginning in mid-1983, and a long article in the *LA Weekly* (6/24/83) focusing on the role of gays in West Hollywood, especially as related to incorporation.

Moos is no doubt correct in his assessment of the political maneuvers during the incorporation campaign (he served as a campaign manager and then as the city's first Director of Rent Stabilization), and is sensitive to "the richness of the social movement that took place within the community" (page 360). He does not, however, analyze in any detail the motivations of the gay community. In his view, rent control advocates and gays operated as equivalent actors in a political coalition, with each group pursuing greater local control, a perspective that misses much of the symbolic struggle that took place. Although Moos believes that the large number of gay candidates allowed (gay) voters to choose among them based on other issues, and was therefore of marginal importance, such a situation was (and for the most part still is) unique. Concentrations of gays exist in many major cities, and gays have achieved a measure of political power in San Francisco, but nowhere do they constitute anything close to a political plurality. In West Hollywood, a city, rather than simply a district of a city, became defined by its connection with gays. This would potentially provide an antidote to the "peculiarly local religiosity" of Los Angeles which suppresses the expression of sexual diversity (Geltmaker, 1992, page 647). The incorporation campaign was a logical step in the creation of gay spaces, because it added the element of legal recognition and power. Since there is a strong tie between gay identity and gay territories, an opportunity to redefine the legal and symbolic character of an area, through incorporation, would also provide the opportunity to redefine an identity for gays. Thus the incorporation movement provided a context in which place became a particularly powerful way to organize the meaning of West Hollywood and gay identity.

**The West Hollywood gay identity**

One might expect that an active local paper should encourage a strong sense of community identity, but this has been difficult to demonstrate in practice. Stamm (1985) makes a relatively comprehensive, but largely unsuccessful attempt. This difficulty in part explains the limited progress in conceptualizing the ties between the local press and community since the work of Park (1922; 1940) and Janowitz (1952). With regard to the link between local newspapers and a sense of community, Stamm writes:

> The difficulty . . . stems not only from the intransigence of the phenomenon, but also from the incompleteness of the model. The most we can hope to produce with such a model is evidence that newspaper reading is associated with community ties some of the time for some kinds of persons (1985, pages 40–41).

The problems associated with developing a *general* model of newspaper–community ties is that there is little reason to think that the press has the same role in different sorts of communities (geographic, ethnic, racial, gender, sexual, etc). Face-to-face interactions and mundane political concerns, rather than the local paper, may have far more influence on a geographically defined community than on a widely dispersed ethnic 'community'.

Cultural studies of mass communication have raised significant questions about how the audience 'reads' newspapers and other texts, arguing that texts always have multiple meanings, that is, they are polyphonic. See Streeter (1989), Bird (1992), and Crang (1992) for a general treatment of this question, and Cohen (1991), for an attempt to analyze the role of 'gay discourse' in the textual interpretations of gay men. Alternatively, one can follow Burgess (1985) and hold that places created by the media are mythical. She argues that like a myth, the meaning of the places created in newspapers is already complete, and does not require interpretation. The problems raised by the polysemy of texts are similar to the ones raised earlier regarding identity. My argument is not that the gay press is representative of a coherent block of gay opinion (it may be), but that as an institution in the gay community, it developed an ideal identity tied to place. It did not seek to portray itself as an objective recorder of events, but actively developed a model of a 'gay city'. Moos (1989, page 354) asserts that established gay activists were largely uninterested in the cityhood movement, and that a local gay paper, *Frontiers*, first made it a 'gay cause' both to boost circulation and to assist incorporation. I do not, however, seek to make any empirical claims about the effect that the gay press had on the gay 'community' during the West Hollywood campaign.[3]

As a relatively accessible public information source, as opposed to private friendship-networks, the gay press can provide an important means to socialize individuals into a gay identity. The empirical extent of this socialization is a somewhat different matter, and it would require an extensive ethnographic study to document. At least one gay publication, however, sees an analogy between its role and the role of public space in the creation of gay culture. The editor and cofounder of *Christopher Street* writes:

> Magazines have been a peculiarly modern device for bringing a public space into existence. Like a town meeting, a magazine enables people to be in each other's company by sharing talk about matters that concern them. It is through talking with each other that most of us start to make some sense out of the world, and begin to discover who we are and what we think . . .
>
> We always thought our task was to open a space, a forum, where the developing gay culture could manifest and experience itself. For people who have been excluded from the social world . . . this access to public space is basic and urgent. (Denneny *et al*, 1983, page 13).

Although this does not address the issue of how gays might 'read' these articles differently, it supports the notion that the gay press – by providing a more stable, reliable, and accessible means for socializing gays – was more likely to encourage geographically centered communities than were informal networks. For individuals unwilling or unable to 'come out', a freely distributed paper also provides anonymity.

In addition, the gay press reported the incorporation campaign quite differently from the nongay ('mainstream' or mass) papers. Both the local paper, *The West Hollywood Post*, and the *Los Angeles Times* portrayed the incorporation campaign primarily as an issue of rent control, whereas the gay press, *The Advocate*, *Edge, Frontiers*, *Update*, and others, emphasized symbolic rather than economic issues (Forest, 1991). The articles also indicate that these latter papers were highly aware of their role, and were adept at both rejecting and incorporating stereotypes about gay men. The gay press was sensitive to incidents or statements which portrayed gays in a negative light during the incorporation campaign. An editorial commenting on several incidents during the campaign states:

> again we are being cast by bigots into the role of irresponsible degenerates and sex-crazed perverts . . . The sincere and earnest efforts of gays and lesbians attempting to run openly in West Hollywood politics are being deliberately twisted and misrepresented. Anti-gays are trying to brand all of us as self-serving and uncaring about anything or anyone but ourselves (*Frontiers* 10/3/84a).

Gay writers did not simply reject stereotypes of gays but rather sought to construct an alternative, positive identity, one that accepted and co-opted some existing stereotypes. The narrative of incorporation in the gay press is reminiscent of Foucault's (1978, page 101) "reverse discourse", that is, a discourse in which "homosexuality began to speak in its own behalf, to demand that its legitimacy or 'naturality' be acknowledged".

The issue of the multiple meaning of texts does raise some difficult questions about my own reading of the gay press, because, if one takes the polysemy of texts seriously, the seven characteristics I have identified may be thought of as merely idiosyncratic, or generated by (unstated) ideological, political, or personal concerns. This is an important issue, but not one which undermines the study. No qualitative analysis can seek to be the only possible interpretation, but rather should be "plausible, useful and allow its own further elaborations and verification" (Strauss, 1987, page 11). Any hermeneutical study depends on a relatively subjective interpretation, but this is not to say that interpretations are arbitrary, or that one 'reading' is as well supported as any other.[4]

With a 'centrist' reading of the press, I emphasized characteristics which are important in the discourse of US civil society. Alexander and Smith (1993) argue that the 'discursive structure' of civil society can be broken down into democratic and counterdemocratic codes. Their 'democratic' codes – rationality, equality, reasonableness, lawfulness, etc – correspond roughly to what I have called the center of US society. Furthermore, I would argue that the same 'codes' that informed my study also provided basic vocabulary for the West Hollywood gay press. Additionally it is important to note that four of the characteristics I have identified – creativity, aesthetic sensibility, an orientation toward entertainment or consumption, and progressiveness – are elements of gay stereotypes, reflected in sources as different as opinion surveys (Taylor, 1983) and literary criticism (Sedgwick, 1985, page 173). Hence the selection of my seven characteristics was not arbitrary, but drew on existing stereotypes. The gay press did not simply reject all stereotypes, but rather it adopted those aspects that it deemed positive. By using the city as a symbol for gays, the gay press could incorporate many diverse characteristics into this identity, some which were closely related, and others which seem contradictory.

### *Creativity*

Creativity was a prevalent theme in the gay press throughout the incorporation campaign. Articles sometimes directly stated that gays are more creative than non-gays: the "motion picture industry . . . has attracted thousands of artistically motivated people . . . [who] include a rather high percentage of folks of the gay persuasion, one of those stereotypes solidly based on fact" (*Frontiers* 5/11/83a). The presence of the Pacific Design Center tended to dominate the discussion of design in West Hollywood, much as it physically dominates the southwestern side of the city. The Center was often tied to gays:

> It's encouraging to find a reference to the gay community's role in the area's development as something other than as wayward stepchildren. West Hollywood is . . . garnering the attention it rightfully deserves as one of the most stylish and

> forward-thrusted spots in the city. The area's current voguish image stems from its emergence as the design center of Los Angeles. The Pacific Design Center . . . stands as a postmodern cathedral for a thoroughly design-conscious community (*Frontiers* 5/11/83b).

It is not just that the Center is physically within the gay community, but that gays have adopted it as a symbol. Design is not simply an occupation, but a cultural praxis deserving a 'cathedral', which lies at the center of the gay community. West Hollywood fully and openly expressed this connection: "The new city could help one of our major industries, situated around the Pacific Design Center, grow . . . we could finally take advantage of our local design talent" (*Frontiers* 5/9/84). The connection between gays and creativity is often linked directly to the cityhood movement, in terms of the benefits the city will be able to draw from its population.

Even preincorporation West Hollywood was "where the gay subculture has the most potential to connect its creative energies to the national media" (*LA Weekly* 6/24/83). Gay cultural activities were important to the city: "It is essential that West Hollywood maintain its leadership as a cultural center and CSW's [Christopher Street West] activities contribute enormously" (*Edge* 5/17/84); and West Hollywood "is already an entertainment center and a design capitol, it's also a gay cultural center, and it could very well become America's gay cultural center" (*Edge* 9/20/84). Incorporation would allow West Hollywood to "take full advantage of the talent and creativity of the gay community" (*Frontiers* 2/1/84). The same article echoed the theme of creativity when Ron Stone (the early leader of the cityhood movement) states that with "residents who manage the music, motion picture, television and design industries, we have an enormous body of creative leadership". Cityhood will make the creativity of gays visible: "For the first time in the U.S., there will be a municipality where the talents of the gay community can be show-cased" (*Edge* 9/20/84). Other articles suggested the future which incorporation will make possible: "It is easy to imagine the city fast becoming a major creative center with numerous cultural events such as city-sponsored festivals, design competitions and sculpture contests" (*Frontiers* 3/14/84); and "The new city of West Hollywood could become an American wonder – a lively urban village that takes a *creative* approach to providing services to people" (Frontiers 5/9/84, emphasis in original). The gay press labeled the future city "West Hollywood, America's Creative Urban Village", featuring "the Harvey Milk Theatre" and "the Stonewall Cinema Center" (*Frontiers* 3/7/84).

It is not so much the 'fact' of having or attracting an artistically talented population, but the symbolic connection between West Hollywood and creativity that is important. Articles used 'creativity' to describe the city in the past, present, and future, suggesting that the relationship between the two is not one to be measured by documentation or empirical fact. The symbolic connection implies that gays in West Hollywood are more creative, innovative, and talented than the nongay population. Creativity is spoken of in a general way, creating the impression that the type of creativity needed to run a city is the same type of creativity used in artistic projects. Such references to creativity comprise part of a 'constitutive narrative' of West Hollywood, a narrative which sought to embody gay identity (Bellah *et al*, 1985).

*Aesthetic sensibility*

Gays' aesthetic sensibility was an attribute closely related to creativity. An article describing public space in the future city of West Hollywood was particularly rich in its use of this characteristic, and it is worth quoting at length:

> Colorful open-air mini-buses take residents from their offices to their homes. The new city's system of 'pocket parks' comes alive with . . . lovers strolling past the fountains and through the gardens . . . . The lights go on along the center strip of West Hollywood Concourse (formerly called Santa Monica Boulevard) which is now lined with tall palm trees, birds of paradise flowers and sculptures . . . The City of West Hollywood has commissioned local architects and designers to create special bus benches, sidewalks and refuse containers for the area. Flower planters now line the entire Melrose district (*Frontiers* 3/7/84).

The emphasis on public, as opposed to private, spaces should be noted. This is a city which is highly conscious of its visual appearance, where even the garbage cans are 'created' by designers. The city would not just 'landscape' the barren gravel-covered median strip along Santa Monica Boulevard (which had been a point of contention with the county for some time), but transform it into a 'concourse', lighted for display. From this description, it seems that flowers, trees, fountains, and sculptures occupy every open space, indicating the high value placed on the aesthetic appeal of the area.

The gay press also used individual candidates to emphasize concern with the aesthetic appearance. In a 'campaign profile', a candidate suggested an overall design plan for the city, including "official city signage and graphics; overall color themes and applications; pedestrian sitting benches; . . . uniform tree plantings; flower planters; . . . 'eyesore-signage buybacks'; . . . seasonal sidewalk decorations; and color applications to transit and service vehicles" (*Frontiers* 8/29/84). Guerrero suggested that the city's sidewalks "fail", because "county ordinances are written in an 'antiseptic manner' " which does not meet the "unique needs of West Hollywood". These antiseptic ordinances may be sufficient for other areas of the county, but not for the more aesthetically demanding West Hollywood.

The Santa Monica median strip is often a focal point for aesthetic improvement. One article described it as an 'eyesore' and states that landscaping it must be a priority of the new city. "This land must be acquired from Southern Pacific and developed as a showcase for the new city, combining landscaping, sculptures and fountains" (*Frontiers* 5/9/84). The article also suggested improvements in the appearance of the city involving telephone poles and parking. The city must not only be efficient, but beautiful: "West Hollywood could quickly act to make itself both more beautiful and functional, and overcome the legacy of ugliness". As a symbol of the gay community, West Hollywood must not simply be a place to live, but must be a showcase, paying constant and careful attention to its appearance.

The gay press sometimes represented beauty as the rightful domain of gays. An article entitled "Beautiful W. Hollywood" (*Update* 5/30/84) described the preincorporation landscaping of the Santa Monica median strip for the 1984 Olympic games in this way: "The problem, say longtime Gay community activists in the area, is that before [Los Angeles County Supervisor] Edelman's tree could be planted, trees put in place four years ago by a group of Gay women and men were bulldozed into oblivion". The suggestion was that gays were the only group truly concerned with the beauty of the area, and that nongay politicians should simply let them take care of the city. In other articles, however, the connection between gays and beauty was less direct, merely suggesting that the coalition of " seniors, gays, business people, homeowners and renters . . . can build a more beautiful, safer city" (*Frontiers* 9/19/84b).

*Entertainment and consumption*

The affinity for entertainment and consumption, particularly centered around performance and bars, often occurred in conjunction with 'creativity', and is evident in several passages noted in that discussion. Gay bars frequently played a critical role as social centers, especially before the gay liberation movement allowed for more open meeting places (Weightman, 1980). Although gays now have numerous meeting centers, including those not directly involved with entertainment and consumption activities, the gay press continued to use this image.

As with creativity, the gay press described West Hollywood in terms that suggested that gays already exercise control over the character of the city. *The Advocate* stated that "West Hollywood . . . is already a popular gay and lesbian tourist center with a number of fashionable gay and lesbian bars, restaurants, clubs, shops and clothing stores" (5/29/84; 9/18/84). An *Advocate* article that ran just after the election used the same phrase to describe the "City Features" of West Hollywood (*The Advocate* 12/11/84). *Frontiers* characterized West Hollywood as a "disco boomtown" and suggested that "only in the strange and wonderful milieu like the gay world could a nightclub be responsible for putting a place on the map" (5/11/83b). (Also see Kilday, 1983.) The gay press was clearly perpetuating the idea that much of gay life centers on entertainment and consumption. In the last passage, the writer seemed to suggest that dance clubs even define West Hollywood for gays. Ron Stone displayed a great deal of insight when he wrote in the *LA Weekly*, "The public image of West Hollywood is that it is a settlement of young men who apparently have nothing better to do than flaunt their good looks while waiting for stop lights, hustling along Santa Monica Blvd., cavorting in discos, and getting drunk in bars" (6/24/83). Stone goes on to suggest that reality does not quite conform to this image, but he nevertheless described West Hollywood as a "desirable, hip place to live" and "very chic". The implication is that gays live in the city, in part, because it is fashionable – that is, they use housing as a means of conspicuous consumption. A *Frontiers* article echoed the connection between gays and consumption for entertainment, where a member of the West Hollywood Incorporation Committee stated that cityhood will attract "high-quality businesses" to the area (*Frontiers* 2/1/84). The context suggests, however, that it is gays, more then cityhood itself, that will attract lucrative retail business.

*Progressiveness*

The gay press portrayed gays, and gays in West Hollywood in particular, as progressive trendsetters, both culturally and politically. The names of gay publications reflect the importance of this: *Frontiers*, *The Edge*, *Update*. These all suggest that gays explore the limits of what is acceptable in society. The importance of gays in design, and thus in determining new clothing, architectural, and theatrical fashions is discussed above. This progressiveness seems to be at the very core of gay identity, because the gay press feared its loss:

> Left to our own direction, it is probable that our specializations will increase, bringing . . . more industries concerned with style and trends and more activities for body and mind. It will be a never-ending struggle for the new city to make certain that we never become so set in our ways that we lose our place as a trendsetter for society as a whole (*Frontiers* 3/14/84).

It is not simply that gays are on the fringe of culture, but are rather on the 'frontier', exploring new ground over which the rest of society will follow.

Gays also expressed this progressiveness politically. In a letter to *Frontiers*, Stone wrote that cityhood, unlike previous efforts in civic affairs, was a 'positive' effort that he compared with five other political movements:

> Too often our motivation in civic affairs is really negative. Stop Anita Bryant. Stop John Briggs. Stop handguns. Stop AIDS. Stop Dan White's parole . . . But the current effort to bring cityhood to West Hollywood is a positive act in which the people get to establish and control their own city government (*Frontiers* 2/15/84).[5]

He does not explicitly categorize any of these as either gay or progressive issues, but one (handgun control) is identified with progressive and/or 'liberal' politics, and the other four have special interest for gays. Stone could have used any number of other political efforts to illustrate the positive versus negative character of the incorporation movement, but this particular combination implied that the cityhood movement was both a progressive and a gay issue. Yet he also affirmed the importance of political involvement. Although cityhood was a progressive, trendsetting cause, it existed well within the realm of acceptable political activity.

Last, many gay politicians presented themselves as part of a progressive program:

> Gay and lesbian politicos . . . believe the council will be afforded an opportunity to build a progressive community which can advance gay and lesbian rights while developing a city within a city that can enhance the quality of life of all its residents (*Frontiers* 5/2/84).

The article illustrates the tension between the symbolic importance of gay council members who "will face intense media attention", and the practical outcomes which cityhood will make possible. Gays were keenly aware that the news media could turn West Hollywood into a symbol of all gays, and that it would be difficult to prevent the mainstream press from focusing on negative or embarrassing events.

Like creativity, 'progressiveness' was used to describe both cultural affairs – fashion, style, etc – and a political orientation. Hence the gay press created a strong symbolic connection between cultural production and politics, but never explicitly documented the material ties between these two kinds of activities. In terms of a community narrative, however, the existence of an actual affiliation between the two was far less important than the symbolic link.

### *Responsibility*

The concern in the gay press for having openly gay elected officials illustrates the characteristic of responsibility. Unlike some of the preceding characteristics, responsibility seemed to have a fairly narrow meaning: the wise use of political power. *The Advocate* suggested that this was the primary concern of the Harvey Milk Democratic Club: " 'We're looking first of all for gay elected officials' " (*The Advocate* 5/29/84). The same article highlighted a quote by Stone, "It [cityhood] would also provide an excellent forum for openly gay candidates to seek and hold public office". Yet these groups do not seek just exposure of gays, but gays who could be trusted to hold office. "According to Stone . . . cityhood would definitely be beneficial to gay people. 'It means that the leaders of the city would be directly accountable to its residents' " (*The Advocate* 5/29/84). The article made the concern for responsibility clear when Valerie Terrigno (identified as a member of the Stonewall Democratic club) stated: "The exciting part . . . is being able to have self-government . . . . I think you have a very responsible group of people living here". Although the paper appeared to mean gays when it wrote about "self-government" and "getting the government closer to the people", it promoted broad definitions of good elected officials. A *Frontiers* article quoted Terrigno, "It's [the cityhood question] about taking responsibility for our lives [and] it's about responsibility for our future" (*Frontiers* 10/25/84a). The concern with this type of conventionally defined responsibility illustrates the essentially assimilationist position of the incorporation campaign. Other *Frontiers* articles also depicted the tie between 'central' US values of responsibility and democracy. The paper quotes Stone in an early article, "If you support cityhood, you trust both yourself and your neighbors to govern responsibly and effectively – in other words – you want to live in a democracy" (*Frontiers* 2/1/84).

### *Maturity*

Many articles closely related maturity and responsibility. The clearest statement appeared in an article by Stone, "Cityhood simply gives us the right to manage our own affairs, just as other cities do . . . . West Hollywood will enter the world as an adult and will be expected to act like one" (*Frontiers* 3/14/84). He used this to suggest that the new city will not embark on radical changes, and that West Hollywood already had the characteristics of a 'real' city. Another gay paper described West Hollywood as "the city that isn't" (*Update* 3/7/84). Both of these passages suggested that areas go through 'developmental' stages, and that West Hollywood had reached a level of maturity deserving city status. This same idea is applied to gays: "West Hollywood represents a great opportunity for the gay community to move on to another phase and elect its own people to city offices" (*The Advocate* 9/18/84), thus establishing the connection between the 'mature' city and 'mature' gays.

The gay press was particularly critical of candidates who suggested that gays were not dealing with the campaign in a mature way. In a report on the destruction of a gay candidate's campaign posters, which *Frontiers* characterized as "attempts to squash cityhood", the paper quoted an anti-cityhood candidate's campaign manager as saying that the poster incidents "are equivalent to little boys and girls playing in the sand box and saying 'it's time to take our marbles and go home' " (*Frontiers* 10/24/84b). The newsmagazine sought to show that the incident in question was indeed part of a larger, important campaign issue, and to deal with the incident in a 'mature' fashion.

*Centrality*

The concern with centrality reflected the 'ethnic strategy' of the gay press. According to Epstein (1987, page 38), like other ethnic groups, "gay ethnicity functions typically through appeals to the professed beliefs of the dominant culture, emphasizing traditional American values such as equality, fairness, and freedom from persecution". (The limits to the analogy of gays as an ethnic group are discussed below.) The gay press promoted the centrality of gays to US society by countering the idea that gays are frivolous, and are only concerned with activities on the fringe of American culture. This attribute is one of the more difficult to distinguish, perhaps because other characteristics (especially the concern with progressiveness, entertainment, and conspicuous consumption) seem to contradict it. This type of opposition, of course, is not unique to gay identity. Tuan (1989), for example, notes that the conflict between morality and imagination is a common theme in Western societies. In an article run on the Fourth of July, the gay press placed the cityhood campaign squarely in the center of the US political tradition, in the spirit of the revolution:

> And so the quest for independence continues. In 1776 and 1984, the founders have put their reputations, money, friendships, business and jobs on the line. *Frontiers* salutes West Hollywood this Independence Day for its inspiring illustration of the power and depth of the American spirit of independence (*Frontiers* 7/4/84).

It is fairly rare to find this sort of rhetoric in the campaign ("no taxation without representation" was common in the early phase of the campaign, but later disappeared), but this article neatly summarized the attempt to show that gays are concerned with issues thought to be fundamentally important in US civil society.

The entire cityhood movement, rather than any particular article, illustrated the strongest attempt to bring gays into a more central position. Entry into political discourse, and the effort to address issues beyond gay rights, (including rent control, taxation, zoning, schools, fire protection, and public services), indicated that gays were not just single-issue candidates. This indicates a practical political plan involving at least two strategies. First, gays did not support cityhood simply as a 'gay' issue, but were concerned with day-to-day matters of running a city. Second, by addressing these more mundane issues, the gay press could demonstrate to nongay society that gays were fully aware of the consequences, problems, and potentials of incorporation. It is possible, however, to identify even more specific evidence that the gay press sought to portray gays as full participants in civil society. An examination of how the qualifications of gay candidates and cityhood leaders were established illustrates this point.

The qualifications of cityhood leaders and candidates were often stated in terms of their involvement or connection with mainstream political parties and leaders. Similarly, the gay press emphasized the formal educational experience of gay candidates. Institutionalized education is a major source of legitimacy, so that a gay candidate with a prestigious diploma could offer convincing evidence of his or her qualifications for political office. Examples of these connections include: Ron Stone's work for the US Senate, and major Democratic political candidates (including 'the Kennedys') (*Frontiers* 2/1/84); Steve Smith's job as assistant to the California Assembly Speaker, and his endorsement by Edward Kennedy and the AFL-CIO (*The Advocate* 9/18/84, 10/30/84, 11/13/84); Valerie Terrigno's membership in the ACLU (*The Advocate* 9/18/84, 12/11/84); and Steve Schulte's involvement with the United Way, National Institutes of Health, California State Senate, and previous local political campaigns, as well as his degrees from Yale and the University of Iowa (*The Advocate* 9/18/84, *Edge* 10/4/84c, *Frontiers* 10/24/84g). Unlike certain kinds of radical identity politics, the knowledge and legitimacy of the candidates depended in part on a relationship to the 'center', and not simply on their membership in the identity group of gays (Scott, 1992). Political and educational qualifications were not, however, the sole basis of support for these candidates. The gay press was even more concerned with candidates' connection with the gay community, documented support of gay rights, work in social services for gays, or other such activities.

It is important to note that the gay press endorsed some nongay candidates, although this was relatively rare. These endorsements, however, always referred to nongay candidates' relationship to gays: "Craig Lawson – his outstanding sensitivity to gay and lesbian issues is particularly remarkable for one not a member of that community" (*Frontiers* 10/31/84d). The gay press, and his own campaign literature, often used Candidate Bud Siegel's relationship with his gay children to illustrate his positive attitude toward gays and his ability to work with the gay community (*Frontiers* 9/5/84a). This shows that, although the gay press sought to portray gays as integrated into mainstream society, it did not want to ignore sexual identity. Candidates could retain their gayness, yet be close to traditional centers of power and legitimacy.

**'Constitutive narratives' and gay communities**

It is tempting to draw parallels between gay and ethnic communities, particularly in light of postmodern treatments of ethnicity that emphasize the unstable and reflective nature of identity. Smith (1992, page 512), for example, defines ethnicity as a "provisional, historically conditioned social construct . . . . A dynamic mode of self-consciousness, a form of selfhood reinterpreted if not reinvented generationally in response to changing historical circumstances". One notes, however, that the transformation of ethnic identity is not a particularly new phenomenon, and that even early urban sociologists observed that immigrant identities are remade from local to ethnic groupings (Park, 1955, pages 157–158). Epstein (1987, page 38) suggests that, in contrast to traditional forms, contemporary "new ethnicities" are all "future oriented" and are concerned with "an instrumental goal of influencing state policy and securing social rewards on behalf of the group", a statement which also aptly describes the goal of many gay advocates.

There are limits, however, to the resemblance between ethnic and sexual identity, and one must not blindly apply concepts developed in the context of ethnic studies to gays (Valentine, 1993c, page 247, note 3). This caution applies to traditional ethnic studies in particular: the use of 'ghetto idea' to study gay neighborhoods, for example, can mask the role gay territories play in the development of gay identity (for example, Levine, 1979). Godfrey (1988, page 215) argues that, although nonconformist communities (gays and lesbians) are more fluid than traditional ethnic communities, the "morphological evidence" indicates that they can be regarded as minorities that cluster in "special-identity neighborhoods". These identifiable neighborhoods "serve as both the symbols and substance of subcultural expression . . . . These places create, maintain, and reinforce group identities" (pages 45–46). For Godfrey the resemblance between gay and ethnic communities is largely a matter of "urban form and spatial structure", but this is hardly a sufficient comparison. A focus on urban morphology may serve to marginalize lesbians further, given Adler and Brenner's (1992) contention that gay women exercise less control over urban space than gay men. Epstein (1987) treats the question of gays as an ethnic group more comprehensively. He argues that, as a strategy, this characterization has a political utility in the context of the US civil rights movement: ethnicity provides social groups with legitimacy.

The issue of citizenship underlies much of West Hollywood's incorporation vis-à-vis gay identity. In the United States after the civil rights movement, ethnicity provides one of the foundations for citizenship claims. Shotter (1993, pages 194–195) argues that the 'topic' of citizenship supplies the basis for a community discussion of identity without imposing an unduly restrictive 'narrative order'. A tradition of argumentation over citizenship creates conditions in which personal issues of belonging and identity can be freely, and reflexively, addressed. This claim is somewhat questionable because the debate over citizenship itself has not been free of constraint and, as Kearns (1992) shows, appeals to 'active citizenship' can serve a variety of political interests. In the United States the struggle for citizenship has focused on efforts to break down barriers to recognition and inclusion in the polity, rather than being "an aspiration to civic participation as a deeply involving activity" (Shklar, 1991, page 3). Smith (1989) believes that the idea of 'citizenship as critique' can measure the degree to which civil, political, and social rights have been achieved, and that this critique is best manifest in localities. Hence it is possible to see the incorporation campaign as an attempt by gays to achieve the entitlements of citizenship by (symbolically) creating themselves as a social group.

The question of whether gays are an ethnicity is unlikely to be settled, because, as Epstein (1987) points out, the answer depends largely on how one defines an ethnic group. I argue that a useful approach is to draw a comparison between ethnic groups and gays using what Bellah (1985) calls the 'constitutive narrative' of communities. This focus on the narratives of identity groups is particularly useful for geographers because it reveals how places come to be morally valued. "Taking language seriously shows, moreover, that the 'quality' of place is more than just aesthetic or affectional, that it also has a moral dimension, . . . for language – ordinary language – is never morally neutral" (Tuan, 1991, page 694). Gay areas like West Hollywood can be described as 'communities of memory' as opposed to 'life-style enclaves' (Bellah *et al*, 1985). Members of a community of memory are tied to both the past and the future of the group; this makes the community 'genuine' or 'real'. "They carry a context of meaning that can allow us to connect our aspirations for ourselves and those closest to us with the aspirations of a larger whole" (Bellah, 1985, page 153). Ethnic and racial communities exemplify this sort of community. There is a limit, of course, to the access one has to narratives of any community. As feminist and postcolonial writers have made clear, the relationships between ethnography, the construction of the 'Other', and power are quite complex (Butler, 1990; Katz, 1992; Keith, 1992; Said, 1979; 1989).

Bellah believes that a community of memory maintains itself by "retelling its story, its constitutive narrative, and in so doing . . . offers examples of the men and women who have embodied and exemplified the meaning of the community" (1985, page 153). The 'constitutive narrative' is not necessarily limited to stories about individuals, because narratives invest places as well as human agents with meaning. Shotter (1993, page 193) discusses the similar notion of a "living tradition" of a community, "a historically extended, socially embodied argument, containing . . . arguments about what should be argued about, and why". Entrikin uses Bellah to discuss the modern relationship between an individual's sense of self, and one's relationship to others. Narratives "center the individual through the attachment to community.

Where such communities are located in particular places or territories, this centering can take on a literal, spatial sense" (199lb, page 66). The narrative the gay press created during the incorporation campaign constructed a gay identity using descriptions of West Hollywood as a place. This served to center gay identity on the city, or more precisely, on an idealized version of the city. The holistic quality of place provides a particularly powerful way to organize segmented facets of identity, a way which relieves the "burden" of modern consciousness, by "offering the fractured ego a nurturing and seemingly harmonious world, a pleasing image of the self" (Tuan, 1982, page 10). Experienced holistically, rather than as a bundle of discrete characteristics, this idealized city served to reconcile contradictory characteristics of gay identity.

## Conclusion

The gay press tried to construct a stable identity for gays by creating a necessary connection between the city of West Hollywood and the 'idea' of gays. In creating this kind of holistic conception of place, the gay press played the role of lay regional geographers, seeking to weave together "nature and culture to create places as areal 'individuals' " (Entrikin, 1991a, page 10). This connection was established through the construction of a 'constitutive narrative' that sought to establish the 'sexual meaning' of West Hollywood, along with elements such as occupations, norms of behavior, clothing styles, political outlooks, and cultural activities. In this sense, the model identity symbolized by West Hollywood is similar to an ethnicity. As an ideal city, West Hollywood incorporated the attributes of creativity, aesthetic sensibility, an affinity with entertainment and consumption, progressiveness, responsibility, centrality, and maturity. By breaking apart old stereotypes and recombining certain components, the gay press could create a new identity that was both complex and closely aligned to 'central' US values. Since the city of West Hollywood would have a tangible existence, it would be a highly visible symbol of this identity. While the descriptions of West Hollywood in the gay press represent an idealized vision of the city, articles consistently portrayed the area as a place that was already a gay territory. Cityhood was a means to embody and reify this quality, not a strategy to initiate radical change.

The coverage of West Hollywood's incorporation in the gay press mirrors an historical trend of gay liberation, where specific "gay territories" have played a major role in the development of particular gay identities and subcultures. In some ways, one can regard the history of gay liberation – and therefore modern gay identity – as a process in which gays self-consciously use place to move themselves into public visibility by claiming and reshaping public spaces. This type of territorial strategy highlights the normative importance of place in the construction of social and personal identities. The close tie between gay identity, space, and place suggests that geographers have an important contribution to make in this area. Although recent work has begun to explore "struggles over sexual representations of, and sexual symbols in, space" (Knopp, 1992, page 666), these studies have maintained an orientation toward political economy, tending to neglect the normative dimension of place as such. The creation of a "gay" civic culture by municipal incorporation is a powerful symbol, as well as a means to affect material social relations. A de jure rather than a de facto recognition of a "gay city" moves gays that much closer to the center of American society, and therefore that much closer to acceptance as a group.

There is an intimate connection between the social process that forms personal and group identities, and the symbolic aspect of place. This connection is now attracting the attention of disciplines such as anthropology (Gupta and Ferguson, 1992). Places, since they are experienced as wholes, organize meaning in such a way that contradictory ideas can be held simultaneously. This quality is particularly important for the component of identity formed by place. Characteristics (of an identity) with conflicting normative values, centrality and progressiveness for example, can be more easily combined if they are embodied in a place rather than in a person. The holistic quality of place also allowed the gay press to create symbolic, "natural" connections between cultural and political activities, when these activities may not be related in practice, for example, artistic creativity and administrative innovation, or political and cultural progressiveness. The connection between West Hollywood and gays tends to disguise the constructed nature of gay identity. Thus the use of place encourages the "common-sense" perception that gays are a social group as natural, and therefore as legitimate, as ethnicities.

The question of identity has received tremendous attention in the past several years, but little consensus has been reached beyond the relatively widespread agreement about its constructed, rather than natural or essential, nature. Studies in geography, as well as other fields, have tended to focus exclusively on the ways in which the "normative ideal" of identity fails to correspond to the particular experience of individuals within that group (Butler, 1990; Sedgwick, 1990). Humanistic geography can complement this perspective by casting the question of identity in terms of place and morally valued ways of life. Place, I have argued, plays a fundamental role in the creation of a particular "normative ideal" of gay identity. Approaches which take note of the perspectival quality of place, and which emphasize the role of language in the creation of place can address the cultural dimensions of these broader relationships of place and identity (Entrikin, 1991b; Tuan, 1991). While the ontological status of identity groups such as gays may remain indeterminate, studies of links between place and identity provide an important means to show how ideal identities are created and perpetuated.

Valentine (1993a), Knopp (1992), and Bell (1991) offer agendas for geographical studies of sexuality generally and of gays and lesbians in particular. This study will be seen, I hope, as complementary to these programs, but as was noted in the introductory section, certain limitations need to be addressed through further research. In particular, humanistic work concerning the role of place in the symbolic construction of lesbian identity needs to be further pursued. If lesbians – or gay men of lower economic status – indeed do not exercise the kind of control over public space that upper-class gay men enjoy, it would be especially important for researchers to heed Bell's (1991, page 328) call to conduct investigation " 'close' to the study group", for example, Valentine (1993b; 1993c). Ethnographic studies and/or surveys of gay men in West Hollywood would also be useful, particularly work that investigates how closely the identity realized by the gay press in 1984 conforms to the self-identity of gay men currently associated with the city.

## Notes

1 It is also important to note that heterosexual identities are as much a social construction as homosexuality (Katz, 1990; Peake, 1993). For example, one can easily envision a study of the role of place in the construction of heterosexual male identity.

2 Except where noted, the historical information on West Hollywood is drawn from Kepner and Williams (1985), Envicom (1986, pages 1.1–1.3), and an interview with Robert Vulcan, president of the West Hollywood Historical Society. For an account of incorporation and annexations in Los Angeles up to 1950, see Bigger and Kitchen (1952); for information after 1950, see Miller (1981).

3 I note, without making a strong claim for their significance, that two relatively recent articles on West Hollywood incorporated the same imagery and symbols that appeared in the gay press: creativity (Waldman, 1988) and maturity (Ellingwood, 1994). The headline of Ellingwood's article is particularly interesting: "The 'Gay Camelot' grows up".

4 Smith (1993, page 292) addresses the problems involved in reading 'polyvocal dialogues' by attempting to make explicit her own position vis-à-vis her subjects or fellow participants, and by invoking a conception of her work as part of an unresolved (and unresolvable) dialogue, rather than as an authoritative statement. This strategy is useful in emphasizing the interpretative nature of ethnographic studies, but is less useful for distinguishing between a well-supported and a less well-supported reading. I think that it would be entirely possible to generate, for example, a radical interpretation of the gay press, and such an exercise would be a valuable complement to this study. I have included a complete listing of articles from the gay press, rather than only the articles cited in this study, in order to facilitate those interested in formulating alternative interpretations of the material.

As for methodology in this study, I first classified each article in one of four categories: (1) focused on gays, (2) focused on rent control, (3) focused on both, or (4) focused on neither. I then examined each passage concerned with gays to identify the most commonly used images, metaphors, words, and phrases. The latter process was not a strict content analysis because I was not only concerned with the absolute frequency of particular words and phrases, but also with their contextual meaning and use.

My methods generally follow the 'rules of thumb' suggested by Strauss (1987, pages 1–20) for qualitative analysis. In particular, there was a long period of reflection, integrating, and revision not only of the conclusions but also of the resource materials and coding categories ('characteristics'). I did not begin with rigid categories, but rather identified the seven characteristics through a reflective reading of many articles. In particular, I isolated the last two characteristics – maturity and centrality – after the other five, and after numerous readings and 'coding' revisions.

5 Anita Bryant, a former Miss America, led a national antigay campaign in the late 1970s. California Senator John Briggs was a conservative leader, who, among other actions, sponsored a proposition banning homosexuals from teaching in public schools in 1978 (Castells, 1983, page 144). In 1978, Dan White murdered San Francisco mayor George Moscone, and city supervisor Harvey Milk, the political leader of the gay community.

## Newspaper references

*The Advocate*

5/29/84, "Move for cityhood gains steam in West Hollywood", Christine S. Shade
9/18/84, "Gays, lesbians caught up in W. Hollywood vote fever", Joan Cort
10/30/84, "Steve Smith for West Hollywood City Council" [advertisement]
11/13/84, "Steve Smith for: He'll do more for West Hollywood" [advertisement]
12/11/84, "First 'Gay City' voted in California", Michael Balter

*Edge*

2/23/84a, "West Hollywood Cityhood Day March 10"
2/23/84b, "Cityhood petition"
4/5/84, "Boystown closer to cityhood"
5/17/84, "CSW endorses Boystown's cityhood"
9/20/84, "Art Guerrero wants to be a city father", Dennis Colby
10/4/84a, "John Mackey campaigns – with a shovel!", Brian John Thorpe
10/4/84b, "Dr. Scott Forbes enters the race", Scott Forbes, O.D.
10/4/84c, "Schultze's new race for a new city", Kevein Bersh
11/1/84, "An urgent appeal . . . "

*Frontiers*

5/11/83a, "West Hollywood's gay roots"
5/11/83b, "West Hollywood: beyond Boystown"
2/1/84, "West Hollywood cityhood"
2/8/84, "Community leaders join cityhood drive"
2/15/84, Letters to editor: "Thanks for cityhood article", Ron Stone; "Cityhood for better representation", Frank Wittenberg
3/7/84, "What would a 'City' of West Hollywood be like?"
3/14/84, "Cityhood should make plenty of dreams come true", Ron Stone
3/28/84, "Feasibility questioned re West Hollywood cityhood"
4/4/84a, "West Hollywood: the time has come for independent jurisdiction", editorial
4/4/84b, "West Hollywood cityhood petition filed"
4/4/84c, "City vote on way to reality?"
4/4/84d, "Club endorses West Hollywood cityhood quest"
4/25/84, "LAFCO approves West Hollywood cityhood petition", Sallie Fiske
5/2/84, "5 West Hollywood council seats up for grabs if cityhood wins", Sallie Fiske
5/9/84, "What lies ahead for the new 'city' of West Hollywood?", Art Guerrero
5/16/84a, "Fiscal analysis for proposed West Hollywood city shows $6.5 million minimum surplus"
5/16/84b, "Christopher Street West, Democrats say yes to W. Hollywood incorporation", Sallie Fiske
6/6/84, "West Hollywood citihood [sic] closer to fruition", Sallie Fiske
7/4/84, "West Hollywood independence: a great American tradition continues"
7/25/84, "This week critical for West Hollywood cityhood"

8/15/84, "Cityhood vote set for Nov. 6"
8/29/84, "Guerrero proposes W. H. 'Sidewalk Transformation' "
9/5/84a, "Bud Siegel: West Hollywood city council candidate"
9/5/84b, "MECLA cityhood breakfast puts West Hollywood in spotlight"
9/19/84a, "WH Candidate denies 'anti-cityhood' charges"
9/19/84b, "Valerie Terrigno runs as qualified candidate for West Hollywood council"
9/26/84, "Don Genhart runs as independent candidate in W. H. Race"
10/3/84a, "West Hollywood incorporation: You can't tell the players without a program"
10/3/84b, "Cityhood foes use survey to stir up anti-gay bias"
10/3/84c, "Homophobia enters West Hollywood cityhood campaign"
10/10/84, "Art Guerrero – an outstanding candidate for city council"
10/17/84a, "Pam Parker wants to put people back in government"
10/17/84b, "John Heilman runs on 'United for West Hollywood' slate"
10/24/84a, "Terrigno fundraiser a circus"
10/24/84b, "Tactics muddy W. H. Campaign issues"
10/24/84c, "Alan Viterbi: wants to end discrimination"
10/24/84d, "John Mackey: avid proponent of rent control"
10/24/84e, "Scott Forbes: wants to end prostitution"
10/24/84f, "Craig Lawson – concerned about rent control"
10/24/84g, "Steve Schulte: he wants a community for all people"
10/31/84a, "The ends do not justify the means"
10/31/84b, "The biggest show in town"
10/31/84c, "Special West Hollywood election section" [endorsements and criticisms of candidates]
10/31/84d, "Gay and lesbian alliance rates highly qualified candidates"
11/7/84, "Cityhood landslide victory"

*LA Weekly*
6/24/83, "Gay pride: will success spoil west Hollywood", Ron Stone
11/2/84a, "West Hollywood cityhood initiative – RR: yes"
11/2/84b, "West Hollywood city council – SS: Stone, Albert, Heilman, Schulte and Terrigno" [endorsements]

*Update*
1/25/84, "West Hollywood seeking cityhood", Richard Labonte
2/22/84, "Cityhood drive is popular", Richard Labonte
3/7/84, "Cityhood day aims at cash and signatures", Richard Labonte
3/21/84, "County opens battle over West Hollywood city budget", Richard Laboote
4/4/84, "Charges fly in cityhood debate", Richard Labonte
4/18/84, "Cityhood movement passes landmark"
5/30/84 "Beautiful W. Hollywood"
6/30/84 "LAFCO Unanimously OK's W. Hollywood Finances", Richard Labonte
8/22/84, "44 vie for new city council", Richard Labonte
10/3/84, "First anti-gay rumble felt in W. Hollywood campaigns", Richard Labonte
10/31/84, "Second guessing gay might in West Hollywood", Richard Labonte
11/14/84a, "New council reflects community", Richard Labonte
11/14/84b, "West Hollywood: gay majority heads council", Richard Labonte

**References**

Adler S, Brenner J, 1992: "Gender and space: lesbians and gay men in the city" *International Journal of Urban and Regional Research* 16, 24–34.

Agnew J, 1989, "The devaluation of place in social science", in *The Power of Place* Eds J A Agnew, J S Duncan (Unwin Hyman, Winchester, MA) pp 9–29.

Agnew J, 1992, "Place and politics in post-war Italy: a cultural geography of local identity in the provinces of Lucca and Pistoia", in *Inventing Places: Studies in Cultural Geography* Eds K Anderson, F Gale (Longman Cheshire, Melbourne) pp 52–71.

Alexander J, Smith P, 1993, "The discourse of American civil society: a new proposal for cultural studies" *Theory and Society* 22, 151–207.

Altman D, 1982 *The Homosexualization of America, the Americanization of the Homosexual* (St Martin's Press, New York).

Anderson K J, 1987, "The idea of Chinatown: the power of place and institutional practice in the making of a racial category" *Annals of the Association of American Geographers* 77, 580–598.

Anderson K J, 1988, "Cultural hegemony and the race-definition process in Chinatown, Vancouver: 1880–1980" *Environment and Planning D: Society and Space* 6, 127–149.

Anderson K, Gale F, 1992 *Inventing Places: Studies in Cultural Geography* (Longman Cheshire, Melbourne).

Barnes T J, Duncan J S, 1992 *Writing Worlds: Discourse, Text and Metaphor in the Representation of Landscape* (Routledge, London).

Bell D J, 1991, "Insignificant others: lesbian and gay geographies" *Area* 23, 323–329.

Bellah R N, Madsen R, Sullivan W M, Swidler A, Tipton S M, 1985 *Habits of the Heart: Individualism and Commitment in American Life* (Harper and Row, New York).

Bigger R, Kitchen J D, 1952 *How the Cities Grew: A Century of Municipal Independence and Expansionism in Metropolitan Los Angeles* (The Haynes Foundation, Los Angeles, CA).

Bird S E, 1992, "Travels in nowhere land: ethnography and the 'impossible audience' " *Critical Studies in Mass Communication* 9, 250–260.

Bondi L, 1992, "Gender symbols and urban landscapes" *Progress in Human Geography* 16, 157–170.

Bourdieu P, 1977 *Outline of a Theory of Practice* (Cambridge University Press, Cambridge).

Burgess J, 1985, "News from nowhere: the press, the riots and the myth of the inner city", in *Geography, the Media, and Popular Culture* Eds J Burgess, J R Gold (Croom Helm, London) pp 192–228.

Butler J, 1990 *Gender Trouble: Feminism and the Subversion of Identity* (Routledge, New York).

Castells M, 1983, "Cultural identity, sexual liberation and urban structure: the gay community in San Francisco", in *The City and the Grassroots* (Edward Arnold, London) pp 138–172.

Christensen T, Gerston L, 1987, "West Hollywood: a city is born" *Cities* 4, 299–303.

Cohen J R, 1991, "The 'relevance' of cultural identity in audiences' interpretation of mass media" *Critical Studies in Mass Communication* 8, 442–454.

Crang P, 1992, "The politics of polyphony: reconfigurations in geographical authority" *Environment and Planning D: Society and Space* 10, 527–549.

Crow D, 1994, "My friends in low places: building identity for place and community" *Environment and Planning D: Society and Space* 12, 403–419.

Davis T H, 1991, " 'Success' and the gay community: reconceptualizations of space and urban social movements", paper presented at the First National Student Conference on Lesbian and Gay Studies, Milwaukee, WI, April; copy available from the author 26 Hemenway Street, #29, Boston, MA 02115.

Denneny M, Ortleb C, Steele T, 1983 *The Christopher Street Reader* (Coward-McCann, New York).

Duggan L, 1992, "Making it perfectly queer" *Socialist Review* 22(1), 11–32.

Ellingwood K, 1994, "The 'Gay Camelot' grows up" *Los Angeles Times* 27 June, page 1.

Entrikin J N, 1991a *The Characterization of Place* Wallace W Atwood Lecture Series No. 5, The Graduate School of Geography, Clark University, Worcester, MA.

Entrikin J N, 1991b *The Betweenness of Place: Toward a Geography of Modernity* (Johns Hopkins University Press, Baltimore, MD).

Envicom, 1986, "City of West Hollywood general plan, technical background report", Envicom Corporation, 28328 Agoura Road, Agoura Hills, CA 91301.

Epstein S, 1987, "Gay politics, ethnic identity: the limits of social construction" *Socialist Review* 17(3–4), 9–54.

Forest B, 1991, "Political territory and symbol: West Hollywood, gays and rent control", unpublished masters thesis, Department of Geography, University of California, Los Angeles, CA.

Foucault M, 1978 *The History of Sexuality: Volume 1* translated by R Hurley (Pantheon, New York).

Geltmaker T, 1992, "The Queer Nation Acts Up: health care, politics, and sexual diversity in the County of Angels" *Enviroment and Planning D: Society and Space* 10, 609–650.

Giddens A, 1991 *Modernity and Self-identity: Self and Society in the Late Modern Age* (Stanford University Press, Stanford, CA).

Godfrey B J, 1988 *Neighborhoods in Transition. The Making of San Francisco's Ethnic and Nonconformist Communities* (University of California Press, Berkeley, CA).

Gupta A, Ferguson J, 1992, "Beyond 'culture': space, identity, and the politics of difference" *Cultural Anthropology* 7, 6–23.

hooks b, 1990, "Choosing the margin as a space of radical openness", in *Yearnings: Race, Gender, Cultural Politics* (South End Press, Boston, MA) pp 145–154.

Jackson P, 1989 *Maps of Meaning* (Unwin Hyman, London).

Janowitz M, 1952 *The Community Press in an Urban Setting* (Free Press, Glencoe, IL).

Katz C, 1992, "All the world is staged: intellectuals and the projects of ethnography" *Environment and Planning D: Society and Space* 10, 495–510.

Katz J N, 1990, "The invention of heterosexuality" *Socialist Review* 20(1), 7–34.

Kearns A J, 1992, "Active citizenship and urban governance" *Transactions of the Institute of British Geographers, New Series* 17, 20–34.

Keith M, 1992, "Angry writing: (re)presenting the unethical world of the ethnographer" *Environment and Planning D: Society and Space* 10, 551–568.

Kepner J, Williams W L, 1985, "West Hollywood incorporation 1924–1985", unpublished manuscript; on file at the International Gay and Lesbian Archives, PO Box 69679, West Hollywood, CA 90069.

Kilday G, 1983, "Hollywood", in *The Christopher Street Reader* (Coward-McCann, New York) pp 54–57.

Knopp L, 1987, "Social theory, social movements and public policy: recent accomplishments of the gay and lesbian movements in Minneapolis, Minnesota" *International Journal of Urban and Regional Research* 11, 243–261.

Knopp L, 1990a, "Some theoretical implications of gay involvement in an urban land market" *Political Geography Quarterly* 9, 337–352.

Knopp L, 1990b, "Exploiting the rent-gap: the theoretical significance of using illegal appraisal schemes to encourage gentrification in New Orleans" *Urban Geography* 11, 48–64.
Knopp L, 1992, "Sexuality and the spatial dynamics of capitalism" *Environment and Planning D: Society and Space* 10, 651–669.
Levine M P, 1979, "Gay ghetto", in *Gay Men: The Sociology of Male Homosexuality* Ed. M P Levine (Harper and Row, New York) pp 182–204.
Lynch F, 1987, "Non-ghetto gays: a sociological study of suburban homosexuals" *Journal of Homosexuality* 13(4), 13–42.
Massey D, 1991, "The political place of locality studies" *Environment and Planning A* 23, 267–281.
Massey D, 1993, "Questions of locality" *Geography* 78(339), 142–149.
Miller G J, 1981 *Cities by Contract: The Politics of Municipal Incorporation* (MIT Press, Cambridge, MA).
Moos A, 1989, "The grassroots in action: gays and seniors capture the local state in West Hollywood, California", in *The Power of Geography* Eds J Wolch, M Dear (Unwin Hyman, Winchester, MA) pp 351–369.
Park R E, 1922 *The Immigrant Press and its Control* (Harper and Brothers, New York).
Park R E, 1940, "News as a form of knowledge: a chapter in thc sociology of knowledge" *American Journal of Sociology* 46, 669–686.
Park R E, 1955, Immigrant community and immigrant press", in *The Collected Papers of Robert Park, Volume 3. Society: Collective Behavior, News and Opinion, Sociology and Modern Society* Eds E Charrington Hughes, L Wirth, R Redfield (The Free Press, Glencoe, IL) pp 152–164.
Peake L, 1993, "'Race and sexuality: challenging the patriarchal structuring of urban social space" *Environment and Planning D: Society and Space* 11, 415–432.
Plummer K (Ed.) 1981 *The Making of the Modern Homosexual* (Hutchinson, London).
Relph E, 1976 *Place and Placelessness* (Pion, London).
Rodman M C, 1992, "Empowering place: multilocality and multivocality" *American Anthropologist* 94, 640–656.
Rutledge L W, 1992 *The Gay Decades: From Stonewall to the Present: The People and Events that Shaped Gay Lives* (Penguin Books, New York).
Sack R, 1981, "Territorial bases of power", in *Political Studies from Spatial Perspectives: Anglo-American Essays on Political Geography* Eds A D Burnett, P J Taylor (John Wiley, New York) pp 53–71.
Sack R, 1986 *Human Territoriality: Its Theory and History* (Cambridge University Press, Cambridge).
Said E, 1979 *Orientalism* (Vintage Books, New York).
Said E, 1989, "Representing the colonized: anthropology's interlocutors" *Critical Inquiry* 15, 205–225.
Scott J W, 1992, "Multiculturalism and the politics of identity" *October* 61, 12–19.
Sedgwick E K, 1985 *Between Men: English Literature and Male Homosocial Desire* (Columbia University Press, New York).
Sedgwick E K, 1990 *Epistemology of the Closet* (University of California Press, Berkeley, CA).
Shils E, 1975 *Center and Periphery: Essays in Macrosociology* (University of Chicago Press, Chicago, IL).
Shklar J N, 1991 *American Citizenship: The Quest for Inclusion* (Harvard University Press, Cambridge, MA).
Shotter J, 1993 *Cultural Politics of Everyday Life* (University of Toronto Press, Toronto).

Smith M P, 1992, "Postmodernism, urban ethnography, and the new social space of ethnic identity" *Theory and Society* 21, 493–593.

Smith S J, 1989, "Society, space and citizenship: a human geography for the 'new times'?" *Transactions of the Insitute of British Geographers, New Series* 14, 144–156.

Smith S J, 1993, "Bounding the Borders: claiming space and making place in rural Scotland" *Transactions of the Institute of British Geographers, New Series* 18, 291–308.

Stamm K R, 1985 *Newspaper Use and Community Ties: Toward a Dynamic Theory* (Ablex Publishing Corporation, Norwood, NJ)

Stein A, 1992, "Sisters and queers: the decentering of lesbian feminism" *Socialist Review* 22(1), 33–55.

Strauss A, 1987 *Qualitative Analysis for Social Scientists* (Cambridge University Press, New York).

Streeter T, 1989, "Polysemy, plurality, and media studies" *Journal of Communication Inquiry* 13, 88–106.

Taylor A, 1983, "Conceptions of masculinity and femininity as a basis for stereotypes of male and female homosexuals", in *Homosexuality and Social Sex Roles* Ed. M Ross (Haworth Press, New York) pp 37–53.

Till K, 1993, "Neotraditional towns and urban villages: the cultural production of a geography of 'otherness' " *Enviroment and Planning D: Society and Space* 11, 709–732.

Tuan Y-F, 1974, "Space and place: a humanistic perspective", in *Progress in Geography: International Reviews of Current Research* Eds C Board, R J Chorley, P Haggett, D Stoddart (Edward Arnold, London) pp 211–252.

Tuan Y-F, 1977 *Space and Place: The Perspective of Experience* (University of Minnesota Press, Minneapolis, MN).

Tuan Y-F, 1980, "Rootedness versus sense of place" *Landscape* 24(1), 3–8.

Tuan Y-F, 1982 *Segmented Worlds and Self: Group Life and Individual Consciousness* (University of Minnesota Press, Minneapolis, MN).

Tuan Y-F, 1984, "In place, out of place" *Geoscience and Man* 24, 3–10.

Tuan Y-F, 1989 *Morality and Imagination: Paradoxes of Progress* (University of Wisconsin Press, Madison, WI).

Tuan Y-F, 1991, "Language and the making of place: a narrative-descriptive approach" *Annals of the Association of American Geographers* 81, 684–696.

Valentine G, 1993a, "(Hetero)sexing space: lesbian perceptions and experiences of everyday spaces" *Environment and Planning D: Society and Space* 11, 395–413.

Valentine G, 1993b, "Desperately seeking Susan: a geography of lesbian friendships" *Area* 25, 109–116.

Valentine G, 1993c, "Negotiating and maintaining multiple sexual identities" *Transactions of the Institute of British Geographers, New series* 18, 237–243.

Waldman T, 1988, "West Hollywood: creative city, diverse constituencies" *California Journal* 19, 541–542.

Weightman B, 1980, "Gay bars as private places" *Landscape* 24(1), 9–16.

Weinberg T S, 1983 *Gay Men, Gay Selves: The Social Construction of Homosexual Identities* (Irvington Publishers, New York).

# SECTION SIX

# *GEOGRAPHIES OF NATIONALISM AND ETHNIC CONFLICT*

**Editor's introduction**

The earliest mention of the word 'nationalism' can be found in a 1774 work of the philosopher Johann Gottfried Herder. But it did not enter into general usage in Europe until the mid-nineteenth century. The multiformity that 'nationalism' has assumed over the years and from place to place has meant that it is hard to provide a general definition that covers all cases (Alter 1989; Woolf 1996). This has not discouraged attempts at doing so. Two approaches have prevailed. One involves seeing nationalism as an ideology in politics that exalts the 'nation' to a central value and in which 'national interests' provide the measure of political thought and action (e.g. Breuilly 1982; Anderson 1983; Hobsbawm 1990). The other views nationalism as an autonomous social 'force' or independent variable in history that, arising first in England and/or Germany, spread through a dual process of elite imitation and mass disaffection with existing identities first into the rest of Europe and then around the world. The understanding of nationalism as a primordial attachment of individuals to a particular people or *volk* goes back to the German philosopher Hegel but the emphasis on elite imitation is a more recent innovation (see, e.g., Nairn 1977; Greenfeld 1992; Huntington 1993).

If the problem with the first approach is that it dissolves nationalism into 'its' particular manifestations and sees it as derivative of other social and political processes, the problem with the second is that it reifies 'the people' and treats nationalism as based on an inheritance from the past, a 'primordial identity' with roots in a misty history, more than a contemporary construct. Each approach to understanding captures only part of the the phenomenon. On the one hand, nationalism is a type of practical politics mobilizing groups by appealing to national interests and identities (Agnew 1989a), but it is also, on the other hand, a set of ideas about the 'nation' as the key or singular reference group for identity that did begin with the vesting of 'sovereignty' in the people as a model of political excellence which then spread under the label of 'self-determination' to groups defining themselves largely on ethnic grounds. The appeal to primordialism, therefore, is the appeal to a mythic common national past (Smith 1991).

The politics of nationalism concern the pursuit of a set of strategies for privileging the presumed shared interests and identity of a population inhabiting a discrete territory (Harris 1990).

From this point of view, therefore, nationalism is not the *natural* expression of nationality. It is a type of politics that depends on claiming non-political legitimacy for political advantage. In other words, nationalism relies on appealing to cultural symbols of identity and collective material interests in pursuit of political goals, prime among which is control over a state. Breuilly (1982, 383–4) puts this case as follows:

> There is no 'natural' basis to politics. There is no cultural or any other non-political unit of humanity which can be regarded as the true basis of legitimate politics. To accept that there is, is to abolish the autonomy and limits of politics. To see that politics does not arise from the nation and that it is a specific and effective form of politics only under certain political conditions can perhaps guard against the idea that there is some natural basis to the legitimate state which lies beyond the public realm.

The articulation of nationalism as a specific form of politics depends on the availability of a state or a political movement which selects and integrates cultural symbols and political claims into a coherent ideological package. Absent this, nationalism will fail to develop a clearly defined and appealing form (Anderson 1986). This is the case even in the presence of widespread and longstanding consciousness of cultural difference or ethnic identity (Armstrong 1982). Ethnic identities must be organized into a nationalist framework for nationalism to result. Nationalism is not an intrinsic feature of ethnic identity itself (Anderson 1983). The appeal of nationalism is still only contingent. If ethnic identity (sense of belonging to a group with shared values and shared history) and the cultural symbols of nationhood have a differential appeal across social strata and places then one can refer to a geography of nationalism (Fitzpatrick 1978; Brustein 1981; Agnew 1989a; Anderson 1989; Cooke 1989).

The major historical change that made it possible to consider the nation as both natural and unitary was the rise of the territorial state (see Section 1). Previous forms of political organization neither defined clear boundaries nor encouraged internal economic integration and cultural homogenization. City-states and empires were poorly integrated as blocs of territory and tolerated a wide variety of cultural practices. Territorial states, in mediating more and more transactions, came to rely on categorical identities to establish eligibility for their political activities and the economic protection they offered. The political revolutions, beginning in England in the 1600s and extending through the American and French ones of the late eighteenth century, added to this a claim for popular sovereignty. Distinctions between who were and who were not 'citizens' became important for the first time since the collapse of the Roman Empire (Spinner 1994; Kofman

1995). This entailed further changes that came to fruition finally in the nineteenth century. Above all, membership in the new common polity required 'mutual communication. This poses an impetus for erasure of differences among the citizens. One of the crucial questions of the modern era is whether meaningful, politically efficacious public discourse can be achieved without this erasure' (Calhoun 1994, 318) The idea of the 'nation-state' at the heart of nationalism rests fundamentally on the creation of either cultural uniformity (as in German-style ethnic nationalism) or a civic 'religion' based on a founding myth and a set of 'special' institutions (as in American-style civic nationalism).

Nationalism is not primarily either geographically integrating or disintegrating (Calhoun 1994, 320). It can be both. The same types of rhetoric can be used to claim unity within one state (British nationalism), across separate states (pan-Slavism, pan-Arabism) or to demand independence for a region within a state (Scottish nationalism, Basque nationalism). What they all share in common is the now longstanding legitimacy of posing claims to autonomy in terms of territorial sovereignty; joining together the people of a specific group within one state. Even with the increased pace of economic globalization this has not lost its attraction. States are still powerful economic regulators and provide the only existing organizational frameworks within which representative and participatory politics can be pursued.

It is difficult to make nationalism appear in an entirely positive light. It is often alleged to have produced some of the darkest moments of the twentieth century. To the political theorist John Dunn (1979, 55), for example, 'Nationalism is the starkest political shame of the twentieth century', associated with the trenches of the First World War and the Nazi seizure of power in Germany in 1933. But it is also 'the common idiom of contemporary political feeling' (Dunn 1979, 56). The inter-communal conflict, 'ethnic cleansing' and use of rape and pillage as weapons of war in the former Yugoslavia and elsewhere around the world stand as a salutary reminder to nationalism's continuing resonance under conditions of political-economic crisis (Glenny 1992; Ignatieff 1993; Denitch 1994; Allen 1996). Indeed, as the Cold War has faded into history since 1991 it is ethnic conflicts that have come into prominence as the main manifestation of competing nationalisms, even though many of these conflicts are not *solely* ethnic in character (see, e.g., de Waal 1994; Keen 1995). A diversity of cultural groups can be found within many existing states. Most states are multiethnic or multinational, despite claims to the contrary. Ethnic groups are not primordial entities but 'collectivities of people who share some pattern of normative behavior, or culture, and who form part of a larger population, interacting within the context of a common social system' (Cohen 1974, 92). The groups are differentiated from one another and integrated internally both by means of constantly recreated and invented ethnic symbols and such political-economic mechanisms as a cultural division of labour and political favouritism (on the former see, e.g., Hobsbawm and Ranger 1983 and Billig 1995; and on the latter see, e.g., Hechter 1974; MacLaughlin and Agnew 1986; Lemarchand 1994; Prunier 1995). The bureaucratization of states, regional economic differences that follow the 'fault lines' of ethnic groups and the 'unfreezing' of the boundaries written into space during the Cold War, have all contributed to the upsurge in communal and ethnic conflicts. These are particularly intractable when they are longstanding and where they involve competing territorial claims, few alternative identities and inter-ethnic economic competition. Violence is important for validating the seriousness of one's claims and encouraging others to accept your inter-ethnic conflict as a zero-sum game. Without the protection of one's own state there are no guarantees against the violence directed at you by the other groups (Agnew 1989b).

The question of how national groups establish the territorial boundaries between one another in which they invest so much meaning and for which they are willing to sacrifice their lives is a contentious one (see, e.g., Bhabha 1990). Until recently it has usually been addressed indirectly within the confines of the two broad approaches to nationalism outlined at the beginning of this Introduction. In his article reproduced here, **Daniele Conversi** (**Reading 16**) provides a spatial-analytic alternative to these two approaches (he calls them *instrumentalism* and *primordialism*, respectively) that views nationalism directly as a process of boundary creation and maintenance. He does so by drawing together three theories – the ethno-symbolist, transactionalist and homeostatic – around the question of boundary definition. He sees nationalism as resting on a process of social categorization in which groups identify themselves in *opposition* to other groups with pre-existing ethnic markers serving to differentiate them one from the other. Internal ethnic 'content' and the spatial segmentation produced by territorial boundaries thus interact to provide the basis upon which a particular nationalism supports and legitimates itself. One may rely more on content, another more on opposition or antagonism across boundaries. In either case: 'nationalism is a struggle over the definition of spatial boundaries, that is, over the control of a particular land or soil' (p. 329).

Conversi is concerned with the socio-spatial process of boundary formation giving rise to nationalism. He gives no attention to the political economy of nationalism, the material conditions under which nationalisms of different types emerge and flourish. This is the focus of the extract from a longer article by **Colin H. Williams** (**Reading 17**). He uses a somewhat revised world-systems theory to frame a discussion of the explosion of ethnic separatist movements (separatist nationalism) around the world since the 1970s. He uses case studies of separatist nationalism in Spain, France and Nigeria to make the claim that the resurgence of nationalism in its separatist guise is the 'playing out of minority aspirations unsatisfied during the critical period of state formation' (p. 340). Changes in the political economy of world capitalism have now created conditions under which it is feasible for those groups ill-digested by existing states to strike out on their own. A variety of local factors also seem to figure in each case (e.g. the aftermath of the Civil War in Spain for Basque nationalism, the hyper-centralization of the French state for Corsican nationalism), suggesting that the material determinants cannot be restricted to the geographical scale of the world-system as a whole. Other political-economic perspectives would be more likely to give greater emphasis to the varying histories of statehood, particularly the degree of *forced* cultural homogenization, economic exploitation and military suppression of subsidiary groups (e.g. Birch 1975; Rokkan and Urwin 1982; Bensel 1990).

How the national or ethnic past from which present identity draws is remembered has become increasingly important for those students of nationalism disillusioned with the typologies (ethnic vs. civic, integral (fascist) vs. unification, etc.) and lists of conditions around which most discussion of nationalism has tended to circulate. This reflects both a concern for the collective memory of national groups and the ways in which memories are represented in nationalist discourses (Boyarin 1994). In this postmodern construction, the production and configuration of national 'images' in literature, cinema, newspapers, monuments and landscape imagery have been seen as ways in which national identities are forged and reshaped (e.g. Gillis 1994). The capacity to place one's self imaginatively within 'your' territory and to develop a 'common sense' about its contemporary shape and history are important features of a 'nationalist imagination' (on 'nature' and 'place' in nationalism, see Deudney 1995). Grasping the nature of your landscape and the heritage of the past contained within it are parts of this imagination (e.g. Lowenthal 1991; Daniels 1993). Some of this has a folk or popular culture component to it, but there are elements of it that are refined and disseminated through official channels (Chatterjee 1993). Attempts at memorializing heroic events in the national past through constructing monuments and organizing celebratory festivals around them have been ways in which the collective past is represented in the everyday lives of people in the present. 'Places of memory' give a concrete reference to the abstract emotions of the national 'imagined community', to use Anderson's (1983) term (Nora 1989; Mosse 1975, 1990).

The sacrifices of war and the heroism of political activists are the most frequent subjects of monumental commemoration. These are examined as to the meanings they convey about national pasts (particularly that of Ireland) in **Nuala Johnson's** article on how landscape forms can contribute to the reproduction of national identity (**Reading 18**). She argues that statues not only are constant reminders of a collective past but also help to 'spatialize' public memory by linking the history of the nation to specific sites in it. This serves to remind everyone, however remote from the battlegrounds of nationhood, of communality with their fellow citizens. The nation itself can also be represented in statue form, frequently as a female: a heroic maiden or doting mother signifying the 'land' for which so much has been sacrificed (also see Agulhon 1981; Hubbs 1988; Nash 1993). Johnson points out, however, that the meaning of statues is not straightforward, even when evidently celebratory or heavily gendered. They are subject to contending interpretations as to what they say about the past. Hence, there are always possible reinterpretations that can lead to new views of the past. Though the statues might be, their *meaning* is not written in stone.

Contemporary students of political geography are often cosmopolitans, welcoming a world in which national identities will no longer exercise the monopoly that they have over so much of the world's population for so long. The irony, of course, as Michael Ignatieff (1993, 13) has so eloquently put it, is that

> Globalism in a post-imperial age permits a post-nationalist consciousness only for those cosmopolitans who are lucky enough to live in the wealthy West. It has brought chaos and violence for the many small peoples too weak to establish defensible states of their own. The Bosnian Muslims are perhaps the most dramatic example of a people who turned in vain to more powerful neighbors to protect them. The people of Sarajevo were true cosmopolitans, fierce believers in ethnic heterogeneity. But they lacked either a reliable imperial protector or a state of their own to guarantee peace among contending ethnicities.

In a world of still-contending nationalisms and ethnicities even cosmopolitans must rely on established national-state boundaries to provide a modicum of security and rights to live cosmopolitan lives. The response of 'If only it could be so for more!' demonstrates the continuing attraction of nation-states as guarantors of security even in a globalizing world. Until we can make sovereignty cosmopolitan (and, thus, global) this is likely to remain the case.

## References

Agnew, J. A. 1989a: Nationalism: autonomous force or practical politics? Place and nationalism in Scotland. In C. H. Williams and E. Kofman (eds) *Community conflict, partition and nationalism*. London: Routledge.

Agnew, J. A. 1989b: Beyond reason: spatial and temporal sources of ethnic conflicts. In L. Kriesberg, T. A. Northrup and S. J. Thorson (eds) *Intractable conflicts and their transformation*. Syracuse NY: Syracuse University Press.

Agulhon, M. 1981: *Marianne into battle: republican imagery and symbolism in France, 1798–1880*. Cambridge: Cambridge University Press.

Allen, B. 1996: *Rape warfare: the hidden genocide in Bosnia-Herzegovina and Croatia*, Minneapolis MN: University of Minnesota Press.

Alter, P. 1989: *Nationalism*. London: Arnold.

Anderson, B. 1983: *Imagined communities: reflections on the origins and spread of nationalism*. London: Verso.

Anderson, J. 1986: Nationalism and geography. In J. Anderson (ed.) *The rise of the modern state*. Brighton: Harvester Press.

Anderson, J. 1989: Ideological variations in Ulster during Ireland's first Home Rule crisis: an analysis of local newspapers. In C. H. Williams and E. Kofman (eds) *Community conflict, partition and nationalism*. London: Routledge.

Armstrong, J. A. 1982: *Nations before nationalism*. Chapel Hill NC: University of North Carolina Press.

Bensel, R. F. 1990: *Yankee leviathan: the origins of central state authority in America, 1859–1877*. Cambridge: Cambridge University Press.

Bhabha, H. 1990: *Nation and narration*. London: Routledge.

Billig, M. 1995: *Banal nationalism*. London: Sage.

Birch, A. H. 1975: *Political integration and disintegration in the British Isles*. London: Allen and Unwin.

Boyarin, J. (ed.) 1994: *Remapping memory: the politics of time space*. Minneapolis MN: University of Minnesota Press.

Breuilly, J. 1982: *Nationalism and the state*. Manchester: Manchester University Press.

Brustein, W. 1981: A regional mode-of-production analysis of political behavior: the cases of Mediterranean and western France. *Politics and Society*, 10: 355–398.

Calhoun, C. 1994: Nationalism and civil society: democracy, diversity and self-determination. In C. Calhoun (ed.) *Social theory and the politics of identity*. Oxford: Blackwell.

Chatterjee, P. 1993: *The nation and its fragments*. Princeton NJ: Princeton University Press.

Cohen, A. 1974: *Two-dimensional man: an essay on the anthropology of power and symbolism in complex society*. Berkeley CA: University of California Press.

Cooke, P. 1989: Ethnicity, economy and civil society: three theories of political regionalism. In C. H. Williams and E. Kotman (eds) *Community conflict, partition and nationalism*. London: Routledge.

Daniels, S. 1993: *Fields of vision: landscape imagery and national identity in England and the United States*. Cambridge: Polity Press.

Denitch, B. 1994: *Ethnic nationalism: the tragic death of Yugoslavia*. Minneapolis MN: University of Minnesota Press.

Deudney, D. 1995: Ground identity: nature, place and space in nationalism. In Y. Lapid and F. Kratochwil (eds) *The return of culture and identity in IR ntheory*. Boulder CO: Lynne Rienner.

de Waal, A. 1994: The genocidal state: Hutu extremism and the origins of the 'final solution' in Rwanda. *Times Literary Supplement*, 1 July: 3–4.

Dunn, J. 1979: *Western political theory in the face of the future*. Cambridge: Cambridge University Press.

Fitzpatrick, D. 1978: The geography of Irish nationalism, 1910–1921. *Past and Present*, 78: 113–44.

Gillis, J. R. (ed.) 1994: *Commemorations: the politics of national identity*. Princeton NJ: Princeton University Press.

Glenny, M. 1992: *The fall of Yugoslavia: the third Balkan war*. London: Penguin.

Greenfeld, L. 1992: *Nationalism: five roads to modernity*. Cambridge MA: Harvard University Press.

Harris, N. 1990: *National liberation*. Reno NV: University of Nevada Press.

Hechter, M. 1974: *Internal colonialism: the Celtic fringe in British national development*. Berkeley CA: University of California Press.
Hobsbawm, E. J. 1990: *Nations and nationalism since 1780: programme, myth and reality*. Cambridge: Cambridge University Press.
Hobsbawm, E. J. and Ranger, T. O. (eds) 1983: *The invention of tradition*. Cambridge: Cambridge University Press.
Hubbs, J. 1988: *Mother Russia: the feminine myth in Russian culture*. Bloomington IN: Indiana University Press.
Huntington, S. P. 1993: The clash of civilizations? *Foreign Affairs*, 72: 22–49.
Ignatieff, M. 1993: *Blood and belonging: journeys into the new nationalism*. New York: Farrar, Straus and Giroux.
Keen, D. 1995: When war itself is privatized: the twisted logic that makes violence worthwhile in Sierra Leone. *Times Literary Supplement*, 29 December: 13–14.
Kofman, E. 1995: Citizenship for some but not for others: spaces of citizenship in contemporary Europe. *Political Geography*, 14: 121–37.
Lemarchand, R. 1994: *Burundi: ethnic conflict and genocide*. Cambridge: Cambridge University Press.
Lowenthal, D. 1991: British national identity and the English landscape. *Rural History*, 2: 205–30.
MacLaughlin, J. G. and Agnew, J. A. 1986: Hegemony and the regional question: the political geography of regional industrial policy in Northern Ireland, 1945–1972. *Annals of the Association of American Geographers*, 76: 247–61.
Mosse, G. L. 1975: *The nationalization of the masses*. New York: Howard Fertig.
Mosse, G. L. 1990: *Fallen soldiers: reshaping the memory of the world wars*. New York: Oxford University Press.
Nairn, T. 1977: *The break-up of Britain*. London: New Left Books.
Nash, C. 1993: Renaming and remapping. *Feminist Review*, 44: 39–57.
Nora, P. 1989: Between memory and history: les lieux de memoire. *Representations*, 26: 7–25.
Prunier, G. 1995: *The Rwanda crisis: history of a genocide*. New York: Columbia University Press.
Rokkan, S. and Urwin, D. (eds) 1982: *The politics of territorial identity: studies in European regionalism*. London: Sage.
Smith, A. D. 1991: *National identity*. London: Penguin.
Spinner, J. 1994: *The boundaries of citizenship: race, ethnicity, and nationality in the liberal state*. Baltimore MD: The Johns Hopkins University Press.
Woolf, S. (ed.) 1996: *Nationalism in Europe, 1815 to the present: a reader*. London: Routledge.

# 16 Daniele Conversi, 'Reassessing Current Theories of Nationalism: Nationalism as Boundary Maintenance and Creation'

Reprinted in full from: *Nationalism and Ethnic Politics* 1(1), 73–85 (1995)

Theories of nationalism have been traditionally divided into two main categories, *instrumentalism* and *primordialism*. The former conceive nationalism as a product of elite manipulation and contend that nations can be fabricated, if not invented. The latter see nationalism as a spontaneous process stemming from a naturally given sense of nationhood. However, as such opposition is reductive and no longer helpful, other models must be added. This article proposes to add three other approaches: ethno-symbolic, homeostatic and transactional.[1] Although all of them will be considered necessary for a fuller understanding of nationalism, the present article focuses on the transactional approach's ability to identify the essential function of nationalism, which is to establish and/or defend boundaries between communities.

## Myths and symbols

Anthony D. Smith's *ethno-symbolic* approach aims at overcoming the distinction between primordialism and instrumentalism by rejecting the axiom that nations may be *ipso facto* invented and that nationalism may be purely a product of elite manipulation.[2] Although nations are a modern phenomenon, they rely on a pre-existing texture of myths, memories, values, and, finally, symbols. The mutual relationship between these remains to be synthesized: myths and memories (including nationalist histories) carry with them the values which shape a common identity. In short, what founds ethnic and then national identity is the power of collective memory. Within this framework, it is possible to reintroduce a non-reductive instrumentalist component by focusing on the role of the intellectuals and, then, of the intelligentsia. Most historical approaches to nationalism focus on these two segments, particularly the intellectuals.[3]

According to one well-known interpretation, the *intellectuals* are the creators, inventors, producers and analysts of ideas which the intelligentsia may then spread.[4] Indeed, the *intelligentsia*, or the professionals, constitute that group which has the power to apply and disseminate the ideas produced by the intellectuals. There are, obviously, overlapping cases, such as individuals who, in their lifetime, have the possibility both to create and disseminate their ideas. However, these are two clearly distinct activities, or 'phases'.[5] Generally, the tendency to be organized in professional categories indicates membership in the intelligentsia.[6] One of its attributes is modernity; intellectuals existed in many epochs, but only with modernization do we encounter large numbers of individuals dedicated exclusively to applying and disseminating ideas, the intelligentsia. Among the intellectuals a decisive role is played by *historians*, whose mission is to provide a legitimizing historical perspective as the basis of the national project. In the late eighteenth and nineteenth centuries, most European historiographers and social philosophers were national 'propagandists'. Extolling the heroic deeds of past national leaders, they provided the political aims of their own leaders with an historicist justification. Nationalist and minority historians are often set apart from mainstream historians. The term *uneven ethno-history* has been suggested by Smith for this purpose.[7]

An *ethno-history* is an imaginary reconstruction of the past. It paints an ideal tableau of what once was. The ethno-historian is thus not concerned with investigating his own claims, but with creating a fiction which is more apt to convey the message of nationalist renaissance. The ethno-historian is nothing more than the modern version of the ancient myth-maker. Yet, there is no value-free history. To paraphrase Marx's dictum, all history is the history of the dominant nations. And borrowing from psychology, we can say that memories are also selective. Turkish history books do not mention the Armenian genocide in which millions perished, yet Turkish mainstream history is not generally classified or referred to as ethno-history. What, in principle, distinguishes professional historians from both 'official' and ethno-historians is that the latter two lack a filter of analytical critique, although their task is precisely to select those materials that are useful for their nation-building purpose.

### State and homeostasis

However, if we focus exclusively on the power of the past and its symbols, we miss two other key features of nationalism: first, its relationship with political power, and particularly with the state; second, its crucial border-generating function.

Nationalism's intrinsic nature is related to the state and no nationalism is conceivable aside from the rise of the modern nation-state. As the state is the most universal mode of political power, nationalism becomes also the most universal ideology in the contemporary world. Nationalism must be seen as both an *attempt to seize control of the state* (through the legitimacy which the ideology of nationalism can bestow), as well as a *reaction against state interference and expansion* (through the negation of such legitimacy). The two phenomena are not contradictory: it is precisely because the existing state structure is rejected that a new state structure is sought after. In the modern world, there is no escape from the nation-state, and from state power in general. Homeostatic approaches focus precisely on the process of reaction against state intrusion. The advent of both modernization and centralized state power has had destructive effects upon traditional lifestyles. People massively uprooted and forced into new endeavours were and are readily available for new forms of mobilization. The term 'new' here includes attempts to reestablish the pre-existing order. As Smith points out, such modes of mobilization cannot be purely arbitrary: a striving for a lost sense of community is what animates most successful nationalist movements.[8]

In more general terms, social movements gather momentum when and where there arises the need to restore a lost equilibrium. In this sense, ethnic insurgency is seen as a movement in defence of a group threatened with extinction by an alien state or by abrupt social change.[9] For instance, Patricia Mayo sees the 'loss of community' as the crucible in the explosion of ethnic conflicts.[10] Their origin lies in the erosion of traditional communitarian structures by the bureaucratic state, with the consequent intrusion of industrial anonymity in every area of modern life. Echoes of Tönnies' distinction between *Gemeinschaft* and *Gesellschaft* reverberate in this approach. Ernest Gellner's theory of social cohesion, which sees industrial society as radically in opposition with agricultural society, is not distant either.[11] We shall consider Gellner's viewpoint in more detail later. But in homeostatic approaches the external variable which causes the nationalist movement to explode is the state.[12] In brief, homeostatic interpretations refer to a primary and spontaneous reaction against state-sponsored bureaucratization and assimilation. However, they do not concern specifically nationalist movements, since a reaction of this kind can be produced through traditionalist, fundamentalist, New-Age, federalist, ecologist and other movements as well. Anti-state reactive movements can, for example, be pursued for religious reasons, which only afterwards may be taken up by the intelligentsia as part of its nationalist platform.[13] Nationalist movements can be distinguished from other reactive movements by their assumption of typical elements of a modern western culture (secularism, appeal to citizenship and equality, a certain degree of centralization and so on).[14]

However, in many historical cases, a homeostatic reaction has been an essential prerequisite for the successive formation of a nationalist movement. This interpretation confirms that the state is the main variable in the formation of nationalism, and that nationalism itself is a product, or a consequence, of the state. This is a theory clearly expressed by John Breuilly. However, Breuilly holds that it is *failure* in state-building that 'gives rise to distinctive nationalist politics'.[15] This can be a rather tautological argument and its results risk reifying the state into the ultimate arbiter of all social processes. The state is a crucial element in my explanation as well, but I consider ethnic nationalism to have been actually reinforced by state intervention, rather than being the result of 'the failure to concentrate sovereignty in particular institutions'.[16] The catalyst of many nationalist upheavals was the state's failure to decentralize its institutions, not to concentrate them. An excess of over-zealous centralism often engendered a homeostatic reaction which, in turn, gave rise to powerful peripheral nationalisms.

A variety of this approach is the *cause–effect model*, which explains ethnonationalist insurgencies as a *direct* response to state repression. This model is often used to explain nationalist violence as a defensive mobilization against the state. The nationalists themselves sometimes have used this explanation in order to justify their recourse to anti-state violence. In political science, the worldwide role of 'politicide', genocide and other forms of state repression in the genesis of ethnic conflicts has been well analysed by Ted Gurr and others.[17]

Some authors go as far as seeing the level of ethnic mobilization as directly proportional to the degree of repression: 'The greater the opposition – economic, political, social, religious, or some combination thereof – perceived by an ethnic group, the greater the degree to which its historical sense of distinctiveness will be aroused, and hence the greater its solidarity or the more intense its movement towards redress.'[18] For instance, state repression was crucial in the crystallization of both Basque and Catalan nationalism, and, if this repression was directed against particular core values (such as the Catalan language), it only served to reinforce the latter's political importance.[19]

Homeostatic approaches can be both primordialist and instrumentalist. That is, anti-state reactions can be seen as either stemming from given pre-existing social identities or as the result of the elite's will to institutionalize a supposedly invented community. Although it is the state which creates the nation, a particular nation may be seen as too artificial, so that a more 'natural' nation can be opposed to it.

**Borders and transactions**

However, if we are to consider the relationship between nations and ethnic groups, or whatever pre-dated them, we must focus on some more essential process, which is the 'spontaneous' construction of ethnic identities as a result of group interaction, a phenomenon which obviously occurs in both the modern and pre-modern world. All forms of interaction need norms and regulations, and borders represent the core of such regulations. Borders indicate a limit which must not be trespassed. Borders are essential to all human processes, both at the individual and the social level. Indeed, all processes of identity construction are simultaneously border-generating and border-deriving.

The Norwegian anthropologist Frederick Barth was the first scientist to examine the crucial role of boundaries in identity construction.[20] Although he mostly studied 'pre-national' societies, his findings are relevant to the study of modern nationalism and nationality formation. *Transactionalism* is the name which I give to the approach of Barth and his disciples: the focus is on human transactions, that is on exchanges and relationship between human groups.

Nationalism is both a process of border maintenance and creation. Hence, it is a process of definition. One of the problems stemming from the lack of a universally acceptable definition of the nation and of nationalism derives precisely from the fact that the nation is itself a tool of definition. As such, it cannot be defined, at least abstractly and extracontextually. If a particular nation is to be defined, it must be bound and delimited, that is, tied to a previously established space.

In the pre-modern world, this 'space' was not necessarily territorial, since many groups intermingled within the same geographical area. Yet, they were allocated separate niches in the same social system and played different functions. Each of them had a separate role. As a consequence, competition was scarce and conflict assumed a more specific, nearly direct cyclic, role. In Ernest Gellner's well-known aphorism, with modernity, 'structure was replaced by culture' and vertical stratification by spatial segmentation.[21] Hence, boundaries shifted from a more internal and all-encompassing level to a more external and territorial one. In contrast with the pre-modern world, the bases of modern nations are eminently geographical. Hence, nationalism is a struggle over the definition of spatial boundaries, that is, over the control of a particular land or soil.[22] Boundaries become even sharper and it becomes theoretically difficult to shift from one national loyalty to another, although it is much easier to trespass class and other cleavages within the national space. Hence, nationalism simultaneously strives at the reinforcement of external border and the elimination of internal borders. The consequence of this is the notorious homogenizing pressure of nationalism, a steam-rolling action which devours and destroys all anti-entropic actors and forces. The reaction to this 'massifying' process is the rise of separatist nationalism and the universalization of ethnic consciousness. Again, homeostatic interpretations can be illuminating:

> Though ethnic consciousness is still to be discovered by much of the world's population, it is expanding very rapidly as outside forces increasingly intrude upon the villagers' former isolation . . . The rapid spread of literacy, the greater mobility of man made possible by dramatic developments in the form and expanse of transportation, and the even more revolutionary strides in communications have rapidly dissipated the possibility of cultural isolation, and, correspondingly, have rapidly propagated national consciousness . . . These developments not only cause the individual to become more aware of alien ethnic groups, but also of those who share his ethnicity.[23]

Indeed, when the process of state centralization is so emphasized that it leads to the erosion of all internal difference, then separatism becomes the only possible option for a human group that wishes to maintain its identity. Indeed, separatism is one of two poles along a continuum that has sadly characterized contemporary nation-building. The other pole is genocide. Genocide occurs when a dominant or expansionist nationalism aspires to the total elimination of an unassimilable group with whatever coercive means at its disposal. Between separatism and genocide other options lie, but these two extreme possibilities are always latent. The tragedy of Bosnia is a reminder of the possibility of the latter 'solution' in the very heart of our 'civilized' world.[24] No multiethnic country is entirely safe from the Bosnian model.

Nationalism can be a process of border creation as well as of border maintenance. The two are difficult to distinguish. When identities slide into each other, borders 'must' be established, although this effort is often presented by nationalist elites as an attempt to maintain a pre-existing or primordial national boundary. Similarly, stateless nations attempt to 'defend' their territory through the reenactment of historically defined ethnic borders, basing themselves either on historical memories of statehood, or on the diffusion of some ethnic marker(s), usually language.

Thus, Catalan nationhood is conceived on the basis of language. This criterion excludes Aragon, to which Catalonia was confederated at the apogee of its imperial splendour, that is, at the peak of its statehood. Similarly, Basqueness is conceived on the basis of several ethnic features, not necessarily language. At the same time, memories of past statehood are grossly exaggerated, since the Basque Country never existed as an independent state.[25]

The crucial mechanism of border conservation is opposition. By definition, borders are oppositional and rely on otherness. In Barth's view, it is the 'other' that defines the 'self', rather than the alleged objective traits through which it is occasionally identified (such as culture, custom, religion and so on). Ethnicity is thus a subjective dimension and the importance of borders overrides the importance of culture. The limit of this approach is that it tends to ignore the internal mechanisms that are needed to maintain boundaries. My approach is, instead, that there are given internal and objective factors that are available and used to demarcate borders without necessarily relying on antagonism and confrontation. These I call *ethnic markers*: some of them can be selected by nationalist elites as the nation's *core values*.[26] Language is normally the most universal of them.[27] Nevertheless, opposition remains the key underlying principle of collective identity formation. This is so because nationalism is first and foremost a process of social categorization, more than a process of cultural creativity.[28] We shall come back later to this concept of social categorization.

**Antagonism and opposition**

According to Barth, ethnic identity cannot be conceived merely as the survival of cultural forms derived from geographic and social isolation.[29] On the contrary, it is the outcome of intense interaction between groups.[30] There is an implicit contradiction between its founding aspects: ethnic groups perceive themselves as independent and autonomous, yet it is precisely their interdependence which is the source of their differential identities and self-perception. 'Groups tend to define themselves, not by reference to their own characteristics, but by exclusion, that is, by comparison to "strangers" '. *Opposition* is then the crucible of ethnic and national identity. According to Lanternari, 'there is always a contradiction in the definition of a group, because it is a group (that is, it has a group identity) only in relation to other groups'.[32]

The central role of opposition in identity formation has been theorized in more detail by Spicer. He defines the *oppositional process* as 'the essential factor in the formation and development of the persistent identity system'.[33] Likewise, Boon sees in this contrast the central feature of cultural groups.[34] They need to produce a sense of themselves only because they encounter and interact with others. Since juxtaposition is the key to ethnic identity and groupness, cultural nuances and details are often grandly exaggerated. It is part of the process which Sigmund Freud called the 'narcissism of small differences'.[35] The less different are the juxtaposed groups, the greater will be their insistence on separateness. Mechanisms of compensation are thus central to groups as well as to individuals. Similarity will be counterbalanced by stress on alleged differentiae. 'Any discourse at some levels alludes to the absences it intrinsically sets in abeyance . . . Every discourse, like every culture, inclines toward what is not: toward an implicit negativity.'[36]

What is important is that diversity in itself cannot generate identity: in order to achieve this, to transform diversity into difference, there must be opposition, an external other is needed. Ethno-genesis is not an endogenous process. In another study, I have analysed what occurs when a cultural void, that is, a lack of shared ethnic markers or common values, makes necessary an emphasis on direct opposition and confrontation.[37] The lack of a common distinctive culture, which can promptly identify or mark a group, is likely to generate violence. The latter works as a form of social cohesion, once a nationalist movement is superimposed on the social structure.

On the other hand, opposition must be distinguished from confrontation and antagonism. Opposition is part of the simpler process of categorization of the world into distinctive units (as engendered by nationalism). Confrontation is a particularly open and direct form of opposition, and one of its most extreme forms is political violence. It is often conterminous with antagonism. An *antagonistic identity* is one constructed essentially through the negative opposition of the in-group to one or more out-groups. All identities are in some way based on opposition, but an antagonistic identity focuses more on the need to define one's own group by exclusion. This border-definition process is carried out by a radical re-evaluation of the positive traits of the in-group and a parallel devaluation of those of the out-group. Borders are stressed rather than content, that is, the group's culture. Culture can also be oppositional, as when it is used in opposition to another culture, generally the dominant one. But when the ethnic culture is weak and ill-defined, the whole group is to be opposed to the out-group, generally the dominant one.

As studied by social psychologists, the process of social categorization is *independent* of real and discrete differences among groups.[38] Nationalism assumes that the world is naturally divided into discrete entities called nations; hence, nationalism is a process of social categorization, both of the 'self' and the 'other'. From here it follows that, although cultural differences may be important in legitimizing nationalism, they are by no means necessary in creating opposition, antagonism and conflict. What is relevant is the capacity of nationalist leaders to impose upon their constituencies the idea that they belong to a united single body called the 'nation'. For many people, to be able to 'imagine' a nation means that a process of social categorization has finally succeeded.[39] The 'art' of imagining and creating a nation is intimately bounded to the 'science' of classifications and categorizations. Imagination and invention concur in the process of founding categories named 'nations'. However, it is highly unlikely that nationalist leaders can manipulate their constituencies at their own discretion, as extreme instrumentalists insist. In order to mobilize the 'masses', the intelligentsia, and before them the intellectuals, must touch some chord, their message must reverberate amongst the people, it must even look familiar to them. As the world is no laboratory, some 'real' element must be present in order for social categorization to become effective. Some pre-existing ethnic markers must facilitate the ascription into a category. However, if the group to be mobilized is too fragmented and assimilated to retain some shared hyphen, social categorization can always be enforced by stimulating borders and opposition, rather than contents and uniqueness.[40] Here is where the role of violence comes in.[41]

## Conclusion

All the three theories presented are crucial for the understanding of nationalism. They interpenetrate, giving a fuller picture of the motives for, and the emergence of, nationalism. This leaves the old instrumentalist/primordialist debate in the background, as all the three approaches can be seen from each of these two standpoints. In general, instrumentalism offers more convincing interpretive keys, especially in so far as the study of elites is concerned. However, transactionalism cannot be reduced to pure instrumentalism. In contrast to Barth's emphasis, this article has described ethnic borders as deeply related to ethnic content.

If such content, that is, the ethnic culture, is weak, boundaries can always be reinforced by enhancing opposition. In cases of global homogenization, total ethnic warfare can be a likely option for proto-nationalist elites wishing to create compact cohesive communities around their goals.

As a form of oppositional social organization, ethnicity may articulate a multiplicity of forms of ecological interdependence between groups, from the exploitation of complementary niches to the maximal competition for specific resources.[42] Ethnicity is defined by *boundaries*. This implies that the cultural and biological content of the group can alter while the boundary mechanism remains unchanged. One of Barth's main contributions is thus the distinction between *ethnic borders* and *ethnic contents*. He observed that the boundaries defining a group's identity may be maintained independently from the culture they enclose. In particular, cultures may change, but the permanence of boundaries themselves is more longstanding. However, Barth's playing down of culture does miss its central importance to the construction of identities: even though identities are often constructed rather than given, they must rely on the pre-existing diffusion of shared symbols and cultural elements as well as on memories of a shared past and myths of a common destiny.[43]

My thesis focuses on the distinction between ethnic boundaries and their content. In doing so, it emphasizes the oppositional dynamics of nationalism (and of ethnic identity in general), without omitting the basis upon which nationalism supports itself and legitimates itself. The latter's legitimacy derives largely from its capacity to absorb, and to delve into, the local culture. The relative weakness or strength of ethnic boundaries will be correlated to the relative weakness or strength of cultural contents.

**Notes**

1 The primordialist/instrumentalist debate focused respectively on the primordiality of nationalism as an irreducible pulsion and on the possibility of 'inventing' nations.

2 Anthony D. Smith, *Theories of Nationalism* (London: Duckworth, 1971); id., *The Ethnic Revival* (Cambridge: Cambridge University Press, 1981); id., 'Ethnic Myths and Ethnic Revivals'. *Archives européennes de Sociologie*, XXV, pp. 238–305, 1984; id., *The Ethnic Origins of Nations* (London: Basil Blackwell, 1986); id., 'The Myth of the "Modern Nation" and the Myths of Nations', *Ethnic and Racial Studies*, Vol. 11, No. 1 (1988), pp. 1–26; id., 'Social and Cultural Conditions of Ethnic Survival', *Journal of Ethnic Studies, Treaties and Documents* (Ljubljana), 21 (1988), pp. 15–26; id., *National Identity* (Harmondsworth: Penguin, 1991).

3 John Breuilly, *Nationalism and the State* (Manchester: Manchester University Press, 1993 [1st edn., 1982]); Miroslav Hroch, *Social Preconditions of National Revival in Europe: A Comparative Analysis of the Social Composition of Patriotic Groups Among Smaller European Nations* (Cambridge: Cambridge University Press, 1985); John Hutchinson, *The Dynamics of Cultural Nationalism. The Gaelic Revival and the Creation of the Irish Nation State* (London: Allen & Unwin, 1987); Elie Kedourie, *Nationalism* (London: Hutchinson, 1993 [1st edn., 1966]); Smith, *The Ethnic Revival.*

4 See Smith, *The Ethnic Revival*, p. 109. Peter Alter speaks of this category as 'the awakener': 'every nation has done its utmost to praise the deeds and merits of the philologists, poets, historians and politicians who substantiated, and in most cases successfully asserted, the nation's claim to independence and self-determination . . . These makers and recreators of states were "great men", individuals powerful enough to have wrought the shape of history'. Peter Alter, *Nationalism* (London: Edward Arnold, 1989), p. 80.

5 See Aleksander Gella (ed.), *The Intelligentsia and the Intellectuals* (Beverly Hills, CA: Sage, 1977); and Smith, *The Ethnic Revival*, p. 109.

6 See Anthony D. Smith, 'Nationalism', *Current Sociology* (Special Issue), Vol. XXI, No. 3 (1973), pp. 1–185, p. 79.

7 Anthony D. Smith, 'A Europe of Nations – or the Nation of Europe?', *Journal of Peace Research*, Vol. 30, No. 2 (1993), pp. 129–35.

8 Smith, 'Social and Cultural Conditions . . .'.

9 In the past, peripheral cultures could feel relatively protected or safe. Change was perceived by their members as mostly acceptable and tied to the constant flux of earthly matters. An Heraclitean conception of *panta rei* may well have been widespread in many cultures (Heraclitus formalized and transcribed a kind of thinking probably present among the wise men of his time, and we have no reason to doubt that a similar vision was current in most places and times). In a perpetual process of adaptation, change was constantly supported by the protective shield of tradition as a steady point of reference, rather than being radical and abrupt.

10 Patricia Mayo, *The Roots of Identity: Three National Movements in Contemporary European Politics. Wales, Euzkadi and Brittany* (London: Allen Lane, 1974).

11 Ernest Gellner, *Nations and Nationalism* (Oxford: Basil Blackwell, 1983).

12 For a more recent and broader theory which sees ethnonationalist movements as a response to both state intervention and the intrusion of technological society, see Gurutz Jaúregui Bereciartu, *The Decline of the Nation State* (Reno, NV: University of Nevada Press, 1994).

13 Alliances between a secularized westernized intelligentsia and religious elements

are a constant in the history of nationalism. In the case of Islam they are certainly decisive: the Algerian example is well known, while the first phase of the Islamic revolution in Iran provides the most indubitable example.

14 Smith appears to be himself a modernist when he distinguishes *nations* from ethnies for the presence of 'western features and qualities: territoriality, citizenship rights, legal codes, . . . political culture . . . [and] social mobility in a unified division of labour' (Smith, *The Ethnic Origins of Nations*, p. 144). Even though the latter are mostly ideal rather than real practices, and even though their origin dates prior to the modern age, they have imbued with their conceptions the contemporary societies which have been forged on the basis of nationalism.

15 John Breuilly, *Nationalism and the State* (Manchester: Manchester University Press, 1993 [1st edn., 1982]), p. 367.

16 Ibid.

17 Ted Robert Gurr, *Minorities at Risk. A Global View of Ethnopolitical Conflicts* (Washington, DC: United States Institute of Peace Press, 1993); Ted Robert Gurr and James R. Scarritt, 'Minorities at Risk: A Global Survey' *Human Rights Quarterly*, Vol. 11, No. 4 (1989), pp. 375–405; Barbara Harff and Ted Robert Gurr, 'Victims of the State: Genocide, Politicide and Group Repression Since 1945', *International Review of Victimology*, Vol. 1, No. 1 (1989), pp. 23–41.

18 George M. Scott, Jr., 'A Resynthesis of the Primordial and Circumstantial Approaches to Ethnic Group Solidarity: Towards an Explanatory Model', *Ethnic and Racial Studies*, Vol. 13, No. 2 (1990), pp. 148–71, p. 164.

19 Daniele Conversi, *The Pen or the Sword. Alternative Paths to Nationalist Mobilization in Spain* (London: Hurst, 1995).

20 Frederick Barth (ed.), *Ethnic Groups and Boundaries. The Social Organization of Cultural Difference* (London: Allen & Unwin, 1969).

21 Ernest Gellner, *Nations and Nationalism* (Oxford: Basil Blackwell, 1983).

22 Gellner's famous pictorial metaphor is still unparalleled in its descriptive capacity: the premodern world resembled an impressionist painting of Kokoschka, where all colours intermingled and faded into each other; the modern world resembles rather a picture of Modigliani, where each colour is sharply separated from the others and no mixture is allowed. Gellner, *Nations and Nationalism*, pp. 139–40.

23 Walker Connor, 'The Politics of Ethnonationalism', *Journal of International Affairs*, Vol. 27, No. 1 (1973), pp. 1–21.

24 Although the international community has recognized Bosnia, it has never upheld its independence and territorial integrity. Indeed, the European Union and the United Nations have first encouraged the Yugoslav government to insist on Yugoslavia's territorial integrity; then, once this integrity could no longer be defended, the Serbs were allowed to carry out a genocide, whilst the Moslems were ostensibly refused the means to defend themselves under the pretext of an arms embargo which, in effect, nullified their sovereignty. For a sweeping critique of mainstream approaches to the conflict in former Yugoslavia, see Stjepan G. Mestrovic, *The Balkanization of the West. The Confluence of Postmodernism and Postcommunism* (London/New York: Routledge, 1994).

25 Of the seven Basque provinces, only Navarre was an independent kingdom until 1512. The other three Basque provinces within Spain enjoyed a considerable degree of autonomy under the regime of the fueros (local charters or special privileges). See Conversi, *The Pen or the Sword . . .*

26 Daniele Conversi, 'Language or Race?: The Choice of Core Values in the Development of Catalan and Basque Nationalisms', *Ethnic and Racial Studies*, Vol. 13, No. 1 (Jan. 1990), pp. 50–70.

27 Daniele Conversi, 'The Influence of Culture on Political Choices: Language Maintenance and its Implications for the Basque and Catalan Nationalist Movements', *History of European Ideas*, Vol. 16, No. 1–3 (1993), pp. 189–200.
28 For the concept of *social categorization*, see Michael Billig, *Social Psychology and Intergroup Relations* (London: Academic Press, 1976).
29 In Barth's words, 'though the naïve assumption that each tribe and people has maintained its culture through a bellicose ignorance of its neighbours is no longer entertained, *the simplistic view that geographical and social isolation have been the critical factors in sustaining culture diversity persists*' (Barth, *Ethnic Groups . . .*, p. 9).
30 The perception that ethnic identities are reinforced through interaction and global communication is not what distinguishes Barth's theory from competing ones. Walker Connor, Michael Hechter, Benedict Anderson, John Armstrong, and several modernists concur that ethnic boundaries are strengthened in response to intense interaction.
31 John A. Armstrong, *Nations before Nationalism* (Chapel Hill, NC: University of North Carolina Press, 1982), p. 5.
32 Vittorio Lanternari, *Identitá e differenza: percorsi storico-antropologici* (Napoli: Liguori, 1986), p. 67.
33 Edward H. Spicer, 'Persistent Cultural Systems. A Comparative Study of Identity Systems That Can Adapt to Contrasting Environments', *Science*, Vol. 174 (19 Nov. 1971), pp. 795–800, p. 797.
34 James A. Boon, *Other Tribes, Other Scribes. Symbolic Anthropology in the Comparative Study of Cultures, Histories, Religions and Text* (Cambridge: Cambridge University Press, 1982).
35 Sigmund Freud, 'Civilization and Its Discontents', in *Civilization, Society and Religion*, Vol. 12 (Harmondsworth: Penguin, 1991 [1st original edn., 1929]), p. 305.
36 Boon, *Other Tribes . . .*, p. 232.
37 Conversi, *The Pen or the Sword . . .*
38 See Billig, *Social Psychology*; Michael Billig and Henri Tajfel, 'Social Categorization and Similarity in Inter-Group Behavior', *European Journal of Social Psychology*, Vol. 3 (1993), pp. 27–52; Henri Tajfel, 'Experiments in Intergroup Discrimination', *Scientific American*, No. 23 (1970), pp. 96–102.
39 For nations as imagined communities, see Benedict Anderson, *Imagined Communities: Reflections on the Origins and Spread of Nationalism* (London: Verso, 1983).
40 'The boundaries are *relational* rather than absolute; that is, they mark the community in *relation* to other communities' (Anthony P. Cohen, *The Symbolic Construction of Community* (London: Routledge, 1985), p. 58.
41 The relationship between violence and *ethno-genesis* has been explored in several case studies. For the Maya in Guatemala, see Richard Wilson, 'Machine Guns and Mountain Spirits: The Cultural Effects of State Repression Among the Q'eqchi' of Guatemala', *Critique of Anthropology*, Vol. II, No. I (1991), pp. 33–61. For the Croats in former Yugoslavia and the Kurds in Turkey, see Daniele Conversi, 'Violence as an Ethnic Border: The Consequence of a Lack of Distinctive Elements in Croatian, Kurdish and Basque Nationalism', in *Proceedings of the International Conference on Nationalism in Europe: Past and Present* (Santiago de Compostela: Santiago de Compostela University Press, 1994).
42 See Barth, *Ethnic Groups . . .*, and Cohen, *The Symbolic Construction . . .* On 'niche theory' and the origins of the term 'niche', see Michael T. Hannan, 'The Dynamics of Ethnic Boundaries in Modern States', in Michael T. Hannan and John

W. Meyer (eds.), *National Development and World System* (Chicago, IL: University of Chicago Press, 1979), pp. 260–64.

43 Smith, *The Ethnic Origins . . .*, and *National Identity*.

# 17 Colin H. Williams, 'The Question of National Congruence'

Excerpt from: R. J. Johnston and P. J. Taylor (eds), *A World in Crisis? Geographical Perspectives,* Chapter 9. Oxford: Blackwell (1989)

The inherent tension between state nationalism and ethnic nationalism within the world system has already contributed to two major catastrophes this century. At times their interaction has created powerful dynamic socio-economic structures. At other times the clash of interests represented by these two forces has produced open conflict reflecting a sustained rivalry between striving participants in the developing state system. There is little reason to believe that violence will be eradicated in this relationship, for 'Despite the attempts of capital to tame and rationalise social relations, to subordinate them to its much more coldly destructive logic, violence will always occur' (Shaw, 1984, p. 4).

The pattern of state formation is abundant testimony to the influence of conflict and warfare in the development of national territories. The size and shape of contemporary states are as much a product of international rivalry as they are reflections of the settlement pattern of constituent 'national' populations. Indeed, the quest for national congruence, defined as the attempt to make both national community and territorial state into coextensive entities, has been a major feature of modern history, particularly in Europe. This 'western model' of state formation has been so influential that Williams and Smith (1983, p. 510) claim that the quest for its constituents – authenticity, legitimacy and equality – 'bedevils interstate and intrastate relations all over the world'. The emergence of the 'territorial–bureaucratic state' has had profound consequences for the political organization and structure of the interstate system, especially in those territories carved out of former colonial dynastic rule. For:

> The central point . . . of the Western experience for contemporary African and Asian social and political change has been the primacy and dominance of the specialised, territorially defined and coercively monopolistic state, operating within a broader system of similar states bent on fulfilling their dual functions of internal regulation and external defence (or aggression). (Smith, 1983, p. 17)

My intention in this chapter is to examine the quest for national congruence within an interdependent world system and to illustrate the manner in which several political movements have sought to change the system of sovereign states so as to effect a more 'representative' distribution of national states.

Taylor (1982) argues that both statism and nationalism, being expressions of the search for ideological legitimacy, are related to specific epochs in modern capitalist development. I want to examine the relationship between these ideologies and the modern world system, taking Wallerstein's work on the effects of the uneven development of capitalism as representative of the central thread of a world-systems argument.

**Expansive capitalism**

Wallerstein's influential analysis of the European-centred world-economy, comprising core, semi-periphery and periphery, is central to the analysis of the modern world-system animated by capitalism. The emerging world-economy encouraged spatial interdependence and a recognizable international division of labour whose profitability was a 'function of the proper functioning of the system as a whole' (Wallerstein, 1979, p. 38). A two-way interaction was initiated between the specialist role of a state's economy within the system and a corresponding set of pressures imposed on domestic political developments by changes within the emerging world-system. Taylor (1982) and Smith (1983) draw attention to the consequences of a state system facilitating the expansion of capitalism. They include the various dynastic and structural changes witnessed in the period 1500–1648, and legitimized in the Treaties of Westphalia, 1648 which established the state system of Europe.

> Since that date, long wars ended by congresses and treaties have become the accepted European norm for state-creation and state-consolidation: witness the treaties of Vienna, Versailles and Yalta. Each new agreement limited the number and extent of new states which could participate in the system; and the later the period, the more did wars and ensuing treaties *create* the recent states. (Smith, 1983, p. 16)

The state system became a framework which facilitated the integrative capacity of capitalism to link previously disparate regions and interest groups into an evolving world-system. Wallerstein's original argument runs as follows:

> Capitalism is based on the constant absorption of economic loss by political entities, while economic gain is distributed to 'private' hands. What I am arguing . . . is that capitalism as an economic mode is based upon the fact that the economic factors operate within an area larger than that which any political entity can totally control. This gives capitalism a freedom of manoeuvre that is structurally based. It has made possible the constant economic expansion of a world-system, albeit a very skewed distribution of its rewards. (Wallerstein, 1974, p. 348)

The details of how capitalism influenced the development of status groups, national bureaucracies, bourgeois ideology and state boundaries are beyond the scope of this chapter (Rich and Wilson, 1967; Anderson, 1974; Wallerstein, 1974). But undoubtedly the evolving state structures owed much to regional and transfrontier economic performance. The early integration of Spain, Portugal, France and southern Britain (Figure 17.1) stemmed from the superior capacity of the local bourgeoisie and state apparatus to control and

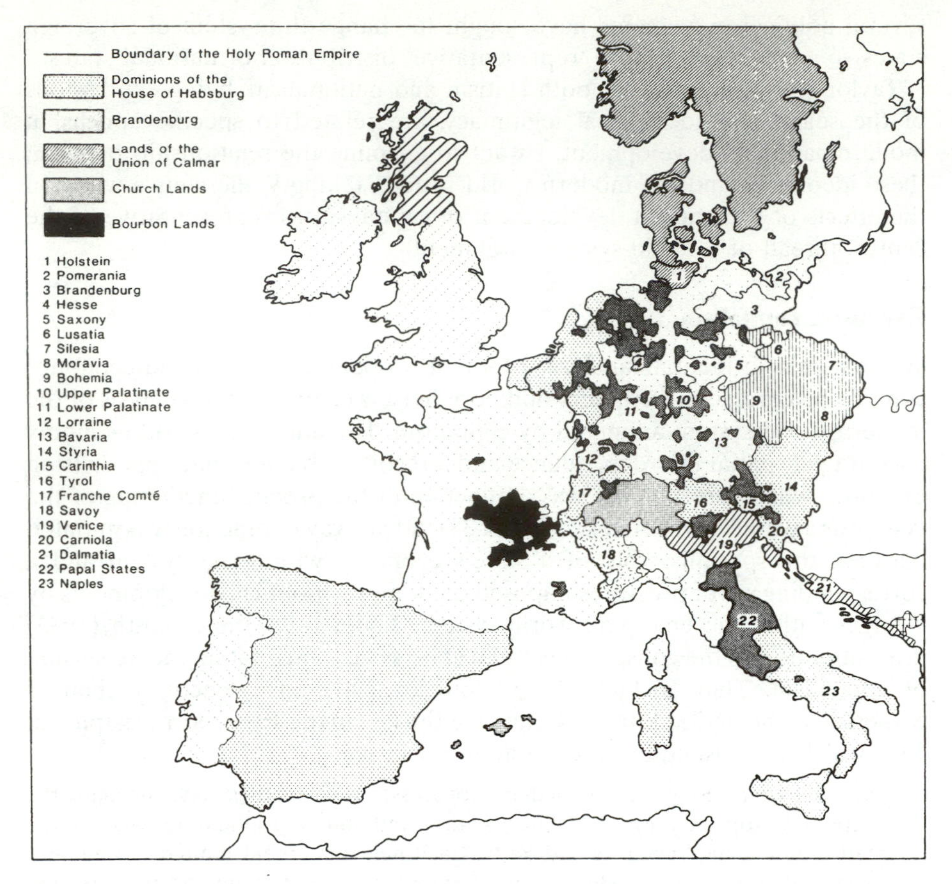


**Fig. 17.1** Europe in 1519 (Treharne and Fullard, 1976, p. 34)

influence internal economic arrangements while also pursuing vigorous foreign trade and revenue campaigns. Core states were thus more able than peripheral states to influence the patterns of commodity flow, and hence to realize a greater share of the surplus value (Wallerstein, 1979, p. 292). Wallerstein argues that this initial advantage was translated into a semi-permanent structure wherein the bourgeoisie of the core states were better placed than were their counterparts within peripheral states. This influenced their specific relationship with the core-state proletariat, a relationship which may have been quite different in kind from that between the peripheral bourgeoisie and proletariat. The argument then turns on the differential character of this relationship, mediated through the emergence of the state as the locus of conflict. He writes: 'Since states are the primary arena of political conflict in a capitalist world-economy, and since the functioning of the world-economy is such that national class composition varies widely, it is easy to perceive why the politics of states differentially located in relation to the world-economy should be so dissimilar' (p. 293).

The state is conceived of as a particular kind of social organization which seeks to perpetuate its advantage through intervention, force and economic manipulation. For Wallerstein, this advantage is institutionalized in the concept of a state's 'sovereignty':

> a notion of the modern world, is the claim to the monopolization (regulation) of the legitimate use of force within its boundaries, and it is in a relatively strong position to interfere effectively with the flow of factors of production. Obviously also it is possible for particular social groups to alter advantage by altering state boundaries; hence both movements for secession (or autonomy) and movements for annexation (or federation). (p. 292)

While individual states and factions may seek to challenge the existing pattern, it is the state system itself which 'entrusts, enforces, and exaggerates the patterns, and it has regularly required the use of state machinery to revise the pattern of the world-wide division of labour' (p. 292).

Though capitalism is encouraged in part by the regulations of a stable state system, Wallerstein argues that the unequal exchange in the appropriation of its surplus value is spatially differentiated, producing regionally variable effects. At its crudest his argument rests on the inherent unevenness of the patterns of exchange such that, in summary:

> Capitalism is a system in which the surplus value of the proletarian is appropriated by the bourgeois. When this proletarian is located in a different country from this bourgeois, one of the mechanisms that has affected the process of appropriation is the manipulation of controlling flows over state boundaries. This results in patterns of 'uneven development' which are *summarized* in the concepts of core, semi-periphery and periphery. (p. 293)

We can accept Wallerstein's characterization of the long sixteenth century as a period within which a multilayered world-system developed, without necessarily accepting his interpretation of this system as a world-empire. Neither need we accept the claim that it was, above all, the economic processes inherent in capitalism which produced this world-system. We should be careful not to promote a historical determinism which argues that, because the division of the world-system into three distinct structural positions is functional to the reproduction of capitalism, the resultant state system was either inevitable or permanent. Wendt (1987) has argued that world-system theorists are prone to interpreting the structure of the world-system as 'given and unproblematic'. In consequence they have tended to reify system structures producing rather static and at times functionalist explanations, primarily because of their earlier preoccupation with the question of structure, treating agencies such as the state or class as 'no more than passive 'bearers' of systemic imperatives' (p. 347). Wendt argues that 'without a recognition of the ontological dependence of system structures on state and class agents, Wallerstein is forced into an explanation of that transition (from feudalism to capitalism) in terms of exogenous shocks and the teleological imperatives of an immanent capitalist mode of production' (Wendt, 1987, p. 348). For our purposes the critical feature of this perspective is its recognition of the European state system as the superstructure whose transformations influenced the differential occurrence and subsequent modification of the process of national congruence. Wallerstein's contribution, in articulating these structural transformations, is well recognized by his critics (Skocpol, 1977; Modelski, 1978; Zolberg, 1981), but they would remind us that he pays too little attention to the 'politico-strategic' linkages between parts of the world-system (Zolberg, 1981, p. 262) and underemphasizes non-economic factors in the genesis of early modern states.

Skocpol (1979) is particularly sensitive to an economic reductionist argument that would 'assume that individual nation-states are instruments used by economically dominant groups to pursue world-market orientated development at home and international economic advantages abroad' (p. 22). She argues that the state system, as 'a transnational structure of military competition was not originally created by capitalism' (p. 22), but rather, quoting Hintze, that 'the affairs of the state and of capitalism are inextricably interrelated . . . they are only two sides, or aspects of one and the same historical development' (Hintze, 1975, p. 452, quoted by Skocpol, 1979, p. 299). Her analysis of both the structures of the capitalist world-economy and of individual national responses to that structure is comparative macro-analysis at its best because it allows for the relative autonomy of several layers or scales of analysis in her work. The international state system, for example, 'represents an analytically autonomous level of transnational reality – *interdependent* in its structure and dynamics with world capitalism, but not reducible to it' (p. 22).

In addition to domestic economic performance and comparative international economic position, she recognizes what Wallerstein underplays, the relevance of factors such as 'state administrative efficiency, political capacities for mass mobilization and international geographical position' (p. 22). In this context the advantage of the world-system approach is that it offers an integrated holistic perspective on an admittedly complex process of global development.

Interstate competition over the past four centuries has animated the ever-fluctuating capitalist system. It has also produced the inexorable integration of diverse culture groups into a 'national population' as part of the process of national congruence. The state has sought to harness the potential of such 'nations' and control their productivity to enhance its own resource base. In consequence, state activity has created a new set of geographies for incorporated peoples, influencing their socio-economic opportunities and political representation. Superordinate 'nations' came to dominate 'unrepresented nationalities' and used the power of the state to buttress their own cultural apparatus as the orthodox legitimized value system. The new opportunities and freedoms were those sanctioned by the state, and woe betide dissident minorities who questioned the verity of state regulations. Indeed, much of the resurgent nationalism of contemporary Europe is but a playing out of minority aspirations unsatisfied during the critical period of state formation.

### *Ethnic resurgence in Western Europe: violence and reform*

Given the overall context of state integration in Europe it is remarkable that current ethnic unrest should be so virulently manifested in the three European states with the longest history of unification and consolidation:

France, Spain and the United Kingdom all possess minorities within minorities who seek greater autonomy. Their style of resisting centralization and assimilation varies from the extra-constitutional activities of violent movements in Ulster, Euskadi and Corsica, through non-violent resistance in Catalonia, Wales and Brittany, to party political opposition in all of the above

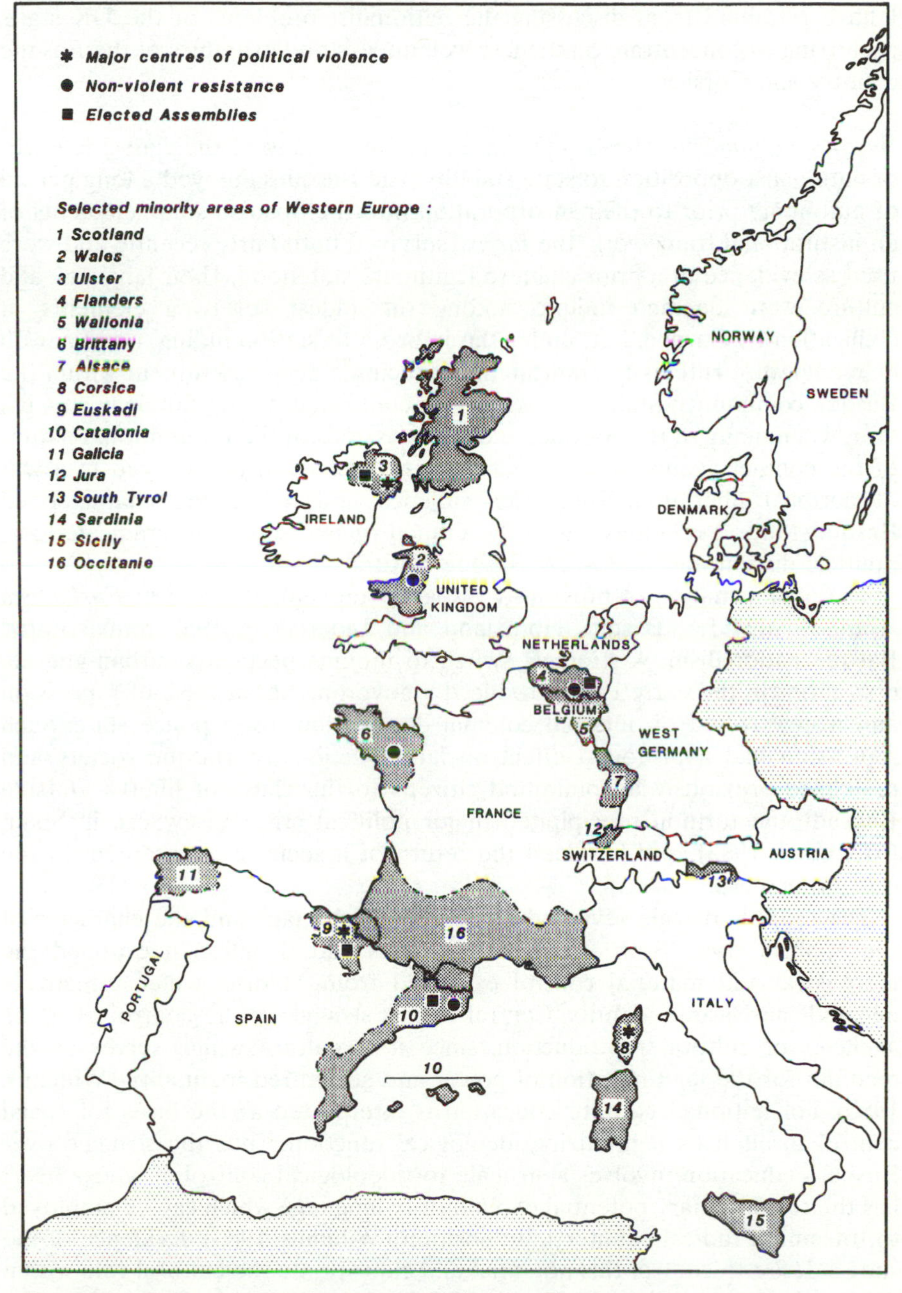


Fig. 17.2 Selected minority areas of Western Europe

plus Scotland, Galicia and Alsace (Figure 17.2). Obviously we should be wary of generalizing about such movements as if they reflected a single concern for decentralization, for a plethora of factors account for their initial emergence and subsequent developments (Anderson, 1978; Foster, 1980; Krejci and Velimsky, 1981; Breuilly, 1982). Here I concentrate on two case studies which reflect much of the ethnic discontent in contemporary Europe. I have refrained from discussing the nationalist problems of the UK state, preferring to concentrate on the less well understood examples of the Basque country and Corsica.

*Basque nationalism.* The Basque case illustrates many of the classic features of nationalist opposition to state stability. The Basques enjoyed a long period of autonomy prior to their incorporation into the Spanish state; elements of an institutional framework, the *fueros*, survived until fairly recently and were used as evidence of a prior claim to legitimate statehood. Their language and culture were deemed unique, among the oldest surviving elements of civilization in Europe, but, under the impress of state-building, non-Spanish elements were ruthlessly eradicated, producing a deep resentment within the Basque community. State oppression was confirmed during the Spanish Civil War, when many of the Basques, though conscious of the international nature of the conflict against fascism, believed themselves to be engaged in a war of national liberation. Thereafter violence and oppression characterized Basque–Madrid relationships as the Franco regime sought to eradicate local political dissent and to destroy Basque cultural identity.

The amalgamation of previously disparate movements to form *Euskadi ta Askatasuna* (ETA; Basque Homeland and Liberty) in 1957 reinvigorated Basque nationalism. A strategic switch to mount spectacular urban-guerilla operations in the early 1960s produced a government backlash of repression and a new round of 'internal colonial domination' by a police state. Such repression had a profound effect on large sections of Basque society and mobilized previously uncommitted citizens to the cause of liberty. Outside Euskadi, the turmoil precipitated major political crises elsewhere in Spain (Medhurst, 1982), and heralded the return of a socialist government in the post-Franco era.

The crucial struggle revolved around the legitimacy and the character of the Spanish state. Both Basque nationalists and socialists questioned the ideological and material control exercised from Madrid, which effectively negated the Basque identity. Central to this struggle is the control over the agencies of cultural reproduction, since it is culture which serves as the medium for the legitimization of power and structured inequality (Williams, 1980). For nationalists, state education is interpreted as the basis for social control through its legitimizing ideological function. Thus the struggle over Euskera education involves a struggle for ideological control at whose heart lies the revolutionary potential of a minority language which can be employed to transmit a radical counter-state position (Williams, 1980). Evidence of the state's desire to control this potential and expropriate the cultural role within the dominant ideology was provided by the twin threats to Basque identity:

the decline of the Basque language, especially among young people, and the post-war influx of workers from other regions. Both trends serve to reduce the ethnic homogeneity of the 'provinces' and have split the nationalist movement in its attempt to devise appropriate measures to counter these deleterious influences. A key result of the large-scale immigration was to strengthen the support base for socialist and communist factions at the expense of the Partido Nacionalista Vasco (PNV).

*Nationalism within France*. Napoleon's inheritance and the Jacobin centralist tradition continue to influence the manner in which the French state negotiates with its constituent 'dissident' nationalities. Previous attempts at determining a specific regional role for areas such as Brittany, Corsica and Alsace in relation to the needs of the French economy have been increasingly questioned since 1945. Clear economic differentials between core and periphery, a failure to devise appropriate regional development policies and a continued stigmatization of 'traditional cultures' are cited as preconditioning grievances for the emergence of reactive nationalist movements. The root of these periodic disturbances is economic exploitation and external control.

In Corsica, after decades of neglect, the French state sought both to colonize the island and to develop its natural resources. Kofman (1982) demonstrates the results of state and capitalist penetration for the period 1950–80. Developments in the three most important employment sectors – agriculture, construction and tourism – produced two forms of marginalization: first, a spatial polarization between coastal development and interior neglect; second, a social polarization consisting of a tripartite stratification. Key positions were reserved for French mainlanders and foreigners, while the Corsicans were squeezed between these spiralists and the influx of North Africans and Iberians imported as semi-skilled labour. Corsicans resented the development of their territory by 'outsiders', and a colonizer–colonized mentality was intensified with the repatriation of some 15,000–17,000 settlers from Algeria and an increase of non-Corsican-born inhabitants from 10 per cent of the population in 1954 to 45 per cent in 1975 (Kofman, 1982, p. 305).

Numerous resistance organizations were formed and re-formed, the strongest of which were the Action Régionaliste Corse dating from 1967 and the more vitriolic Front de Libération Nationale de la Corse formed in 1976. The socialist government's *statut particulier* (a devolutionist reform recognizing Corsica as a territory with a regionally elected assembly), announced by Gaston Defferre, the Socialist Minister of the Interior and 'Decentralization', in August 1981, has not assuaged the sporadic violence associated with Corsican autonomists (Kofman, 1982, pp. 309–10). But the new interdependent structure of councils, the state–regional employment committee and new agencies for transport, hydroelectric power, agriculture and regional development do provide an innovative institutional framework wherein grievances can be voiced. However, as Mény has demonstrated, if the reforms are limited to institutional changes they will be ineffective; the real transformation would accompany 'the establishment of the rule of law and a respect for universal suffrage' (*Le Monde*, 20 November 1983, quoted by Mény, 1984, p. 74). The Defferre reforms seek to increase local democracy and involvement and have gone a long way to stifling the opposition cries that a diminution of centralist control would threaten the viability of the French state.

Government pronouncements that its decentralist measures will reduce the source of core–periphery conflict in France have been judged premature and overoptimistic. Mény (p. 75) observes that the very creation of new organs of government and the establishment of new relationships may create problems, for they 'constitute a disruption of the system'. But in resolving such problems attention will have been lifted from the activities of the central government to competition at the local level for the exercise of regional power (Hirsch, 1981; Mény, 1984). After reviewing the experience of Mitterrand's regionalist accommodation programme, Mény concludes that in offering territorial minorities 'a right to roots' as well as a 'right to *options*' (Urwin, 1985), 'the State has laid the groundwork for a consensual integration of far greater efficacy than earlier, more authoritarian attempts' (Mény, 1987, p. 60). Such reforms may very well strengthen the central state apparatus in time, by making it an arbiter of local and regional conflicts and by distancing it from events in Brittany, Corsica and Occitania, so 'objectifying' and 'depersonalizing' the role of the state that its legitimacy increases.

*The break-up of the state*? The French and Spanish examples were selected to reflect the wider structural discontent in contemporary Europe. The expansion of the national territorial state ideal, in the past two centuries in particular, has strained the basic resources which constitute the building blocks of nation formation. Yet while materialist approaches often view the state as the unwitting midwife of global capitalism, unable to control the very processes it brought into being, idealist approaches root much of the threat to global stability in the aggrandizing process inherent in the competitive state system itself. Idealist alternatives focus on a return to small historical communities as a panacea for the problems created by the emergence of a world-economy and its concomitant state system.

### *National congruence in former colonial territories*

In contrast with the long experience of state formation and national congruence in Europe, Third World societies convinced of the value of the European-derived 'nation-state' have a seemingly impossible task of reconciling divergent interests in the pursuit of state stability and economic development. In many respects nationalism was more virulent in the colonial context, even though the traditional factors which were conducive to nationhood were often but a pale reflection of their European origins. Indeed, the very notion of a nation itself, linked intimately to its own state structure for full political expression, was an alien concept to all but a few of the western-educated elite. Let us illustrate the interplay of nationalism, conflict and state formation by reference to the Nigerian experience, conscious, of course, that we are selecting but one of a large number of possible cases.

History permits us to interpret colonialism as a fascinating example of nineteenth-century liberal notions superimposed on the *anciens régimes* of Africa, Asia and Latin America. High ideals, and even higher profit levels, determined the extension of European influence over the newly conquered territories of the far-flung empire, regulating the interaction between a world-system of core, semi-periphery and peripheries. The 'accidents' of imperial Balkanization produced a diverse pattern of multi-ethnic colonies whose transition to independence is one of the most squalid and tragic episodes in world history. Not that the call to liberty was in itself a tragedy, but the passage from servitude to statehood was so often marred by widespread conflict, warfare and subsequently new forms of domination that their effects largely determined the future role and direction of many newly independent states.

The basic political crisis is legitimacy. Principles of national unity are employed in the most unpromising circumstances to shape a population largely incorporated forcibly into a nation served by a strong state apparatus. Under the drive of nationalism, post-independence developments have strengthened the process of filling power vacuums, extending the reaches of the bureaucratic–military elite to the furthest periphery. Central to this European-style pattern of nation-building is identity formation. As we have seen, national integration is a difficult enough process in Europe; it is well-nigh impossible in many African contexts. The basic building blocks are absent from the national construction of a unified society. Reflect on Azikiwe's statement in 1945: 'Nigeria is not a nation. It is a mere geographic expression. There are no Nigerians in the same sense as there are English, Welsh or French. The word "Nigerian" is merely a distinctive appellation to distinguish those who live within the boundaries of Nigeria from those who do not' (Sklar, 1963, p. 233, quoted by Oberschall, 1973, p. 91).

The interdependent relationship between the professional intelligentsia and varieties of nationalism is crucial in the post-colonial state (Smith, 1983, pp. 90–4). In the search for national pride, economic development, state unity and international recognition, the intelligentsia are uniquely placed to interpret local and global events, and to analyse their effects on the 'nation' which they have helped to forge into a self-conscious political community. However, as Smith demonstrates: 'it follows that the chief political struggles in Africa today, including ethnic ones, are at root factional conflicts within the intelligentsia – civilian versus military, liberal versus marxist, regional or ethnic conflicts – and that any involvement on the part of the other strata or classes is at the invitation or behest of one or other faction within the ruling stratum of the intelligentsia' (p. 90). In Nigeria, the scale of the intelligentsia's problems in transmitting these various schemes for the state's future was daunting – a task compounded abroad by the constraints of neo-colonialism and at home by the vast ethnic and regional disparities inherited on independence.

It is now commonplace to attribute to Nigeria's ethnic diversity the seeds of the eventual civil war. But many commentators on the period 1950–66 have argued that certain historical factors, not inevitable elsewhere, have exacerbated these differences. Three are of prime importance: the legacy of colonial administration – direct rule in the south and indirect rule in the north; the internal dynamics of the Nigerian military and of civil–military relations; federal–regional rivalry with each of the major political parties trying to outmanoeuvre its opponents in gaining access to power, patronage and privilege.

## Selected references

Anderson, M. 1978: The renaissance of territorial minorities in western Europe. *West European Politics* 1, 128–43.

Anderson, P. 1974: *Lineages of the Absolutist State*. London: New Left Books.

Breuilly, J. 1982: *Nationalism and the State*. Manchester: Manchester University Press.

Foster, C. R. (ed.) 1980: *Nations Without a State*. New York: Praeger.

Hintze, O. 1975: Economics and politics in the age of modern capitalism. In Gilbert (ed.), *The Historical Essays of Otto Hintze*. New York: Oxford University Press.

Hirsch, J. 1981: The apparatus of the state, the reproduction of capital and urban conflicts. In M. Dear and A. J. Scott (eds) *Urbanisation and Urban Planning in Capitalist Society*. London: Methuen, 593–607.

Kofman, E. 1982: Differential modernisation, social conflicts and ethno-regionalism in Corsica. *Ethnic and Racial Studies*, 5, 300–13.

Krejci, J. and Velimsky, V. 1981: *Ethnic and Political Nations in Europe*. London: Croom Helm.

Medhurst, K. 1982: Basques and Basque nationalism. In C. H. Williams (ed.), *National Separatism*. Cardiff: University of Wales Press, 235–61.

Mény, Y. 1984: Decentralisation in socialist France: the politics of pragmatism. *West European Politics*, 7, 66–79.

Mény, Y. 1987: France: the construction and reconstruction of the centre, 1945–86. *West European Politics*, 10, 52–69.

Modelski, G. 1978: The long cycle of global politics and the nation state. *Comparative Studies in Society and History*, 20, 214–35.

Nairn, T. 1981:*The Break-Up of Britain*. London: Verso.

Oberschall, A. 1973: *Social Conflict and Social Movements*. Englewood Cliffs, NJ: Prentice-Hall.

Rich, E. E. and Wilson, C. H. (eds) 1967: *The Economy of Expanding Europe in the 16th and 17th Centuries*. Cambridge: Cambridge University Press.

Shaw, M. 1984: War: the end of the dialectic? *Journal of Area Studies*, 9, 3–6.

Sklar, R. 1963: *Nigerian Political Parties*. Princeton, NJ: Princeton University Press.

Skocpol, T. 1977: Wallerstein's world capitalist system: a theoretical and historical critique. *American Journal of Sociology*, 82, 1075–90.

Skocpol, T. 1979: *States and Social Revolutions*. Cambridge: Cambridge University Press.

Smith, A. D. 1983: *State and Nation in the Third World*. Brighton: Wheatsheaf Books.

Taylor, P. J. 1982: A materialist framework for political geography. *Transactions, Institute of British Geographers*, NS 7, 15–34.

Urwin, D. 1985: The price of a kingdom: territory, identity and the centre–periphery dimension in Western Europe. In Y. Mény and V. Wright (eds), *Centre–Periphery Relations in Western Europe*. London: Allen & Unwin, 151–70.

Wallerstein, I. 1974: *The Modern World System*. New York: Academic Press.

Wallerstein, I. 1979: *The Capitalist World Economy*. Cambridge University Press.

Wendt, A. E. 1987: The agent–structure problem in international relations theory. *International Organization*, 41, 334–70.
Williams, C. H. and Kofman, E. (eds) 1988: *Community Conflict, Partition and Nationalism*. London: Routledge.
Williams, C. H. and Smith, A. D. 1983: The national construction of social space. *Progress in Human Geography*, 7, 502–18.
Williams, G. 1980: Review of E. Allardt's Implications of the Ethnic Revival in Modern Industrial Society. *Journal of Multilingual and Multicultural Development*, 1, 363–70.
Zolberg, A. R. 1981: Origins of the modern world system: a missing link. *World Politics*, 33, 253–81.

# 18 Nuala Johnson, 'Cast in Stone: Monuments, Geography, and Nationalism'

Reprinted in full from: *Environment and Planning D: Society and Space* 13, 51–65 (1995)

## Introduction

The building and unveiling of a statue to Sir Arthur 'Bomber' Harris in London in June 1992 aroused a small but vocal protest in London. Commemoration of the loss of 55 573 aircrew during the Second World War was the impulse for erecting the public monument. The commemoration, however, witnessed a protest from the Peace Pledge Union in London and from the mayors of Köln and Dresden, the two cities most affected by the saturation raids initiated by the commander (MacKinnon, 1992). A visit by Queen Elizabeth to the city of Dresden was greeted with a demonstration by local people for the role Harris played in the carpet-bombing of their city.

The changing political organisation of eastern Europe has precipitated the mass removal of public statuary that celebrated leaders of communist rule. In Budapest the city council has removed in excess of twenty monuments including those of Marx and Engels. Veterans of the 1956 uprising were among those seeking their removal. The Red Army Monument, however, has been retained in one of the city's main squares, but it is under constant police protection. According to the deputy Mayor of the city "History should not be re-written again. Despite what happened later, the Russian army played a very important role in the Second World War and actually did free this city from the Nazis. The Russian soldiers who died – and many died – deserve a monument" (quoted in Dent, 1992, page 10). The city plans to build a statue park to house the monuments that have been removed. No longer to adorn public space these monuments will now enter more explicit 'heritage space' and will be subject to the tourist gaze (see Urry, 1990). Ironically then, the

Hungarian past will be used to generate foreign revenue through an open-air museum display, yet that heritage is rejected in the civic landscape of the city. The space which these monuments occupy is not just an incidental material backdrop but in fact inscribes the statues with meaning. A civic square is an altogether different space from a specialist theme park.

Particularly since the nineteenth century public monuments have been the foci for collective participation in the politics and public life of towns, cities, and states (Agulhon, 1981; Mosse, 1975; Warner, 1985). Despite their location in public space and their role as sites of shared unity or protest, geographers, in general, have underutilised public monuments as a vehicle for conceptualising the nation-building process. In this paper I seek to highlight the usefulness of public monuments as a source for unravelling the geographies of political and cultural identity especially as they relate to conceptions of national identity.

## Geogaphy and the study of national identity

> – A nation? says Bloom. A nation is the same people living in the same place.
> – By God, then, says Ned, laughing, if that's so I'm a nation for I'm living in the same place for the past five years.
>
> So of course everyone had the laugh at Bloom and says he, trying to muck out of it:
> – Or also living in different places.
> – That covers my case, says Joe.
> – What is your nation if I may ask? says the Citizen.
> – Ireland, says Bloom. I was born here. Ireland.
>
> (James Joyce *Ulysses* 1922)

The above extract from Joyce's *Ulysses* amply captures some of the principal issues surrounding the vexed question of defining the nation and concomitantly national cultural identity. Drawing on the temporal (over time), ethnic (same people), and geographical (same place) elements implicit in a commonsense knowledge of the constituent features of a nation, Joyce manages to parody and render complex Bloom's seemingly straightforward definition of a form of political organisation that dominates the global map. Today we have a huge academic literature proposing a variety of definitions of the nation (for an overview see Hobsbawm, 1990) and Anderson's (1983) assertion that it is an 'imagined community' has been generally received as one of the most authoritative accounts. For geographers, however, Watts (1992) warns that "imagination and territoriality are employed often quite loosely, as though individuals cook up some sort of ideal world out of thin air" (page 125). An examination of public statuary, I argue, highlights some of the ways in which the material bases for nationalist imaginings emerge and are structured symbolically.

Cultural geography has recently witnessed an increased concern with the articulation, constitution, and representation of identity – be it social, ethnic, or gender identities (Cosgrove, 1990; Jackson, 1989), yet much of this literature has been somewhat hesitant in addressing the ways in which national cultural identity at the *popular* level is constructed, maintained, or challenged. Thus we have studies which look at how particular landscape images are constructed, largely through the imaginings of what Gramsci (1971) would have considered an intellectual elite (that is, spiritual or political leaders, artists, writers, architects, critics) (see Cosgrove and Daniels, 1988; Duncan, 1990), but we have a comparatively sparse literature on how these sorts of images are popularised, consumed, or resisted by groups within the state. As Jackson (1989) has pointed out, there is an increasing emphasis on the fluid and fragmented nature of political or cultural identity, yet the empirical focus has largely been on minority cultures within states.

One could peruse the geographical literature over the past decade, for instance, and find precious few in-depth studies which address the conflict of identity in Northern Ireland. Yet, in a British context, it is here in particular that conceptions of national political culture are contested and where the popular imagination is highly territorialised at a variety of geographical scales. Research, outside of geography, on the location and semiotics of murals, the routes of marches or parades, and the spatialisation of prison life (Jarman, 1992; Rolston, 1987; 1988; 1991) all confirm Cosgrove's assertion (1989) that "geography is everywhere". It is the absence of sustained geographical research on these topics that requires some attention. Recent suggestions that Northern Irish Protestants have no 'intrinsic' identity outside of an oppositional politics antithetical to the Republic of Ireland, or no collective landscape imagery (Graham, 1994) could be furthered through empirical investigation. Indeed Jackson (1992) in an analysis of unionist myths in Northern Ireland claims that "the cult of 1912 14 . . . is central to the historical consciousness of modern Unionism . . . [and] the contemporary memory of 1912–14 has been tailored by dead partisans to a degree unusual even in twentieth-century Ireland" (pages 183–184).

Historians have paid considerable attention to the processes involved in the articulation of a heroic version of the past and the invention of traditions (Hobsbawm and Ranger, 1983) which are subsequently popularly consumed within a nation-building framework. This process has been taken on board by others interested in nation-building (McCrone, 1992; A D Smith, 1986; 1991). Geographers too have analysed the invention of traditions, especially in an American context. A special issue of *Journal of Historical Geography* (1992) offers insights into how some discourses about American identity emerged historically and the specific role of particular actors or interests in the promotion of certain types of landscape image (for example, the Great American desert, a New England colonial village). The persistent power of some of these myths, as Watts (1992) has shown, was exposed in the uproar precipitated by the 1991 exhibition "The West as America: Reinterpreting Images of the Frontier 1820–1920". The exhibition evoked comments such as "perverse" (Senator Ted Stevens, Alaska) or "historically inaccurate, destructive exhibition" (Daniel Boorstin, former Librarian of Congress) and resulted in the cancellation of some of the venues for the exhibition (all quoted in Watts, 1992).

Thus, although the invented nature of some traditions associated with nations' histories have been exposed, their persistence in the popular consciousness is rather less understood. Nation-building is, as Smith (1986) has observed, an ongoing historical process – whose myths prevail at particular moments is the crucial question. The connections between elite and popular 'imagined communities', where subaltern voices are not always assumed to be epiphenomenal to identity formation, is crucial to an investigation of nationalism. I now wish to provide a brief overview of some of the literature on public monuments.

**Monumental studies or the study of monuments?**

The transformation of urban space through monumental architecture and statuary has been explored by urban historians, art historians, and some geographers. Schorske's (1979) compelling investigation of the redesign of the Ringstrasse in nineteenth-century Vienna under the liberals "as a visual expression of the values of a social class" (page 25) meshes an analysis of the economic, political, and aesthetic values of Vienna's 'triumphant middle class' in the reconceptualisation of late nineteenth-century urban form. Although Schorske (1979) makes some reference to individual statues, the overall focus is placed on specific buildings or street blocks constructed at a monumental scale.

Whereas Schorske's study focuses on the rise of a particular social class in Vienna rather than on the evolution of Austrian nationalism, Mosse's (1975) investigation is of the 'new politics' which "attempted to draw the people into active participation in the national mystique through rites and festivals, myths and symbols which gave concrete expression to the general will" (page 2). Through a broad-ranging analysis of monuments, architecture, theatre, and public festivals Mosse traces the various ways in which the masses were nationalised in Germany from the Napoleonic wars to National Socialism. Deviating from studies which consider the economic and political evolution of the nation-state, Mosse is concerned with the aesthetics and symbolism central to the new politics of the nineteenth century and convincingly argues that "the reality of nationalism and of National Socialism represented itself to many, perhaps most people, through a highly stylized politics, and in this way managed to form them into a movement" (page 214).

It was beyond the remit of Mosse's study to examine the ways in which the new politics of style was challenged both from the right and from the left, nor are the connections between the geographical constitution of identities (Agnew, 1987; Agnew and Duncan, 1989; Watts, 1992) discussed in any detail. Although Cosgrove (1990, page 564) rightly suggests that the nation-state sought to "promote a single identity within the bounds of its territory", it is important to stress that this process has been strongly spatialised and frequently resisted in particular regions of the state where subaltern symbols and alternative versions of history prevail. The reluctance of Ireland to cohere with the Union of Britain and Ireland, for instance, reinforces this contention and resistance was inscribed visually through monuments erected on the island (Johnson, 1994).

Harvey's (1979) brilliant study of the Basilica of the Sacre Coeur in Paris highlights the contested political meaning of the site at Montmartre, where both conservatives and communards could lay symbolic claim. The eventual alliance of monarchists and intransigent Catholics guaranteed the building of a monument dedicated to the cult of the Sacred Heart. Uncovering the politics underlying the development of the Sacre Coeur not only reveals the deep fissures informing late nineteenth-century Parisian and national politics but also opens up the ways in which the building can be read which "rescue that rich experience from the deathly silence of the tomb and transform it into the noisy beginnings of the cradle" (Harvey, 1979, page 381).

**War memorials**

Memory and commemoration have become an area of increased academic interest especially among historians. Although political geographers have long been concerned with studies of war or peace, geopolitical discourse, and geostrategy, they rarely examine the symbolic fallout associated with national or international conflict (see Reynolds, 1992). Similarly cultural geographers concerned with landscape interpretation and iconography have largely ignored war sculpture in their analyses of the relationship between politics and culture (for instance, Agnew and Duncan, 1989; Cosgrove and Daniels, 1988; Duncan, 1990). Yet war memorials are of special significance because they offer insights into the ways in which national cultures conceive of their pasts and mourn the large-scale destruction of life. Wagner-Pacifini and Schwartz (1991) posit that "Memorial devices are not self-created; they are conceived and built by those who wish to bring to consciousness the events and people that others are inclined to forget" (page 382).

Historians have begun examining the relationship between memory and history. Nora (1989) claims that with the demise of peasant society sites of memory have replaced real environments of memory – "true memory, which has taken refuge in gestures and habits, in skills passed down by unspoken traditions, in the body's inherent self-knowledge, in unstudied reflexes and ingrained memories" (page 13) has been replaced by modern memory which is self-conscious, historical, individual, and archival. This distinction between true and modern memory becomes more persuasive when connected with the style of politics associated with the rise of the national state, where extralocal memories are intrinsic to creating 'imagined communities' and new memories necessitate collective forgetting or amnesia (Anderson, 1983; Hobsbawm, 1990).

Space or more particularly territory is as intrinsic to memory as historical consciousness in the definition of a national identity. These new sites of memory are not simply arbitrary assignations of historical referents in space but are consciously situated to connect or compete with existing nodes of collective remembering. Thus the claim that "Statues or monuments to the dead . . . owe their meaning to their intrinsic existence . . . [and] one could justify relocating them without altering their meaning" (Nora, 1989, page 22), warrants some revision. The meaning of the Vietnam Veterans' Memorial in Washington DC, for instance, is partly defined by its location in the capital of the USA (MacCannell, 1992) linking it to US foreign policy and geopolitical discourses (Sturken, 1991). It could be interpreted rather differently in Ho Chi Minh City.

Zelinsky (1988) has commented on the paucity of literature on American monuments in general, but Civil War statuary (especially Confederate statues) have been considered in some detail both by geographers and by historians (Davis, 1982; Foster, 1987; Gulley, 1993; Winberry, 1982; 1983). Although the location of Civil War statues reflects the geographical division of allegiances, the historian Savage (1994) demonstrates how statues which ought to have reflected a serious divide between antislavery and proslavery lobbies in America gradually became transformed and "Americans perceived this kind of monument building as part of a healthy process of sectional reconciliation – a process that everyone knew but no one said was for and between whites" (Savage, 1994, page 132). The difference between those supporting the Union and those supporting the Confederates was disguised through a racial politics which, each in its own way, denied black memory. The South's defence of slavery became blurred. In the commemorative statue of Lee, for instance, he was depicted as an American hero who fought out of loyalty to his home state. In the North, memorials generally omitted cultural representation of blacks in the war effort. Only three Civil War statues represent blacks, the most famous Shaw memorial designed by Augustus Saint-Gaudens "facilitates opposing readings of its commemorative intent" (Savage, 1994, page 136). Overviews of Union monuments have received far less scholarly attention and the connections between the two traditions are only beginning to be explored (Foster, 1987; Savage, 1994). Yet even preliminary investigations of Civil War statuary highlight the ways in which the physical memorialisation of war adds fresh interpretation to the events themselves. The treatment of civil war in geographical scholarship is sparse, yet it is precisely at these points of fracture that some of the particularly interesting questions regarding the spatialisation of historical imagination which Watts (1992) has commented on, come to the fore.

There has been a recent upsurge of interest in collective memory, especially as it relates to the experience of two world wars. Historians have begun to trace the origins of a new, more democratic, style of war memorial, which iconographically moved away from the commemoration of generals or rulers to the acknowledgement of the role of ordinary soldiers and armies in the war effort (Fussell, 1975; Mosse, 1990). Sherman (1994) in his discussion of World War I memorials in France identifies two types of tension inherent in the commemoration of that event. He claims that "the decision to construct a monument implicated a community in several kinds of latent contestation" (page 188), one centring on the secular or religious question and the second centred on "the negotiation of local and national claims to memory of the dead" (page 188). This negotiation produced numerous types of debate and discrete outcomes in different national contexts. In France the government agreed, where possible, to pay for the return home of the bodies of dead soldiers and frequently the local memorials named the individuals killed (Sherman, 1994). This practice differentiated it from commemorations where memorials named the leader but very rarely the rank and file war dead (Laqueur, 1994). Britain decided to bury the bodies of its dead along the Western Front, resisting any attempt to return them to their families, but "the state poured enormous human, financial, administrative, artistic and diplomatic resources into preserving and remembering the names of individual common soldiers" (Laqueur, 1994, page 155). The national commemorative activity centred on the building of the catafalque at Whitehall and the burying of the unknown soldier in Westminster Abbey: "the unknown warrior becomes in his universality the cipher that can mean anything, the bones that represent any and all bones equally well or badly" (Laqueur, 1994, page 158). The unknown soldier has become a common motif for the sacrifice of the rank and file in Europe and beyond (Inglis, 1993).

Although seen by some as a legitimate form of collective commemoration these tombs have also been subject to parody. When passing the Arc de Triomphe in 1920 and asked how long he thought the eternal flame would burn, James Joyce caustically replied "Until the Unknown Soldier gets up in disgust and blows it out" (quoted in Ellmann, 1959, page 486). As a pacifist he, for one, found such public monuments offensive (Fairhall, 1993). The extent to which the public supported or dissented from such commemorative activity has yet to be documented. Geographers have just begun to examine some of the debates surrounding the Great War and the landscapes of remembrance produced (Heffernan, 1995), but this work is still in its infancy.

In the case of the United States' commemoration of the Great War the populace refused to be treated solely as servants of the state. In a fascinating analysis Piehler (1994) examines the ways in which women articulated their response to the war, picking up from existing studies on gender relations and World War I (Higonnet et al, 1987). Despite the War Department's wishes to bury US soldiers in cemeteries in Europe and inscribe the United States' role in the world political order on the fields of France and Belgium, women demanded that their sons be repatriated for burial at home; eventually over 70% of soldiers were returned. The ways in which discourses of citizenship, motherhood, and peace were rearticulated in the interwar years in the United States provide preliminary insights on how the spaces of women were redefined (Piehler, 1994).

With the recent success of Steven Spielberg's film epic *Schindler's List*, memory of the Holocaust in the popular consciousness has been rekindled. Indeed the place where the movie was filmed is starting to become a tourist attraction as increasing numbers of people visit the site and make connections between their viewing of the film and the real landscape (Borger, 1994). Yet thousands of monuments to the Holocaust were erected after World War II and amongst them the Warsaw Ghetto monument has received much of the attention. On the consumption of this public monument and its role in popular memorialisation, Young (1989) provides a seminal analysis. Prefacing his discussion with a critique of the narrow conceptual framework adopted by art historians and the commentators of 'high' public art, Young suggests that "it may be just this public popularity that finally constitutes the monument's aesthetic performance" (page 99).

Although views differ on the ability of monuments in general, and figurative ones in particular, to engage the viewer reflexively with the past or future, and the antimonument movement seeks to reclaim memory as part of everyday life (Gillis, 1994), Young's study suggests that monumental figurative statuary continues to engage the viewer. The iconographic effect and public popularity of Maya Lin's quasi-abstract Vietnam Veterans' Memorial in Washington confirms the continuing appeal of public statues (Sturken, 1991). Geographers attempting to conceptualise how the public memory works and how national imaginings unfold could do well to move beyond examining the elite landscapes of the ruling classes, and begin to broaden their own imaginative remit to the popular mind. Unlike creative literature or painting, the production and consumption of public monuments is more firmly part of a collective process – "Sculpture . . . is, so to speak, more democratic than painting because it is simpler and more solemn, more appropriate to the public square, to huge dimensions, and to emblematic figures that are both a product of and a stimulus to the imagination" (Republican leader, Godefroy Cavaignac, 1834; cited in Agulhon, 1981, page 4). Statuary offers a way of understanding nation-building which moves beyond top-down structural analyses to more dialectical conceptualisations (for a critique of structural analyses of nationalism see MacLaughlin, 1986).

**Gender and monuments**

The relationship between gender and national iconography has become an area of increasing interest in studies of statuary. Warner (1985) provides seminal insights into the use of the female body in statues and other allegorical representations. She claims that "The body is still the map on which we mark our meanings; it is chief among metaphors used to see and present ourselves, and in the contemporary profusion of imagery, from news photography to advertising to fanzines to pornography, the female body recurs more frequently than any other: men often appear as themselves, as individuals, but women attest the identity and value of someone or something else" (page 331). In national commemoration the role of women is largely allegorical, and, although states use women as symbols of identity such as the Figures of Liberty or Marianne (Agulhon, 1981), women rarely appear in sculpture as political or cultural leaders. Outside the Reichsrat (parliament) in the Viennese Ringstrasse, Athena was chosen as the symbol to adorn the new building, the Austrian parliamentarians avoided Liberty because of her association with a revolutionary past – "Athena, protectrix of the polis, goddess of wisdom, was a safer symbol. She was an appropriate deity . . . to represent the liberal unity of politics and rational culture" (Schorske, 1979, page 43). Allegorical figures cannot be simply subsumed under a single model of gender relations but must be disaggregated in terms of the version of history being promoted (by men) in specific political contexts. The use and reading of female allegories are not fixed, as can be seen in the case of Hibernia in a late nineteenth century Irish context (see Johnson, 1994).

The female body in contemporary monuments in the city continues to reveal the ways in which 'gender performance' is articulated. Smith's (A Smith, 1991) fascinating interpretation of the monument Anna Livia Plurabella, erected in Dublin in 1987, highlights the fragility of women's position in the city under patriarchy. The statue, whose inspiration was drawn from Joyce's character, Anna Livia Plurabella, a representation of womanhood, of the city of Dublin, and its river Liffey in *Finnegan's Wake*, was proposed by the city's civic government, financed through private capital, and located in the city's principal street. After its unveiling the statue underwent a series of renamings – 'the floozie in the jacuzzi', 'the whore in the sewer', 'the skivvy in the sink' – a strategy by Dubliners to deflate the high-art pretensions of the monument itself, to cut it down to size so to speak; but also a strategy which reveals a "male role-shift from that of Slave to Master" in a postcolonial context (Smith, 1991, page 11). The female figure in this instance does not represent the virtues alluded to by Warner (1985) but invokes gender-coded stereotypes of woman in public space as whore, temptress, pollutant, and scaled to virtual anorexic proportions as she bathes in the waters of the city (Figure 18.1). Although allegorical figures of woman as 'motherland' and protector of the private sphere of home and family enjoy acceptance in nationalist discourse (Nash, 1993), in the city woman's role in public space, as suggested by the renaming of Anna Livia Plurabella, is confined to that of prostitute or seductress strolling streets normally occupied by men. Joyce's character may be more complex than her representation on

**Fig. 18.1** 'The floozie in the jacuzzi' – Anna Livia Plurabella monument, O'Connell Street, Dublin

O'Connell Street, but the monument, I suspect, has far more viewers than *Finnegan's Wake* has readers!

Women not only feature in monuments themselves but recent research has been emphasising their role in the organisation of commemorative activity and in the debates surrounding the articulation of public memory (Gulley, 1993). Women have been active in mobilising support for statue building, most notably in the context of war dead. As mothers, wives, and sisters of soldiers women have been an active, if underrepresented, grass-roots lobby for the repatriation of killed men and in postwar peace movements (Gillis, 1994). Occasionally they have also been the sculptors (the Vietnam memorial is the best-known example). By considering the role of women in this context we may begin to reveal and challenge how the imaginary unfolds in the discursive practices of identity formation.

## The sculptural mapping of Dublin – the Parnell monument

Nineteenth-century Dublin, like other European capitals, had its streets and parks decorated with public and private statues. Public monuments were listed in the catalogue for visitors to the Great Industrial Exhibition of 1853. It was noted in 1856, however, that "No public statue of an illustrious Irishman has ever graced the Irish capital" (quoted in Murphy, 1994, page 202). Until the middle of the century there were two principal types of statue. The first were royal monuments such as the King William III statue erected in 1701 in College Green or the equestrian statue of George I in the gardens of the Mansion House (the Mayor of Dublin's official residence). The second type were those erected to commemorate the prowess of military leaders. Unlike war memorials commemorating the efforts of ordinary soldiers (Gillis, 1994) these statues reserved public reference to military (imperial) leaders. The two most significant in Dublin were the Wellington and the Nelson monuments. Designed as an obelisk and column, respectively, each was "rising to a soaring height and visible from some distance, [they] employed scale in an aggressive manner" (Murphy, 1994, page 203). Rather than soliciting collective public memory of war casualties, these monuments functioned in iconography, location, and subject to solicit public knowledge of heroes in British military campaigns. Unlike civil war memorials in the USA or memorials to the Great War in Europe, these statues did not attempt to evoke collective reconciliation with the past or mourning for 'national' sacrifice, but further inscribed Dublin as a provincial capital within a Union whose centre was London.

The first attempts to counterpoise the sculptural mapping of Dublin with Irish people came with the erection of statues to honour literary figures. The credentials of the city as a literary capital are now well documented as any glance at current tourist brochures attests (Lincoln, 1993; Titley, 1990). Yet before Dublin was publicising its 'production' of three winners of the Nobel Prize for Literature, the city was erecting statues to Oliver Goldsmith, Thomas Moore, and Edmund Burke in the 1850s and 1860s. As respectable figures, pedestalised outside their alma mater Trinity College, these Protestant men of letters were comparatively uncontroversial and could comfortably coexist with the more imperial statues gracing the city.

It was in the second part of the nineteenth century that Dublin began to celebrate overtly nationalist leaders, with the O'Connell monument on the city's main street being the only statue of a Catholic erected in the nineteenth century (Murphy, 1994). Its location at the head of O'Connell Street competed with the Nelson column situated centrally along the thoroughfare. Designed as a figurative statue O'Connell is perched at the apex of the pedestal, surrounded below by a series of figures supporting Catholic emancipation and pivoted by the female allegory Hibernia; the base is crowned with four winged victories. The design combines classical motifs with Celtic iconography (Figure 18.2). Unlike Sri Lanka where statues associated with empire were removed and replaced with statues of nationalist leaders after independence (Duncan, 1990), in Dublin both types of statue existed simultaneously; the confrontational nature of Irish politics at the time was reflected in the statue-building on the streets of the capital and elsewhere (Johnson, 1994). Collective memories were being consciously aroused in stone and bronze. Statues did not necessarily merely reflect the values of a particular social class as in the

**Fig. 18.2** The O'Connell monument, O'Connell Street, Dublin (courtesy of the Lawrence Collection, National Library of Ireland)

case of Vienna (Schorske, 1979) (indeed many were heavily criticised on artistic grounds), their imposition was a means of negotiating and contesting popular nationalist politics. Their design, funding, location, and unveiling were well publicised events reported in the popular press (Murphy, 1994) and in some instances they created great controversy (O'Keefe, 1988). The ability to shape public commemoration and negotiate the geography of public ritual had important implications for the various political interests of the city.

The Parnell monument was first proposed in 1898, the year of the centenary celebrations of the 1798 rebellion (see Johnson, 1994). As he was one of the most contentious political leaders in modern Ireland, the proposal to erect a monument to him generated considerable discussion. Dublin Corporation were of the view "that no statue should be erected in Dublin in honor of any Englishman until at least the Irish people have raised a fitting monument to the memory of Charles Stuart Parnell" (quoted in Murphy, 1994, page 206). The Irish people, however, were not of one mind when it came to assigning Parnell's role in the public memory. The Irish Parliamentary Party, which he effectively led before the revelation of his adulterous affair, split into a number of factions, with supporters of Parnell coalescing under the leadership of John Redmond (Foster, 1988). The Parnell episode marked a period of intense public and private debate conducted "with all the venom of a fratricidal feud" (O'Keefe, 1984, page 7). A powerful worm's-eye view of the controversy is found in the Christmas dinner argument in Joyce's *A Portrait of the Artist as a Young Man*. The commemoration of the leader, therefore, would not necessarily be a peaceful affair, although the primary organiser of the monument, Redmond, hoped that it would heal the wounds of conflict between 'Irishmen'. Unlike the centenary of the 1798 rebellion where the memory of the rebellion could be reinterpreted by influential parties such as the Catholic Church (Johnson, 1994; O'Keefe, 1988; 1992; Turpin, 1991), in the case of Parnell, the recency of his death and the widespread press coverage of his 'affair' rendered revisionist interpretations much more difficult.

At the laying of the foundation stone at the northern end of O'Connell Street in 1899, it was clear that unanimity of opinion was far from achieved. The absences at the unveiling ceremony were significant. From over eighty elected members of the Irish Parliamentary Party only a handful attended, no Catholic clergy participated, which contrasts with the 1798 celebrations, and other civic leaders such as city and county magistrates were thin on the ground (O'Keefe, 1984). The occasion was marred with incident – hecklers interrupted speeches, fist fights broke out, and in general it was a disorderly affair. The Irish Republican Brotherhood saw the event as "a direct challenge to their own campaign to honor the father of Irish republicanism, Wolfe Tone, who had died more than a century before" (O'Keefe, 1984, page 8). According to Gillis (1994) modern memory stems from, amongst other things, "an intense awareness of the conflicting representations of the past and the effort of each group to make its version the basis of national identity" (page 8). Redmond's hegemonic influence in the face of republican-minded nationalists certainly prevailed in the first decade of the twentieth century.

Amidst a siege of criticism and dissent, Redmond decided to raise funds for the monument in the United States, fearing that the Irish public might be unwilling to foot the bill. Parnell, whose mother was American, was popular in the USA and Redmond broadened the fund-raising activities to include the building of a headstone on Parnell's grave and the acquisition of the ancestral home in Avondale, County Wicklow. Although the effort was not wholly successful, Redmond did manage to raise some funds especially from Richard Croker, boss of Tammany Hall. The second objective in the United States was to find a sculptor, preferably with connections to Ireland, to design the monument, and reinforce Parnell's lineal connection with the USA (O'Keefe, 1984). The geographical base of Irish nationalism went far beyond the shores of the island, and was cultivated periodically in the process of memory making.

The sculptor chosen was Augustus Saint-Gaudens, a well-established name in American sculpture (for example, the Shaw memorial). His credentials were ideal, born of an Irish mother and French father who emigrated to the

**Fig. 18.3** The Parnell monument, O'Connell Street, Dublin (courtesy of the Lawrence Collection, National Library of Ireland)

USA during the famine (O'Keefe, 1984). In the midst of various financial crises and personal difficulties Saint-Gaudens completed the monument which was finally unveiled in October 1911. The figure in bronze of Parnell (Figure 18.3), clothed in a frock coat, arm outstretched as if he is in the act of speaking, and standing beside a table draped in a flag of Ireland, was designed according to Saint-Gaudens "as simple, impressive and austere as possible, in keeping with the character of the Irish cause as well as of Parnell" (quoted in O'Keefe, 1984, page 17). The backdrop to the statue was a triangular shaft, constructed of Shantalla granite and inscribed with a brief passage, deliberately chosen by Redmond, from one of Parnell's more extreme versions of his political aims. It reads "No man has a right to fix the boundary to the march of a nation. No man has a right to say to his country – thus far shalt thou go and no further. We have never attempted to fix the ne plus ultra to the progress of Ireland's nationhood and we never shall". For those reading the inscription Parnell's nationalism appears unequivocal. The use of this quotation might be seen as an attempt to reconcile the public's distaste for his personal life with their admiration of his public one. As Savage (1994) noted in the context of the American civil war monument, a selective reading and representation of the past served to heal the wounds of a society divided; the same was hoped for in the context of Parnell. The use of swags and ox heads on the monument attempted to make it visually compatible with the motifs decorating the nearby Rotunda. Statue design was not an isolated activity separated from the context of an existing streetscape (Schorske, 1979).

By 1911 Redmond's popularity had risen since the days of laying the foundation stone. Over fifty MPs and some Catholic clergymen attended the unveiling ceremony and the public disorder of a decade earlier had all but disappeared. Parnell, at last, it would seem, had been embraced by Dubliners, but only in the shadow of a rising militancy among the working class (Rumpf and Hepburn, 1977). A railroad strike and a bakers' strike were held during the day of the unveiling, a precedent to the well-documented 1913 lockout in Dublin. A small but significant socialist movement was taking hold in the capital (Foster, 1988; Lyons, 1979). The monument at the northern end of O'Connell Street completed what Yeats referred to as the triumvirate of 'old rascals' – O'Connell, Nelson, and Parnell. Ironically O'Connell Street's General Post Office, overlooked by Nelson would be the pivotal node for the 1916 uprising, an expression of republican nationalism, from which would emerge a new coterie of heroes, whose sacrifice would later be commemorated in stone and bronze.

O'Connell Street today has lost its Nelson pillar, dynamited by republicans in the 1960s, but it retains its O'Connell and Parnell, supplemented by other statues including Anna Livia Plurabella. New memories have been aroused since the heady first decades of this century, new definitions of national identity have been articulated.

## Conclusion

In this paper I have sought to highlight the usefulness of public monuments as a source for understanding the emergence and articulation of a nationalist political discourse. The location of statuary reveals the ways in which monuments serve as the focal point for the expression of social action and a collectivist politic, and the iconography of statues exposes how class, 'race', and gender differences are negotiated in public space. Although historians have devoted considerable attention to the role of public statuary especially in the context of war memory, geographers are just beginning to examine the relationships between the memorialisation of the past and the spatialisation of public memory. Using Anderson's (1983) thesis that nations are 'imagined political communities' as a guiding principle, I have argued that an examination of public statues enables the researcher to gain some insights into how the public imagination is aroused and developed in the context of the ongoing task of nation-building. Statues, as part of the cityscape or rural landscape, act not only as concentrated nodes but also as circuits of memory where individual elements can be jettisoned from popular consciousness. Their role in the geography of the city as points of physical and ideological orientation requires much further research. In the case of Belfast, anthropologists and communications theorists have powerfully elucidated some of the ways in which divided identities are structured and maintained in the geographies of everyday life through analyses of public art, popular parade, and ritual. In so doing they have elided the tendency to treat nationalism as an ideology and a practice born out of the preoccupations of a privileged intelligentsia and have instead treated it as one of the dominant discourses to emerge from a complex web of political, cultural, and economic inequalities.

## References

Agnew J. 1987: *Place and Politics*: *The Geographical Mediation of State and Society* (Allen and Unwin, London).

Agnew J., Duncan J. (eds), 1989: *The Power of Place: Bringing Together Geographical and Sociological Imaginations* (Unwin Hyman, London).

Agulhon M. 1981: *Marianne in Battle: Republican Imagery and Symbolism in France, 1798–1880* (Cambridge University Press, Cambridge).

Anderson B. 1983: *Imagined Communities* (Verso, London).

Borger J. 1994: "Site of Schindler's story repackaged as city of the film" *The Guardian* 16 May, page 24.

Cosgrove D. 1989: "Geography is everywhere: culture and symbolism in human landscapes", in *Horizons in Human Geography* eds D Gregory, R Walford (Macmillan, London) pp 118–135.

Cosgrove D. 1990: ". . . Then we take Berlin: cultural geography 1989–90" *Progress in Human Geography* 14, 560–568.

Cosgrove D., Daniels S. (eds), 1988: *The Iconography of Landscape* (Cambridge University Press, Cambridge).

Davis S. 1982: "Empty eyes, marble hand: the Confederate monument and the South" *Journal of Popular Culture* 16, 2–21.

Dent B. 1992: "The Hiss in history" *The Guardian* 13 November, page 10.

Duncan J. 1990: *The City as Text: The Politics of Landscape Interpretation in the Kandyan Kingdom* (Cambridge University Press, Cambridge).

Ellmann R. 1959: *James Joyce* (Galaxy, New York).

Fairhall J. 1993: *James Joyce and the Questions of History* (Cambridge University Press, Cambridge).

Foster G. M. 1987: *Ghosts of the Confederacy: Defeat, the Lost Cause, and the Emergence of the New South* (Oxford University Press, New York).

Foster R. F. 1988: *Modern Ireland 1600–1972* (Allen Lane, London).

Fussell P. 1975: *The Great War and Modern Memory* (Oxford University Press, Oxford).

Gillis R. 1994: "Memory and identity: the history of a relationship", in *Commemorations: The Politics of National Identity* ed. R. Gillis (Princeton University Press, Princeton, NJ) pp. 3–24.

Graham B. 1994: "No place of mind: contested Protestant representations of Ulster" *Ecumene* 257–281.

Gramsci A. 1971: *Selections from the Prison Notebooks* translated by Q. Hoare, G. N. Smith (International Publishers, New York).

Gulley H. E. 1993: "Women and the lost cause: preserving Confederate identity in the American Deep South" *Journal of Historical Geography* 19, 125–141.

Harvey D. 1979: "Monument and myth" *Annals of the Association of American Geographers* 69, 362–381.

Heffernan M. 1995: "For ever England: the Western Front and the politics of remembrance in Britain" *Ecumene* 2, 293–324.

Higonnet M. R., Jenson J., Michel S. and Weitz M. C. 1987: *Behind the Lines: Gender and the Two World Wars* (Yale University Press, New Haven, CT).

Hobsbawm E. 1990: *Nations and Nationalism since 1780; Programme, Myth, Reality* (Cambridge University Press, Cambridge).

Hobsbawm E. and Ranger T. (eds) 1983: *The Invention of Tradition* (Cambridge University Press, Cambridge).

Inglis K. S. 1993: "Entombing unknown soldiers: from London and Paris to Baghdad" *History and Memory* 5(2), 7–49.

Jackson A. 1992: "Unionist myths 1912–1985" *Past and Present* number 136, 164–185.

Jackson P. 1989: *Maps of Meaning* (Unwin Hyman, London).

Jarman N. 1992: "Troubled images" *Critique of Anthropology* 12, 179–191.

Johnson N. C. 1994: "Sculpting heroic histories: celebrating the centenary of the 1798 rebellion in Ireland. *Transactions of the Institute of British Geographers* 19, 78–93.

*Journal of Historical Geography* 1992: Special issue on "Creation of myth, invention of tradition in America" ed. J. L. Allen 18, 1–138.

Joyce J. 1922: *Ulysses* eds Hans Walter Gabler et al (Garland, London, 1984).

Laqueur T W, 1994, "Memory and naming in the Great War", in *Commemorations: The Politics of National Identity* ed. R. Gillis (Princeton University Press, Princeton, NJ) pp. 150–167.

Lincoln C. 1993: "City of culture: Dublin and the discovery of urban heritage", in *Tourism in Ireland: A Critical Analysis* eds B. O'Connor and M. Cronin (Cork University Press, Cork) pp. 233–257.

Lyons F. S. L. 1979: *Culture and Anarchy in Ireland 1890–1939* (Oxford University Press, Oxford).

MacCannell D. 1992: "The Vietnam memorial in Washington, DC", in *Empty Meeting Grounds* ed. D. MacCannell (Routledge, London) pp. 280–282.

McCrone D. 1992: *Understanding Scotland: The Sociology of a Stateless Nation*

(Routledge, London).

MacKinnon I. 1992: "'Bomber' Harris protest confronts Queen Mother" *The Independent* 1 June, page 1.

MacLaughlin J. 1986: "The political geography of nation-building and nationalism in social sciences: structural vs dialectical accounts" *Political Geography Quarterly* 5, 299–329.

Mosse G. L. 1975: *The Nationalization of the Masses* (Howard Fertig, New York).

Mosse G. L. 1990: *Fallen Soldiers: Reshaping the Memory of the World Wars* (Oxford University Press, Oxford).

Murphy P. 1994: "The politics of the street monument" *Irish Arts Review Yearbook* (Guinness Peat, Aviation, Dublin) pp. 202 –208.

Nash C. 1993: "Renaming and remapping" *Feminist Review* 44, 39–57.

Nora P. 1989: "Between memory and history: les lieux de memoire" *Representations* 26, 7–25.

O'Keefe T. J. 1984: "The art and politics of the Parnell monument" *Eire-Ireland* 19, 6–25.

O'Keefe T. J. 1988: "The 1898 efforts to celebrate the United Irishmen: the '98 centennial" *Eire-Ireland* 23, 67–91.

O'Keefe T. J. 1992: "Who fears to speak of '98: the rhetoric and rituals of the United Irishmen centennial, 1898" *Eire-Ireland* 27, 67–91.

Piehler G. K. 1994: "The war dead and the gold star: American commemoration of the First World War", in *Commemorations: The Politics of National Identity* ed. R. Gillis (Princeton University Press, Princeton, NJ) pp. 168–185.

Reynolds D. R. 1992: "Political geography: thinking globally and locally" *Progress in Human Geography* 16, 393–405.

Rolston B. 1987: "Politics painting and popular culture: the political wall murals of Northern Ireland" *Media, Culture and Society* 9, 5–28.

Rolston B. 1988: "Contemporary political wall murals in the North of Ireland: drawing support" *Eire-Ireland* 23, 3–18.

Rolston B. 1991: *Politics and Painting: Murals and Conflict in Northern Ireland* (Fairleigh Dickinson University Press, Toronto).

Rumpf E. and Hepburn C. 1977: *Nationalism and Socialism in Twentieth Century Ireland* (Liverpool University Press, Liverpool).

Savage K. 1994: "The politics of memory: black emancipation and the Civil War monument", in *Commemoration: The Politics of National Identity* ed. R. Gillis (Princeton University Press, Princeton, NJ) pp. 127–149.

Schorske C. E. 1979: *Fin-de-siècle Vienna: Politics and Culture* (Weidenfeld and Nicolson, London).

Sherman D. J. 1994: "Art, commerce and the production of memory in France after World War I", in *Commemorations: The Politics of National Identity* ed. R. Gillis (Princeton University Press, Princeton, NJ) pp. 186–211.

Smith A. 1991: "The floozie in the jacuzzi" *Feminist Studies* 1, 6–28.

Smith A. D. 1986: *The Ethnic Origin of Nations* (Basil Blackwell, Oxford).

Smith A. D. 1991: *National Identity* (Penguin Books, Harmondsworth, Middx).

Sturken M. 1991: "The wall, the screen and the image: the Vietnam Veterans Memorial" *Representations* 35, 118–142.

Titley A. 1990: "The city of words", in *Dublin and Dubliners* eds J. Kelly and U. Mac Gearailt (Helicon, Dublin) pp. 127–146.

Turpin J. 1991: "Oliver Sheppard's 1798 memorials" *Irish Arts Review* pp. 71–90.

Urry J. 1990: *The Tourist Gaze* (Sage, London).

Wagner-Pacifini R. and Schwartz B. 1991: "The Vietnam Veterans memorial:

commemorating a difficult past" *American Journal of Sociology* 97, 376–420.

Warner M. 1985: *Monuments and Maidens: The Allegory of the Female Form* (Picador, London).

Watts M. J. 1992: "Space for everything (a commentary)" *Cultural Anthropology* 7, 115–129.

Winberry J. J. 1982: "Symbols in the landscape: the Confederate memorial" *Pioneer America Society Transactions* 5, 9–15.

Winberry J. J. 1983: " 'Lest we forget': the Confederate monument and the southern townscape" *Southeastern Geographer* 23, 107–121.

Young E. 1989: "The biography of a memorial icon. Nathan Rapoport's Warsaw Ghetto monument" *Representations* 26, 69–106.

Zelinsky W. 1988: *Nation into State: The Shifting Symbolic Foundations of American Nationalism* (University of North Carolina Press, Chapel Hill, NC).

# INDEX


ACT-UP (AIDS Coalition to Unleash Power) 292
Adams, John Quincy 102
Adler, S. 305
*Advocate, The* 294, 296, 300, 302, 303, 304, 310
aesthetic sensibility, and gay identity 299–300
Africa 24, 112, 204, 345
  economy 128–9, 136, 137, 200
Afro-American community 250, 251
  in US ghettos 269–87
Agar, H. *The City of Man: A Declaration of World Democracy* 106
Agnew, J. A. 221, 222, 289
agriculture
  agrarian societies and state power 73–5, 79
  agricultural location theory 11
  commercialization of, Baliapal 227–8
  US rural labour 178
Al-Sabah family 144
Alavi, H. 202
Alexander, J. 297
Alexander the Great 74
Alter, Peter 333n
America *see* Latin America, United States
American Federation of Labor 173, 192, 193–4, 195
American Revolution 32, 100, 318
Anderson, B. 348, 361
Anderson, P. 33
Anna Livia Plurabella statue 355–6, 360
antagonism, and identity formation 331–2
Arab conquests, eighth century 42
Arab-Israeli War (1973) 125
Arab oil embargo 114
Archer, J. C. 207, 211
areal differentiation, and post-modernism 25–30
Aristotle 32
Arnheim, R. 107–8
Austria 46, 350, 354
  multilingual empire 53–4
authoritarian states 63–4
autonomous power of the state 33, 58–81
Azikiwe, Nnamdi 345

Baker, James 150, 151
Baliapal movement, India 219–47
Ball, G. 113
Bandung Conference (1955) 111
Bangladesh, border with India 86, 87–8
Bangla Desh Rifles (BDR) 87, 88
Barth, Frederick 328, 330, 332, 335n
Basque provinces 328, 330, 334n, 342–3
Baudrillard, Jean 161n
Bauman, Z. 159, 160
Behera, Purushottam 234–5
Belgium, linguistic regionalism in 53, 252, 256–69
Bell, Daniel 173
Bell, D. J. 289, 308, 309
Bellah, R. N. 306
Bennett, Sari 167, 172–98
Benton, Thomas Hart 102–3
Bernstein, Richard 90n
Bharat, Akhand 84
Bharatiya Janata Party (BJP), India 84, 88, 90n, 92n, 238
*Bheeta Maati* (our soil), Baliapal 233, 240
bilingualism 262, 263, 264
Bismarck, Otto 55
Black community
  and Civil War statuary 352
  US ghettos 252–3, 269–87
Bonaparte, Louis 68–9
Bonaparte, Napoleon *see* Napoleon Bonaparte
Bonn Constitution (1949) 55
Boon, J. A. 330
Bopal disaster 224
Border Security Force (BSF), India 87, 88
borders *see* boundaries and borders
Bosnia 322, 329, 334n
boundaries and borders
  choosing political 22–3
  identifying 19–20
  and identity construction 251–2, 320, 328–30, 331–2
  Indian 34, 82–3, 83–6, 87–9
  linguistic, as interaction barriers 259–61
  peripheral interfaces 47–50
  physical 20, 83
Bourdieu, P. 89
Bowman, I. 106
Brenner, I. 305
Bretton Woods monetary system 124, 125, 132

Breuilly, John 318, 327
Britain 93, 113, 144, 196, 201, 205, 210, 337
  economic hegemony 123, 132, 200
  Falklands expedition 71
  war memorials 347, 353
Brussels 252, 256, 258, 261–3
  linguistic capabilities 264
Buchanan, James 13–14
Budapest, public monuments in, 347–8
Budd, Edward 176
Bundesbank 137
bureaucratic states 63
Burgess, J. 295
Bush administration 142–3, 145, 146–53
Bush, President George 141, 143, 146, 147, 148–9, 151, 152, 154–5, 161n, 162n
Busia, K. A. 205–6, 207, 212, 213, 214, 217n, 218n
business, internationalization of 94–5, 122–40
*Business Week* 149
Butler, J. 291

Calhoun, C. 249
Canada 5, 22
  politico-cultural landscapes, Vancouver 26
capitalism 33, 224
  geopolitics of 95
  golden age of 124
  industrial 75–6
  and socialism 173, 175, 181, 187, 188, 206
  and territorial centralization 77–8
  in world-systems framework 199–200, 321, 337–40
capitalist democracies, state power in 62–3, 64, 76
Caribbean (greater) region 21–2, 23–4
Carter, President Jimmy 98, 100, 127, 154
cartographic anxiety in postcolonial India 34, 81–92
cash crops, Baliapal 227–8
caste structure, Baliapal 227, 229–30, 237–8, 239–40
Catalan nationalism 328, 330
Catholic Church/Catholics 45, 46, 351, 357, 358
  nation-building 52–3, 54
Cayton, Horace R. 281
Celtic expansion 41
centralist politics, Ghana 205–6, 207–14, 217n, 218n
  mobilization 213–14
centrality, and gay identity 303–5
centralization
  in India 223
  and separatism 329, 341–2
  territorial 69–72, 73, 74, 77–80
centre-periphery model
  positioning of United States 94, 98–121
  Western European development 37–58
Charlier, Jacques 259
Chase-Dunn, C. K. 199–200
Chatterjee, P. 229, 233
Chicago, South Side ghettos 252–3, 275–81
Chicago School 10
China 45, 61, 84
  relationship with US 113–14, 116n
Christianity 44
*Christopher Street* 296
chronopolitics 141, 153–6
Churchill, Winston 152
cities
  size, and failure of US socialism 174, 187, 190, 192, 195, 196
  urban wage ratios 167, 190–5
  US ghettos 252–3, 269–87
  and Western European development 46, 52
'Citizens for a Free Kuwait' 144, 145
citizenship 32, 82, 290, 305–6, 318–19
cityhood campaign, West Hollywood 287–8, 294–305, 307
Civil War, US 182, 184
  statuary 352
class
  of gay men 289, 308–9
  inequality, India 223–4
  in nineteenth century Vienna 350
  and state power 60
  structure, Baliapal 229–30, 237–8, 239–40
  in 'transitional states' 68–9
  in US ghettos 252–3 270, 275, 278–9, 280, 284n
  in world-systems framework 201, 202, 203, 338, 339
Clastres, P. 73
Clay, Henry 102
Clayton Act (US, 1914) 194
cocoa-growing economy, Ghana 206–7, 216n
Cohen, I. R. 295
Cohen, J. L. 240, 242
Cold War 5, 93, 132
  effects of end 94, 95, 122, 142–3, 149–50, 319, 320
Columbus, Christopher 100
communication
  effects of new technologies 12–13
  and state power 65, 68
Communist Party of India (Marxist) 88
'compulsory co-operation' 73
conceptual map of Europe 50–4
Congress (I) Party, India 84, 88, 90n, 210, 238
Connor, Walker 257
consecutive mobilizations, Ghana 211–14
consumption, and gay identity 300–1
'containment' policy 93, 110–11
  'flexible' 154
Convention People's Party (CPP), Ghana 205, 212
Conversi, Daniele 320, 325–36
Corbridge, Stuart 94–5, 122–40
core/semi-periphery/periphery structure 18, 20–1, 168, 200, 337–9
  in greater Caribbean region 21–2
  Hall's modification of 19–20
  in US 22

Corsica 343
Cosgrove, D. 349, 350
cotton textile industry, US 177
Cox, Kevin R. 6, 10–17
creativity, and gay identity 297–9
Cressey, G. B. 109
Creveld, M. van 74
Cuban missile crisis 113
culture/cultural
  and attachment to land 233, 240, 241
  and Basque identity 342–3
  codes 169, 241
  dimensions, and Western European development 38–41, 44, 45, 52–4
  ethnic, and borders 330, 331, 332
  exaggeration of difference 330–1
  expressions of resistance 235, 241
  and nationalism 351
  popular 348–9
  significance of war memorials 351
  variables, and unification vs federalization 54–8

Davis, T. H. 292
De Certeau, M. 28, 87, 89
De Vriendt, Sera 263
Debs, Eugene 188, 192, 195
decentralization 135–6
  in France 343–4
Declaration of Independence, US 100
deconstruction 28–9, 95
Defferre, Gaston 343
DeLeon, Daniel 188, 192
democracy 32, 169
  and responsibility 302
  *see also* capitalist democracies
Democratic Party, US 188, 201, 207
Denmark, interface peripheries 49–50
Der Derian, J. 158
despotic state power 33, 61, 63–4, 72, 78, 79
Detente 113, 115
deterritorialization 134–5, 143
Deudney, D. H. 32
'difference', post-modern insistence on 28
'Diffusion of Power' thesis 115
Dikshit, R. D. 210
*Dispatch* 146, 147, 148–9, 150, 151, 152, 153
Drake, St Clair 280, 281
Dublin, public monuments in 355–6, 356–60
Dunn, John 319
Dye, T. R. 16

Eagleton, Terry *Against the Grain* 25, 29
Earle, Carville 167, 172–98
Eastern Europe 135, 136, 137, 347
Easton, D. *The Political System* 16
Eban, Abba 114
economy/economic
  of Baliapal 227–9
  concerns, vs military 93, 94–5
  dimensions, and Western European development 38–41, 44–5, 50–2
  effects of development, India 224
  effects of Iraqi invasion of Kuwait 144
  expansion, and territorial centralization 77–9
  externalities 6, 13–15
  and geopolitical world order 94, 106, 122–40
  labour, in industrializing US 175–81
  location theory 11
  power, compared with state 69–70
  redistribution, as state function 68, 72
  variables, and unification vs federalization 54–8
  *see also* political-economic perspectives; world-economy
*Edge* 296, 298, 301, 304, 310
Eisenstadt, S. M. 74
elections 6
  Bengal, village level 87–8
  Ghana (1954–1979) 168, 198–219
  India (1989) 84, 238
  US (1892–1920) 167, 188–90, 192–5
Elizabeth II 347
Elshtain, Jean 86
Emergency period, India 224
Engel, D. W. 74
Engels, Frederick 181, 188, 347
England 46, 317, 318
  monetization 52
  relationship with America 100–1
  wage ratios 179
Enlightenment 26
entertainment, and gay identity 300–1
Entrikin, J. N. 289, 290, 306–7
Epstein, S. 303, 305, 306
Eratosthenes 100
ethnic conflict 258, 319–21, 327–8, 329, 340–6
ethnic-linguistic landscape of early Western Europe 41–4
ethnicity
  content (culture) and borders 330, 331, 332
  gay 303, 305–6
  geoethnic variables and Western European development 38, 41
  and Ghanaian elections 168, 213, 214, 216, 216n
  in peripheral states 204
  and regionalism 257–8
ethno-symbolist theory of nationalism 320, 325–6
'euphoria', post-Cold War 143, 160–1n
Eurocurrency 124–5
Europe
  and Arab oil embargo 114
  relationship with US 100, 102, 104–5, 110, 113–14, 117n
  territorial centralization 77–8
  *see also* Eastern Europe; Western Europe
European Currency Unit (ECU) 137
European Economic Community (EC) 95, 122, 124, 132, 133, 137, 166, 263
*Euskadi ta Askatasuna* (ETA) 342

externalities 6, 13–15
Exxon Corporation 114–15

Fahd, King 145
Falklands expedition 71
federalism
  vs centralization in Ghana 205–6
  and fragmentation 75,
  vs unitary structures in Western Europe 54–8
Fernand Braudel Center, New York 18
feudal states 63, 66–7
  rise of 44–5
Finland, interface peripheries 50
First World War 49, 105, 106, 110, 123, 157, 173, 319
  memorials 352–3
Fischer, E. *The Passing of the European Age* 110
fishing, Baliapal 228
Flanders, Flemings 252, 256, 258
  and Fourons/Voeren controversy 265–6
  interaction with Walloons 259–61
  linguistic capabilities 263–4
  relationship with Brussels 262
Ford Motor Corporation 126–7
Fordism 252
Forest, Benjamin 253–4, 287–315
Foucault, M. 90n, 221, 297
Fourons/Voeren controversy 264–6
Frampton, K. 26
France 113, 337
  linguistic standardization 53
  monetization 52
  separatist nationalism in 320, 321, 341, 343–4
  territorial history 46–7, 47–8
  war memorials 352–3
French Revolution 32, 46–7, 318
Freud, Sigmund 330–1
Friesian interface peripheries 49–50
'frontier thesis' 175–6
*Frontiers* 296, 297–8, 299, 300–1, 302, 303, 304, 310–11
functional vs institutional definitions of state 60–1
Fussell, P. 152

Gamble, A. 201
Gandhi, Rajiv 81
Gates, Robert 149
GATT talks 132
gay identity 253–4, 287–315
Gbedemah, K. A. 217n
Geertz, C. 28
Gellner, Ernest 327, 329
Geltmaker, T. 294
gender
  identity 290–1
  and monuments 354–6
  structure, Baliapal 229, 237–8, 239–40
General Motors 70
genocide 329
geopolitics
  global positioning of US 94, 98–121
  of Gulf crisis and War 95–6, 140–64
  internationalization of business 94–5, 122–40
  overview 93–7
German Empire 46, 48
German-Roman Empire 48, 54–5
Germany 144, 317
  economy 123, 129, 132, 134, 137, 200
  interface peripheries 49–50
  invasions by, fourth and fifth century 41
  linguistic borders 53
  militarist tradition 59
  nationalization 350
  Nazi regime 64, 93, 152, 159, 319
  post-twelfth century expansion 42
  relationship with US 142, 143, 149, 150
  unification 55–6
Ghanaian elections (1954–1979) 168, 198–219
Giddens, Anthony 60, 64, 290
Gilded Age, US 174, 181, 185–7, 188–90
Gillis, R. 358
Gilpin, William 103
Giri, Gadhagar 229, 230, 238
globalization 166
  and nationalism 322
  of world economy 123–31
Godfrey, B. J. 305
Goldberg, M. 22
Golden Bull (1356) 47
Gompers, Samuel 184
Goody, J. 45
Gorbachev, Mikhail 142, 149
grain economy, US 178, 195–6
Gramsci, A. 219, 223, 349
Gray, C. 94
Great Britain *see* Britain
Gregory, Derek 8, 25–30
Guerrero, 299
Guha, Ranajit 220, 223
Gulf crisis and War (Iraq War) 95–6, 134, 136, 140–64
Gutenberg, J. G. 45, 52
Gutman, Herbert 191

Habakkuk, H. J. 176
Habermas, J. 26, 27, 271
Habsburgs 46, 47, 49
Hägerstrand, T. 28
Hall, T. D. 19–20
Hamilton, Alexander *Report on Manufactures* 176–7
Hamilton, N. 202
Handler, R. 250–1
Happart, Josè 265, 266
Harley, J. B. 83
Harris, Sir Arthur 'Bomber' 347
Harrison, Richard Edes 107
Hartshorne, R. 20, 28

*The Nature of Geography* 28
Harvey, David 28, 135, 221, 222, 253, 351
Hassner, P. 116
'Heartland' theory 94, 109
Heeger, C. A. 203, 204
Hegel, G. W. F. 317
Henrikson, Alan K. 94, 98–121
Herder, Johan Gottfried 317
Hettne, B. 204, 206, 207
Hinduism 84, 233, 234
Hintze, Otto 60, 340
Hirschman, A. O. 45
historical regions 21–2
history
 and memory 351
 and nationalism 326, 349–50
Hitler, Adolf 152
Holland *see* Netherlands
Holocaust 159, 353
homeostatic theory of nationalism 320, 326–8, 329
hooks, bell 253, 272, 288
Hoover, President Herbert 106
Hopkins, T. K. 19
human ecology, spatial approach to 10–11
humanistic perspectives 288, 308
 post-modern objection to 28
Humboldt, Alexander van 102, 103
Hungary
 interface peripheries 49
 public monuments, Budapest 347–8
Hussein, Saddam 141, 143, 145, 148, 152
hyperghettos 271, 274–81

IBM Europe 124–5
identity/identities 8
 Basque 342–3
 Black, in US ghettos 252–3, 269–87
 ethnolinguistic, in Belgium 251–2, 256–69
 gay, in West Hollywood 253–4, 287–315
 national and ethnic 53, 81–2, 318, 321–2, 325, 348–50
 overview 249–55
 postcolonial 345
 role of boundaries 251–2, 328–30, 332
 role of opposition 330–1
identity cards 81–2, 86, 87–8
identity-orientated theories of social movements 168, 219, 220–1, 222, 242
ideology
 of African governments 200
 and centrality 74
 power of, compared with state 70
 Woodrow Wilson's 105
Ignatieff, Michael 322
imperial states 63
India
 Baliapal movement 168–9, 219–47
 cartographic anxiety in 34, 81–92
 Punjab elections (1955–1977) 210
Indian National Congress 224
industrial capitalism, and state power 75–6
industrial location theory 11
Industrial Revolution 75
industrialization
 and population redistribution 13
 in US 175–81, 190, 191–2
infrastructural state power 33, 62, 63–4, 64–6, 72, 77–80
institutional vs functional definitions of state 60–1
instrumentalism 317, 320, 325, 328, 332
intellectuals 325–6, 332, 349
intelligentsia 325–6, 331–2, 345
international debt crisis 128–9, 133–4
International Monetary Fund (IMF) 124, 129, 133, 135, 136–7
internationalization of business 94–5, 122–40
Iran-Iraq War 145
Iraq War *see* Gulf crisis and War
Ireland 349
 public monuments in 350–1, 355–6, 356–60
Irish Parliamentary Party 358
Irish Republican Brotherhood 358
'Isothermal Zodiac' 103
Italy 252
 interface peripheries 49
 unification 48, 55–6

Jackson, A. 349
Jackson, P. 349
Jameson, Frederic 26
Japan 144
 economy 95, 122, 123, 124, 125, 129, 132, 134, 137
 relationship with US 142, 143, 149, 150
Jefferson, President Thomas 102
Johnson, President Lyndon B. 270
Johnson, Nuala 321–2, 347–64
Johnson, Robert 148
Joshi, Murli Manohar 84
*Journal of Historical Geography* 349
Joyce, James 353
 *Finnegan's Wake* 355
 *Portrait of the Artist as a Young Man*, A 358
 *Ulysses* 348

Kannan, Brigadier R. S. 225
Kashmir 84
Kearns, A. 306
Kemeny, J. G. 11
Kennan, G. F. 110
Kennedy, President John F. 125
Kerner Commission Report (1989) 271, 282
Keynes, John Maynard 124, 133
Kissinger, Henry 113, 143, 151, 160–1n
Knights of Labor 181, 182, 184, 185, 186, 187, 188, 191, 192, 193–4
Knopp, L. 288, 291–2, 307, 308
knowledge, post-modern critique of 7–8, 26
Kofman, E. 343
Kolko, G. 173
Korean war 124, 280

Krishna, Sankaran 34, 81–92
Kuhn, T. S. 27, 99, 116n
Kuwait, and Gulf crisis 96, 140, 141, 143–5, 146–7, 150

*LA Weekly* 294, 298, 300, 311
labour
  market, in industrializing US 175–81
  movement, and failure of US socialism (1865–1920) 181–95
Labour Party 210
language
  and Basque identity 342
  of enclave and interface peripheries 48–9
  and nationalism 53–4, 330
  linguistic regionalism 252, 256–69
  moral dimension of 306
  vernacularization 45
  *see also* ethnic-linguistic landscape of early Western Europe
Lanternari, V. 330
Laqueur, T. W. 353
'late development' response 72
Latin America 102, 281
  economy 112, 126, 128, 129, 132, 135
Lattimore, Owen 71
Lawson, Craig 304
*Le Soir* 264
Lee, General Robert E. 352
Lefebvre, Henri 92n, 221
lesbians 288, 289, 292, 294, 305, 308–9
Levinas, Emmanuel 159, 160
Lewisohn, Sam 196
Ley, David 26
liberal democracies 201–2
liberal politics, Ghana 205–6, 207–14, 217n, 218n
  mobilization 212–13
*Life* magazine 152
Limman, Hilla 218n
Lippmann, W. 109, 110
Lipset, Seymour Martin 173
Liska, G. 113
literacy, and state power 64, 65–6
locale 169, 222, 227–9, 229–30, 236, 239–40
location 169, 222, 225–7, 227–9, 236, 239
locational approaches to power and conflict 10–17
Lotharingia 47
Louis XIV 47
Luce, Henry *The American Century* 106, 110, 151
Luke, T. 96
Luxembourg 53
Lyotard, J-F. 26, 28

Machiavelli, N. *The Prince* 34
Mackinder, Halford 10, 93, 94, 104–5, 109
Mahan, Alfred Thayer *The Influence of Sea Power upon History* 103–4
Malraux, André 110
Mann, Michael 33, 58–81, 159
Mann, Patricia 158–9
maps 99
  conceptual, of Europe 50–4
  positioning of US 100–1, 103, 106–9, 112–13, 117n, 118n
  as texts 83–6
*Maran Sena* (suicide squad), Baliapal 232, 234, 235, 240
Marx, Karl 68–9, 326, 347
Marxist theory 7, 59, 200
Massey, D. 289
maturity, and gay identity 302–3
Mayo, Patricia 327
McNaught, Kenneth 173
media
  gay press and West Hollywood gay identity 253, 287–8, 295–305, 307, 309n
  Gulf War coverage 96, 156–7, 158–9
Melucci, 221
Mény, Y. 343–4
Mercator projection 107, 112
Mercer, J. 22
meridians, positioning of 101
Mesopotamia 65–6
Mexican Border Industrialization Program 126
Mexico
  boundary with US 22
  economy 126, 128, 129
Middle East
  oil production 114
  *see also* Gulf crisis
militarist state theory 59, 73
military
  concerns, vs economic 93, 94–5
  coups and governments, Africa 204, 206, 212, 213, 217n
  force, and state power 72, 73–5, 77
  functions of the state 68
  interests, Baliapal 225–7, 239
  might of US 110–11, 113, 134
  power, compared with state 70–1
  *see also* Gulf crisis and War
military-administrative dimensions, and Western European development 38–41, 44
Mishan, E. J. 14
Mistry, Kagen 88
Mitterand, François 344
Monroe Doctrine 102
Montgomery, David 184
Montgomery, James 176
monuments, public 321–2, 347–64
Moore, B. 45, 52
Moos, A. 293, 294, 295–6
moral narratives 290
moral proximity 159–60, 166
Mosse, G. L. 350
multilateralization 136–7
multinational corporations (MNCs) 124, 125, 126
Murphy, Alexander B. 252, 256–69

Mussolini, Benito 49

Napoleon III 48
Napoleon Bonaparte 47, 55
Napoleonic Wars 22
Narmada River Valley Project 224
nation-states 78, 82, 90–1n, 257, 319, 322, 326, 344, 350
nationalism
  as boundary maintenance and creation 320, 325–36
  Indian 82
  national congruence 320–1, 336–47
  overview 317–24
  and public monuments 321–2, 347–64
  and regionalism 257–8
nationalist mass parties, Ghana 203–4
  mobilization 212
Nehru, Jawaharlal
  *Autobiography* 83
  *Discovery of India* 83
Nelson monument, Dublin 356, 357, 360
Netherlands/United Netherlands 46, 55
New Mexico 19
'New World Order' 98, 123, 135, 137, 138, 149–51
*New York Times* 144
Nigeria, separatist nationalism in 320, 344–6
Nixon, President Richard 93, 114
Nkrumah, Kwame 205–6, 207, 208, 212, 213, 214, 217n
Nora, P. 351
'normal vote' 207, 209–10
North/South relationship 18, 96, 135
  and Western European development 50–4
  and cartography 111–13
Nunn, Sam 155

Ó Tuathail, Gearoid 95–6, 140–64, 166
O'Brien, D. C. 203
O'Brien, Richard 124–5
O'Connell monument, Dublin 357, 360
October War (1973) 114
oil
  and Gulf crisis 146–9
  price rises 125, 132
  world movements 114–15
Oliver, Melvin 283n
Olson, M. 166
OPEC (Organization of Petroleum Exporting Countries) 125, 132, 133, 148
opposition, and identity formation 330–1
order, state maintenance of 67–8
Osei-Kwame, Peter 168, 198–219
Osofsky, Gilbert 271
Ottoman Empire 54

*paan* (betel nut) cultivation, Baliapal 227, 228
Pacific Design Centre 297–8
Paine, Thomas *Common Sense* 100–1
Pakistan 83
  border disputes with India 85–6
Panama 143, 154
paradigm 99
  post-modern rejection of 26–7
Park, R. E. 10
Parnell monument, Dublin 358–60
Patra, Gannanath 229, 238
peripheral interfaces 47–50
peripheral states, paradoxes in 202–5
Peters, Arno 112
physical boundaries 20, 83
Piehler, G. K. 353
place/places
  dematerialization of 96, 156–9
  effacement of 159–60
  and identity 250, 251–2, 254, 287–315
  and social movements 219, 221–3, 239–40, 241
  understanding 20–1
'Politechnic Zone' idea 115
political-administrative variables, and unification vs federalization 54–8
political and social movements 5, 360
  Baliapal movement, India 168–9, 219–47
  failure of US socialism 167, 172–98
  Ghanaian elections (1954–1979) 168, 198–219
  and identity 249
  and nationalist movements 327
  overview 165–71
political-economic perspectives 2, 7, 33, 95, 168, 200, 216, 251, 253, 288, 320, 321
political parties
  and social movements 166–7, 220, 224–5, 238, 241
  in world-systems framework 168, 201, 203–4
population
  decline in Chicago ghettos 278
  redistribution 13
  stability in world-system zones 21
  *see also* size *under* cities
Portes, Alejandro 281
Portugal 21, 337
postmodern perspectives 2, 7–8, 25–30, 251, 321
postcolonialism 98, 168
  cartographic anxiety in India 34, 81–92
  separatist nationalism in Nigeria 344–6
Poulantzas, N. 69, 77
poverty, in US ghettos 270, 271, 274, 279, 280, 282n
power 3
  autonomous, of the state 33, 58–81
  and conflict, locational approaches to 6, 10–17
  loci of 221–2
  politics of 201–2
PRIDE (gay group) 294
primordialism 317, 320, 325, 328, 332
Progressive Era, US 174, 181, 188–90
progressiveness, and gay identity 301–2

Protestant Church/Protestants 349, 356–7
nation-building 52–3, 54
Prussia
monetization 52
state power in 69
strategy for control 53–4
territorial history 46, 47
public goods 14–15
public space
and gays 291–2, 299
monuments in 347–8

Queer Nation (gay group) 292

Ratzel, Friedrich 10, 93
Reagan, President Ronald 93, 145, 270
Reagonomics 129, 142
redistributive states 68, 72
Redmond, John 358, 359, 360
reductionist state theory 59
Reformation 45, 52
regional geography and world-systems analysis 17–25
regionalism, linguistic 252, 256–69
religion 70
and attachment to land 233, 240, 241
and nationality 82
Relph, E. 26
Renner, George 109
Republican Party, US 143, 201
resource-mobilization theories of social movements 168, 219, 220, 222, 242
responsibility, and gay identity 302
retail location theory 11
'Retour à Liège' 265, 266
Reynolds, David R. 6, 10–17
Riker, W. H. 54
Rokkan, Stein 33, 37–58
Roman Empire 32, 55
conquests 41
disintegration of 38, 40, 44–7
state power of 61, 72
Roosevelt, President Franklin D. 106, 107
Roosevelt, President Theodore 105, 151
Routledge, Paul 168–9, 219–47
Russia
monetization 52
relationship with US 108–9
*see also* Soviet Union
Russian Empire 42
Rutledge, L. W. *The Gay Decades* 287

Sack, R. 290
Sacre Coeur, Paris 351
Said, Edward 159
Saint-Gaudens, Augustus 352, 359–60
Sardar, Chitta 88
Saudi Arabia 145, 147
Savage, K. 352, 360
Scandinavia 46
Schaefer, 28
Schattschneider, E. E. 201
Schick, A. 15–16
*Schindler's List* 353
Schorske, C. E. 350, 354
Schulte, Steve 304
Schwartz, B. 351
Schwartzkopf, General 145, 155, 158
Scowcroft, Brent 145, 154
script, ideographic and alphabetic 45
Second World War 5, 64, 93, 106, 110, 123
memorials 347, 353–4
mythic narratives of 151–2, 153
Sedgwick, E. K. 290–1
sense of place 169, 222, 233–5, 236, 240
separatist nationalism 320–1, 329, 340–6
Shahal, Balbir Singh 87–8
Sharma, J. C. 210
Sharp, G. 220, 222
Shaw memorial 352, 359
Sheikh, Atiar 88
Sherman, D. J. 352
Sherry, M. 157
Shotter, J. 305–6
Siegel, Bud 304–5
Simmel, George 251
Singh, V. P. 238
Singrauli development scheme 224
Skocpol, Theda: *States and Social Revolutions* 60, 340
Slav interface peripheries 49
Smith, A. 355
Smith, Anthony D. 325, 326, 327, 334n, 336, 337, 345, 350
Smith, M. P. 305
Smith, P. 297
Smith, S. J. 290, 306, 309n
Smith, Steve 304
social categorization 320, 331–2
social construction
of sexual identity 291
of space 256–69
social movements *see* political and social movements
socialism 306
failure in US 167, 172–98
in Ghana 206
Socialist Labor Party, US 188, 192
Socialist Party of America 181, 188–90, 191, 192–5
society and state 80
Soja, E. 29
Sombart, Werner 173
'soul', and Black identity 272–3
Soviet Union 93, 137
Armenian-Azerbaidzhani conflict 258
collapse of 122, 134
relationship with US 98, 110–11, 113, 116n, 132, 142–3, 150
state power in 61, 64, 66, 76
space, social construction of 256–69
*see also* public space
'Spaceship Earth' doctrine 115–16
Spain 21, 46, 144, 337

separatist nationalism in 53, 320, 321, 341, 342–3
Spanish Civil War 342
Spanish-American War 104
spatial-analytic perspectives 2, 6, 33, 94,167–8, 251, 320
spatiality of the Gulf crisis 140–64
Spencer, H. 73
Spicer, E. H. 330
Spykman, N. J. 109
Stamm, K. R. 295
Stanislaw I 48
state/states
autonomous power of 33, 58–81
defining 60–6
development in Western Europe 33, 37–58, 337–40
and fate of US ghettos 252–3, 269–87
as focus of social movements 165, 166, 169, 239, 240–2, 327
and geoeconomic competition 94–5
Marxist theory of 200
militarized 157–9
mobilization, Ghana 212
and nation 34, 52–4, 81–92, 318–19, 326–8, 336–47
paradoxes in peripheral 202–5
politics of power and politics of support 201–2
preoccupation with 257
as producer of statistics 18
and region 20, 22–3
repression 173, 222–3, 224, 236–7, 342
spatiality of, overview 31–6
statistics, critiquing 18–20
Stone, Ron 298, 300, 301, 302–3, 304
strikes
Ireland 360
US (1790–1880) 182–7, 190, 191
structural interpretation of failure of socialism in US 172–3
structuralism, post-modern opposition to 27–8
Suez crisis 113
support, politics of 201–2
Sweden 46, 52,
interface peripheries 50
unitary structure 56
Swiss Confederation 46, 55
Switzerland
federal structure 56
language 53, 55
'systems' perspective 6, 15–17

tactical interpretation of failure of socialism in US 173
Taylor, Peter J. 7, 17–25, 168, 198–219, 337
technology
communication 12–13
in industrializing US 176, 177, 178–9
military 96, 153–4, 157–8
printing 45
television 118n
Gulf War coverage 96, 156–7, 158–9
Temin, Peter 176
'terrains of resistance' 169, 219, 222, 235–6, 238–42
Terrigno, Valerie 302, 304
territory/ territorial
approach to language rights 265–6
centralization 69–72, 73, 74, 77–80
and memory 351–2
and state 31–3
strategy, and identity formation 290, 307–8
Western European development 33, 37–58
Thatcher, Margaret 145, 161n
Third World 135, 138, 344
cartographic representation 111–13
economic development 126–7
elections in 199
US perception of 143, 150–1
Thirty Years War 47
Thomson, J. 32
Tilly, Charles 33, 60
*Time* 106
Tocqueville, A. de 165, 169
Tone, Wolfe 358
totalization, post-modern rejection of 27
trade unions, US 173, 174–5, 182–4, 192–4, 195, 196
transactionalist theory of nationalism 320, 328–30
'transitional states' 68–9
Treaties of Westphalia (1648) 337
Tuan Y.-F. 289, 290, 303, 306, 307
Turkey 326
Turner, Frederick Jackson 175–6

'underclass' theory 253, 270, 272–3, 273–4, 281–2, 283n
unemployment, in Chicago ghettos 278, 279
unitary vs federal structures 54–8
United Gold Coast Convention (UGCC), Ghana 205
United Kingdom 131, 341
United Nations (UN) 135, 136, 140, 141, 150
United Party, Ghana 212
United States (US) 5, 15, 93, 169
American identity 349
Black ghettos in 252–3, 269–87
boundaries in 22–3
citizenship in 305–6
failure of socialism in 167, 172–98
and geopolitical economy 95, 106, 122–40, 200
global positioning of 94, 98–121
and greater Caribbean region 21–2, 23–4
and Gulf crisis and War 95–6, 134, 136, 140–64
and Parnell monument 359
politics of power and politics of support in 201
war memorials in 352, 353, 354

*Update* 296, 300, 301, 303, 311
urbanization, and externalities 13–14
  *see also* cities

Valentine, G. 292, 308
Van de Craen, Pete 263
Van Wolferen, K. 149
Vancouver, politico-cultural landscapes 26
vernacularization 45
*Vichar* institution Baliapal 230, 238, 240
Vienna 350, 354
Vietnam war 98, 113, 152–3, 155
  Veterans Memorial 352, 354, 356
Viking raids 42
violence
  inter-ethnic 320, 331, 332, 335n
  in US ghettos 276–8, 283–4n
Virilio, Paul 141, 153–4, 155–6, 157
*Voelkerwanderungen* 41, 53

Wacquant, Loïc J. D. 252–3, 269–87
wage differentials in US labour force 174, 177–81, 190–5, 195–6
Wagner-Pacifini, R. 351
Waldseemuller, Martin 100
Wallerstein, Immanuel 7, 18, 19, 20–1, 23, 33, 199–200, 337–40
  *Historical Capitalism* 19
Wallonia, Walloons 252, 256, 258
  and Fourons/Voeren controversy 265–6
  interaction with Flemings 259–61
  linguistic capabilities 264
  relationship with Brussels 262
war memorials 351–4
Warner, M. 354
Washington, President George 102
Watts, M. J. 348, 349, 352
Weber, Max 60, 69, 79
Webster, William 145
Weimar Constitution (1919) 55
Weinstein, James 173
Wellington monument, Dublin 356
Wendt, A. E. 339
West Hollywood, gay identity in 253–4, 287–315
West, Cornel 273
West-East axis, in modern Western Europe 50–4
Western Europe 5, 143, 182
  economy 124, 125, 132, 133
  ethnic conflict in 340–4
  feudal states in 44–5, 63, 66–7
  and greater Caribbean region 21–2
  state formation in 33, 37–58, 337–40
  territorial centralization in 77–8
Whirney, Eli 178
Wiebe, Robert 191
Williams, Colin H. 320–1, 336–47
Wilson, President Woodrow 105–6, 110, 117n, 149, 194
Wittgenstein, Ludwig 25
Wohlstetter, A. 113
Wolf, E. R. 19
women
  role in Baliapal movement 230, 232, 233, 234, 237
  and war dead 353, 356
  *see also* gender
Woodward, B. 154
World Bank 124, 127, 131, 133, 135, 136–7
World War I *see* First World War
World War II *see* Second World War
world-economy 17–25, 168, 199–200, 204–5, 337–40
  and Ghanaian elections 206–7, 214–16
world-systems framework 7, 168, 199–205, 320, 337–40
  and regional geography 17–25

Yeats, William Butler 360
Young, C. 200
Young, E. 353–4

Zelinsky, W. 352